Data Science Concepts and Techniques with Applications

Usman Qamar • Muhammad Summair Raza

Data Science Concepts and Techniques with Applications

Second Edition

 Springer

Usman Qamar
College of Electrical and Mechanical
Engineering
National University of Sciences
and Technology
Islamabad, Pakistan

Muhammad Summair Raza
Department of Computer Science
Virtual University of Pakistan
Lahore, Pakistan

ISBN 978-3-031-17441-4   ISBN 978-3-031-17442-1   (eBook)
https://doi.org/10.1007/978-3-031-17442-1

This Springer imprint is published by the registered company Springer Nature Switzerland AG
The registered company address is: Gewerbestrasse 11, 6330 Cham, Switzerland

To our lovely daughters.

Foreword

In the recent years, data science has gained a truly remarkable visibility and has attracted a great deal of interest coming from academe and industry. We have been witnessing a wealth of architectures, paradigm, and learning methods materialized through concepts of deep learning cutting across challenging areas of computer vision and natural language processing, among others. With an omnipresence of floods of data, data science has impacted almost all endeavors of human life. With the rapidly evolving area and waves of new algorithms, there is a genuine and timely need for a systematic exposure of the fundamentals of data science to help the audience to build a coherent and systematic picture of the discipline and facilitate sifting through the meanders of current technology and understand the essence and applicability of the developed methods.

Professors Qamar and Raza brought forward a truly timely, prudently organized, and authoritatively presented material on the recent state of the art of data science. The book serves as an accessible textbook with the aim of presenting an introduction as well as covering advanced concepts of the emerging and interdisciplinary field of data science.

The textbook systematically covers all aspects of data science, including the central topics of data analytics, data preprocessing, machine learning, pattern mining, natural language processing, deep learning, and regression analysis.

The material is organized in a coherent manner, and this definitely helps the reader fully acquire fundamental knowledge. The presentation is completed in a top-down fashion.

The book is structured into three main parts. The first one offers a focused introduction to data science. Starting from the basic concepts, the book highlights the types of data, their usage, their importance, and issues that are normally faced in data analytics, followed by discussion on a wide range of applications of data science and widely used techniques in data science. The second part is devoted to the techniques of data science. It involves data preprocessing, feature selection, classification, clustering, deep learning, frequent pattern mining, regression analysis, as well as an introduction to text mining and opining mining. Finally, the third part of the book covers two programming languages commonly used for data science projects, i.e., Python and R, as well as provides a comprehensive tutorial of the popular data science tool WEKA.

The textbook illustrates concepts with simple and intuitive examples, accompanied by a step-by-step explanation. It provides numerous examples using real-world data, supports understanding through hands-on experience of solving data science problems using Python and R programming language, reviews a range of applications of data science, as well as provides supplementary code resources and data.

The book is suitable for both undergraduate and postgraduate students as well as those involved in research in data science. It can be used as a textbook for undergraduate students in computer science, engineering, and mathematics. It can also be accessible to undergraduate students from other disciplines. The more advanced chapters can be used by researchers intending to gain a solid theoretical understanding. The book can also be used to develop new courses in data science, evaluate existing modules and courses, draft job descriptions, and plan and design efficient data-intensive research teams across scientific disciplines. Although this book primarily serves as a textbook, it will also appeal to practitioners and researchers.

In sum, the book delivers a systematic, carefully thoughtful material on data science. Professors Qamar and Raza should be congratulated on producing this timely and very much needed book.

Edmonton, Canada   Witold Pedrycz
February 2023

Preface

What is *data science*? This is a very difficult question to answer. For us, data science is a cross-disciplinary set of skills. We will say it is composed of three aspects: *the statistician* who can model and summarize the data, *the data scientist* who designs and uses the algorithms to present the knowledge extracted from the data, and finally *the domain expert* who will validate the knowledge.

Organization of the Book

In *Chap. 1*, we will discuss the data analytics process. Starting from the basic concepts, we will highlight the types of data, their use, their importance, and issues that are normally faced in data analytics. Efforts have been made to present the concepts in the most simple possible way, as conceptual clarity before studying the advanced concepts of data science and related techniques is very much necessary. *Chapter 2* discusses various applications of data science and their corresponding benefits. Before discussing common techniques used in data science applications, we first will try to explain three types of learning, which include supervised, unsupervised, and reinforcement learning. This is explained in *Chap. 3*.

In *Chap. 4*, data preprocessing is introduced. It is essential to extract useful knowledge from data for decision-making. However, the entire data is not always processing ready. It may contain noise, missing values, redundant attributes, etc., so data preprocessing is one of the most important steps to make data ready for final processing. Feature selection is an important task used for data preprocessing. It helps reduce the noise and redundant and misleading features. Classification is the process of grouping objects and entities based on available information, which is explained in detail in *Chap. 5*. Classification is the process of grouping objects and entities based on available information. It is an important step that forms the core of the data analytics and machine learning activities. In this chapter, we will discuss the concepts of the classification process. We will provide a comprehensive overview of classification techniques including decision tree, naïve Bayes, K-nearest neighbor, support vector machines, and artificial neural networks. Finally the concept of ensemble is introduced along with different methods for model evaluation.

Chapter 6 explains clustering, which is the process of dividing objects and entities into meaningful and logically related groups. In contrast with classification where we have already labeled classes in data, clustering involves unsupervised learning, i.e., we do not have any prior classes. We just collect similar objects in similar groups. In *Chap. 7*, text mining is introduced. These days a large portion of the data exists in the form of text. This motivates us to have the methods for processing this data so that meaningful information could be derived from it. Based on the knowledge discovery perspective, this is similar to knowledge discovery in the database. Text is the natural form of storing and transferring information, and the Internet is one of the major examples where a large portion of information exists in the form of textual data.

Chapter 8 provides a comprehensive overview of deep learning. Deep learning comprises various other domains including machine learning, neural networks, artificial intelligence, etc. Neutral networks form a core of deep learning. In the last few years, there has been an unprecedented increase in applications of deep learning. Products like Siri, self-driving cars, and virtual assistants are part of daily life now, and their use is increasing on a daily basis. In *Chap. 9*, frequent pattern mining is explained as well as its applications. Frequent pattern mining deals with identifying the common trends and behaviors in datasets. These trends represent information that may not be available otherwise. Identification of these patterns can help organizations to improve their decision-making and policy-related matters. *Chapter 10* discusses regression analysis. Regression is the process of learning relationships between inputs and continuous outputs from data. The chapter covers different methods and techniques of regression like simple (single-variant), multiple (multivariate), polynomial, and logistic regressions as well as some advanced methods of regression.

In *Chap. 11*, we discuss two programming languages commonly used for data science projects, i.e., Python and R programming language. First Python will be discussed and in the later part, we will discuss the R programming language. *Chapter 12* is an introduction to WEKA. WEKA stands for Waikato Environment for Knowledge Analysis. It is an open-source tool for performing different machine learning and data mining tasks as part of a data scientist's daily activities.

To the Instructor

The textbook is designed to give a broad, yet detailed, overview of the data science field. The textbook comprehensively covers all aspects of data science, including data analytics, data preprocessing, machine learning, pattern mining, natural language processing, deep learning, regression analysis, and several other related topics. The book is suitable for both undergraduate and postgraduate students as well as those carrying out research in data science. It can be used as a textbook for both undergraduate and postgraduate students in computer science, engineering, and mathematics. It can also be accessible to students from other areas with adequate backgrounds. The more advanced chapters can be used by postgraduate researchers in the field of data science.

The book was specifically written to enable the teaching of both the basic data science and advanced data science courses from a single book. Specifically, the courses that could be offered with various chapters are as follows:

- *Basic data science course:* If you are teaching an introductory course to data science, then *Chaps. 1, 2, 3, 4, 5, 6, and 11* are a good place to start with. They comprehensively cover the core concepts of data science.
- *Advanced course:* For more advanced courses, where the students already have the basic knowledge of data analytics, *Chaps. 4, 5, 6, 7, 8, 9, 10, and 12* can be taught. Such a course would cover advanced topics on data science.

If you are teaching a course that comprises lab work, then the last two chapters of the book, i.e., *Chaps. 11 and 12*, focus on two programming languages commonly used for data science projects, i.e., Python and R programming language, as well as a comprehensive tutorial of the data science tool "WEKA."

Individual chapters in the textbook can also be used for tutorials. Also each chapter ends with a set of exercises.

To the Student

Data science is an umbrella term that encompasses data analytics, data mining, deep learning, and several other related disciplines, so contents have been devised keeping in mind this perspective. An attempt has been made to keep the book as self-contained as possible. The textbook illustrates concepts with simple and intuitive examples, along with stepwise explanations. It provides numerous examples using real-world data and supports understanding through hands-on experience of solving data science problems using Python and R programming language. It is suitable for many undergrad and postgrad courses such as Data Science, Machine Learning, Data Mining, Artificial Neural Networks, Text Mining, Decision Sciences, and Data Engineering. The textbook is written in a simple style to make it accessible even if you have a limited mathematical background. However, you should have some programming experience.

To the Professional

Although this book primarily serves as a textbook, it will also appeal to industrial practitioners and researchers due to its focus on applications. The book covers a wide range of topics in data science and thus can be an excellent handbook. Each chapter has been written as self-contained; you can choose to read those chapters that are of interest to you.

Islamabad, Pakistan

Lahore, Pakistan

February 2023

Usman Qamar

Muhammad Summair Raza

Contents

About the Authors

Usman Qamar has over 15 years of experience in data engineering and decision sciences both in academia and industry. He has a Masters in Computer Systems Design from the University of Manchester Institute of Science and Technology (UMIST), UK. His MPhil in Computer Systems was a joint degree between UMIST and the University of Manchester which focused on feature selection in big data. In 2008/2009, he was awarded a PhD from the University of Manchester, UK. His PhD specialization is in Data Engineering, Knowledge Discovery, and Decision Science. His post-PhD work at the University of Manchester involved various research projects including hybrid mechanisms for statistical disclosure (feature selection merged with outlier analysis) for Office of National Statistics (ONS), London, UK; churn prediction for Vodafone UK; and customer profile analysis for shopping with the University of Ghent, Belgium. He is currently Tenured Professor of Data Sciences at the National University of Sciences and Technology (NUST), Pakistan, and Director of Knowledge and Data Science Research Centre, a Centre of Excellence at NUST, Pakistan. He has authored nearly 200 peer-reviewed publications which include 27 book chapters, 45+ impact factor journals, and over 100 conference publications. He has two patents with Intellectual Property Office, UK. He has also written four books which have all been published by Springer & Co, including a textbook on data science. Because of his extensive publications, he is a member of the Elsevier Advisory Panel and Associate Editor of various journals, including *Information Sciences, Applied Soft Computing, Computers in Biology and Medicine, ACM Transactions on Asian and Low-Resource Language Information Processing, Neural Computing and Applications, Informatics in Medicine Unlocked, PLOS One*, and *Array*. He has successfully supervised 5 PhD students and over 100 master's students. He has received multiple research awards, including Best Book Award 2017/2018 by Higher Education Commission (HEC), Pakistan, and Best Researcher of Pakistan 2015/2016 by Higher Education Commission (HEC), Pakistan. He was also awarded a gold medal in the Research & Development category by Pakistan Software Houses Association (P@SHA) ICT Awards 2013 & 2017 and silver medal by APICTA (Asia Pacific ICT Alliance

Awards) 2013 in the category of R&D hosted by Hong Kong. Finally, he is a visiting fellow at Cranfield University, UK.

Muhammad Summair Raza has a PhD specialization in software engineering from the National University of Sciences and Technology (NUST), Pakistan. He completed his MS at International Islamic University, Pakistan, in 2009. He is currently associated with the Virtual University of Pakistan as an Assistant Professor. He has published various papers in international-level journals and conferences with a focus on rough set theory. His research interests include feature selection, rough set theory, trend analysis, software design, software architecture, and nonfunctional requirements.

Chapter 1
Introduction

1.1 Data

Data is an essential need in all domains of life. From the research community to business markets, data is always required for analysis and decision-making purposes. However, the emerging developments in the technology of data storage, processing, and transmission have changed the entire scenario. Now the bulk of data is produced on daily basis. Whenever, you type a message, upload a picture, browse the web, or type a social media message, you are producing data that is being stored somewhere and available online for processing. Just couple this with the development of advance software applications and inexpensive hardware. With the emergence of the concepts like Internet of Things (IoTs), where the focus is on connected data, the scenario has worsened further. From writing something on paper to online distributed storage, data is everywhere.

Each second, the amount of data is increasing by the rate of hundreds of thousands of tons. By 2020, the overall amount of data is predicted to be 44 Zetta-bytes and just to give an idea, 1.0 Zettabyte is equal to 1.0 Trillion gigabytes. With such huge volumes of data apart from the challenges and the issues like the curse of dimensionality, we also have various opportunities to dig deep into these volumes and extract useful information and knowledge for the good of society, academia, and business.

Figure 1.1 shows a few representations of data.

1.2 Analytics

In the previous section, we have discussed huge volumes of data that are produced on daily basis. The data is not useful until we have some mechanism to use it to extract knowledge and make decisions. This is where the data analytics process steps

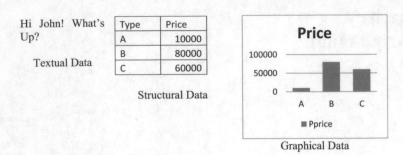

Fig. 1.1 Few data representations

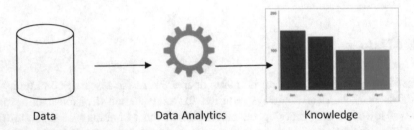

Fig. 1.2 Data analytics process, a broad overview

in. There are various definitions of data analytics. We will use a simple one. Data analytics is the process to use the data (in any form), processing it through various tools and techniques and then extracting useful knowledge from this data. The knowledge will ultimately help in decision-making.

Overall data analytics is the process of generalizing knowledge from raw data which includes various steps from storage to final knowledge extractions. Apart from this, the process involves concepts from various other domains of science and computing. Starting from basic statistic measures, e.g., means, medians, variances, etc., to advance data mining and machine learning techniques, each step transforms data to extract the knowledge.

This process of data analytics has also opened the doors to new ideas, e.g., how to mimic the human brain with a computer so that the tasks performed by a human could be performed by the machines at the same level. Artificial Neural Network was an important development in this regard.

With the advent of such advance tools and techniques for data analytics, several other problems have also emerged, e.g., how to efficiently use the computing resources and enhance the efficiency and how to deal with the various data-related problems, e.g., inaccuracy, huge volumes, anomalies, etc. Figure 1.2 shows the data analytics process at a high level.

1.3 Big Data vs Small Data

The nature of the data that applications need now is changed. Starting from basic databases which used to store the daily transactions data, the distributed connected data has become reality. This change has impacted all the aspects related to data including the storage mechanisms, processing approaches, and knowledge extractions. Here we will present a little comparison of the "small" data and "big" data in Table 1.1.

As discussed above, the data has grown from a few Gigabytes to Zettabytes; the change is not only in size; it has changed the entire scenario, e.g., how to process distributed connected data. How to ensure the security of your data that you don't know where it is stored on the cloud? how to make the maximum out of it with limited resources? The challenges have opened new windows of opportunities. Grid computing, clouds, fog computing, etc. are the results of such challenges.

The mere expansion of the resources is not sufficient. Strong support from software is also essential because conventional software applications are not sufficient to cope with the size and nature of big data. For example, a simple software application that performs data distribution in a single server environment will not be effective for distributed data. Similarly, an application that just collects data from a single server as a result of a query will not be able to extract distributed data from different distributed nodes. Various other factors should also be considered, e.g., how to integrate such data, where to fetch data efficiently in case of a query, should data be replicated, etc.

Even the abovementioned issues are simple, and we come across more complex issues, e.g., in the case of cloud computing, your data is stored on a cloud node, you don't have any idea, so, how to ensure the sufficient security of your data and availability?

One of the techniques common techniques to deal with such big data is a MapReduce model. Hadoop, based on MapReduce, is one of the common platforms

Table 1.1 Comparison between small data and big data

Small data	Big data
Small volumes of data mostly organizational data	Large volumes including the text, images, videos, etc.
Stored on single systems	Data is distributed, i.e., various geographical locations
Standalone storage devices, e.g., local servers	Data is connected, e.g., Clouds, IoTs
Homogeneous data, i.e., data has the same format	Heterogeneous data, i.e., data may be in different formats and shapes
Mostly structured data	May include both structured, unstructured, and semi-structured data
Simple computing resources are sufficient to process the data	May require most computing resources including cloud computing and grid computing, etc.
Processing such data is less challenging	Processing such data is more challenging especially when it comes to accuracy and efficiency of processing

for processing and managing such data. Data is stored on different systems as per needs and the processed results are then integrated.

If we consider the algorithmic nature of the analytics process, data analytics for small data seems to be basic as we have data only in a single structured format, e.g., we may have data where you are provided with a certain number of objects where each object is defined through well-precised features and class labels. However, when it comes to big data, various issues need to be considered. Talking simply about the feature selection process, a simple feature selection algorithm for homogeneous data will be different from the one dealing with data having heterogeneous nature.

Similarly, the small data in the majority of the cases is structured, i.e., you have well-defined schemas of the data; however, in big data you may have both structured and unstructured data. So, an algorithm working on a simple structure of data will be far simple as compared to the one working on different structures.

So, as compared to simple small data, big data is characterized by four features as follows:

Volume: We deal with Petabytes, Zettabytes, and Exabytes of data.
Velocity: Data is generated at an immense rate.
Veracity: Refers to bias and anomalies in big data.
Variety: The number of types of data.

As compared to big data, small data is far simple and less complex with respect to big data.

1.4 Role of Data Analytics

The job of data analysts includes knowledge from domains like statistics, mathematics, Artificial Intelligence, and machine learning. The ultimate intention is to extract knowledge for the success of the business. This is done by extracting the patterns from the data.

It involves a complete interaction with data throughout the entire analysis process. So, a data analyst works with data in various ways. It may include data storage, data cleansing, data mining for knowledge extraction, and finally presenting the knowledge through some measures and figures.

Data mining forms the core of the entire data analytics process. It may include extraction of the data from heterogeneous sources including texts, videos, numbers, and figures. The data is extracted from the sources and transformed into some form that can easily process, and finally we load the data so that we could perform the required processing. Overall, the process is called the Extract, Transform, and Load (ETL) process. However, note that the entire process is time-consuming and requires a lot of resources, so, one of the ultimate goals is to perform the entire process with efficiency.

Statistics and machine learning are a few of the major components of data analytics. They help in the analysis and extraction of knowledge from the data. We input data and use statistics and machine learning techniques to develop models. Later, these models are then used for prediction analysis and prediction purposes. Nowadays, lots of tools and libraries are available for this purpose including R and Python, etc.

The final phase in data analytics is data presentation. Data presentation involves visual representations of the results for the concept of the customer. As the customer is the intended audience of data representation, the techniques used should be simple, formative, and as per his requirements.

Talking about the applications of data analytics, the major role of data analytics is to enhance the performance and efficiency of businesses and organizations.

One of the major roles data analytics plays is in the banking sector, where you can find out credit scores, predict potential customers for a certain policy and detect outliers, etc.

However, apart from its role in finance, data analytics plays a critical role in security management, healthcare, emergency management, etc.

1.5   Types of Data Analytics

Data analytics is a broad domain. It has four types, i.e., descriptive analytics, diagnostic analytics, predictive analytics, and prescriptive analytics. Each has its nature and the type of tasks performed in it. Here we will provide a brief description of each.

- Descriptive analytics helps us find "what happened" or "what is happening." In simple words, these techniques take the raw data as input and summarize it in the form of knowledge useful for customers, e.g., it may help find out the total time spent on each customer by the company or total sales done by each region in a certain season. So, the descriptive analytics process comprises data input, processing, and the results generations. Results generated are in visual forms for a better understanding of the customer.
- Diagnostics analytics: Taking the analytics process a step ahead, diagnostic analytics help in analyzing "Why it happened?". Performing analysis of historical and current data, we may get details of why a certain event happened at a certain period in time. For example, we can find out the reasons for a certain drop in sales over the third quarter of the year. Similarly, we can find out the reasons behind the low crop yield in agriculture. Special measures and metrics can be defined for this purpose, e.g., yield per quarter, profit per 6 months, etc. Overall, the process is completed in three steps:

  - Data collection
  - Anomaly detection
  - Data analysis and identification of the reasons

- Predictive analytics: Predictive analytics as the name indicates helps in making predictions. It helps in finding "what may happen." Predictive analytics helps us find patterns and trends by using statistical and machine learning techniques and tries to predict whether the same circumstances may happen in the future. Various machine learning techniques like an Artificial Neural Network, classification algorithms, etc. may be used. The overall process comprises of following steps:

  - Data collection
  - Anomaly detection
  - Application of machine learning techniques to predict patterns

- Prescriptive analytics: Prescriptive analytics, as the name implies, prescribes the necessary actions that need to be taken in case of a certain predicted event, e.g., what should be done to increase the predicted low yield in the last quarter of the year? What measures should be taken to increase the sales in the off-season?

So, the different types of analytics help at certain stages for the good of the business and organizations. One thing common in all types of analytics is the data required for applying the analytics process. The better the quality of data, the better the decisions and results. Figure 1.3 shows the scope of different types of data analytics.

1.6   Challenges of Data Analytics

Although data analytics has been widely adopted by organizations and a lot of research is underway in this domain, still there are lots of challenges that need to be catered to. Here, we will discuss a few of these challenges.

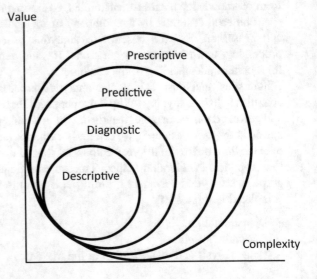

Fig. 1.3 Scope of different data analytics types

1.6.1   Large Volumes of Data

In this data-driven era, organizations are collecting a large amount of data on daily basis, sufficient to pose significant challenges for the analytics process. The volumes of data available these days require a significant amount of resources for storage and processing. Although analytics processes have come up with various solutions to solve the issues, still lot of work needs to be done to solve the problem.

1.6.2   Processing Real-Time Data

In the majority of the cases, the significance of the data remains valid only for a certain period. Coupling the problem with a large amount of data, it becomes impossible to capture such huge volumes in real time and process them for meaningful insights. The problem becomes more critical in case we do not have sufficient resources to collect real-time data and process it. This is another research domain where data analytics processes deal with huge volumes of real-time data and process it in real time to give meaningful information so that its significance remains intact.

1.6.3   Visual Representation of Data

For the clear understudying of the customer and organizations, data should be presented in simple and easy ways. This is where visual representations like graphs, charts, etc. may help. However, simply presenting the information is not an easy task especially when the complexity of the data increases, so we need to present information at various levels. Just inputting the data into the tools and getting knowledge representations may seem to be simple, but it involves a large amount of processing in real time.

1.6.4   Data from Multiple Sources

Another issue that needs to be catered to is distributed data, i.e., the data stored at different geographical locations. This may create problems as manually the job seems very cumbersome. However, regarding the data analytics process, data distribution may have many issues which may include data integration, mismatch in data formats, different data semantics, etc.

The problem lies not only in data distribution but in the resources required to process this data. For example, processing distributed data in real-time may require expensive devices to ensure high-speed connectivity.

1.6.5   Inaccessible Data

For an effective data analytics process, data should be accessible 24/7. Accessibility of data is another issue that needs to be addressed. Backup storages and communication devices need to be purchased to ensure the data is available whenever required. Because even if you have data but it is not accessible due to any reason, the analytics process will not be significant.

1.6.6   Poor-Quality Data

Data quality lies at the heart of the data analytics process. Incorrect and inaccurate data means inaccurate results. It is common to have anomalies in data. Anomalies may include missing values, incorrect values, irrelevant values, etc. Anomalies may occur due to various reasons, e.g., defects in sensors that may result in incorrect data collections, users who are not willing to enter correct values, etc.

Dealing with this poor-quality data is a big challenge for data analytics algorithms. Various preprocessing techniques are already available, but dealing with anomalies is still a challenge especially when it comes to large volumes and distributed data with unknown data sources.

1.6.7   Higher Management Pressure

As the results of data analytics are realized and the benefits become evident, the higher management demands more results which ultimately increase pressure on the data analysts. So, working under pressure always has its negatives.

1.6.8   Lack of Support

Lack of support from higher management and peers is another issue to deal with. Data analytics is not useful if higher management is not supportive and does not give authority to take action after extracting the knowledge from the results through the analytics process. Similarly, if peers, e.g., other departments are not willing to provide data, the analytics process will not be much useful.

1.6.9   Budget

Budget is one of the core issues to deal with. The data analytics process requires expensive systems having the capacity to deal with large volumes of data, hiring consultants whenever needed, purchasing data and tools, etc. which involves a significant amount of budget. Until the required budget is not provided and organizations are not willing to spend on the data analytics process, it is not possible to get the fruits of the data analytics.

1.6.10   Shortage of Skills

Data analytics is a rich field involving skillsets from various domains like mathematics, statistics, artificial intelligence, machine learning, etc. So, it becomes an issue to find such experts having knowledge and experience in all of these domains. So, finding the right people for the right job is also an issue that organizations and businesses are facing so far.

1.7   Top Tools in Data Analytics

With the increasing trend in data analytics, various tools have already been developed to create data analytics systems. Here we will discuss a few tools that are most commonly used for this purpose.

R Programming R is one of the leading tools for analytics and data modeling. It has compatible versions for Microsoft Windows, Mac OS, Unix, etc. Along with this, it has many available libraries for different scientific tasks.

Python Python is another programming language that is most widely used for writing programs related to data analytics. This open-source and object-oriented language have many libraries from high-profile developers for performing different data analytics-related tasks. A few of the most common libraries used in Python are NTLK, Numpy, Scipy, Sci-Kit, etc.

Tableau Public It is free software that can connect to any data source and create visualizations including graphs, charts, and maps in real time.

QlikView QlikView offers in proc data processing thus enhancing efficiency and performance. It also offers data association and data visualization with compressed data.

SAS SAS is another data analytics-related tool that can analyze and process data from any source.

Microsoft Excel Microsoft Excel is one of the most common tools used for organizational data processing and visualizations. The tool is developed by Microsoft and is part of the Microsoft Office suite. The tool integrates many mathematical and statistical functions.

RapidMiner RapidMiner is mostly used for predictive analytics; the tool can be integrated with any data source including Excel, Oracle, SQL server, etc.

KNIME KNIME is an open-source platform that lets you analyze and model data. Through its modular data pipeline concepts, KNIME provides a platform for reporting and integration of data.

OpenRefine As per their official statement, OpenRefine (previously Google Refine) is a powerful tool for working with messy data: cleaning it, transforming it from one format into another, and extending it with web services and external data.

Apache Spark Apache server is one of the largest large-scale data processing tools. The tool executes applications in Hadoop clusters 100 times faster in memory and 10 times faster on disk.

Splunk Splunk is a specialized tool to search, analyze, and manage machine-generated data. Splunk collects, indexes, and analyzes the real-time data into a repository from which it generates the information visualizations as per requirements.

Talend Talend is a powerful tool for automating big data integration. Talend uses native code generation and helps you run your data pipelines across all cloud providers to get optimized performance on all platforms.

Splice Machine Splice Machine is a scalable SQL database that lets you modernize your legacy and custom applications to be agile, data-rich, and intelligent without modifications. It lets you unify the machine learning and analytics consequently reducing the ETL costs.

LUMIFY LUMIFY is a powerful big data tool to develop actionable intelligence. It helps find the complex relationships in data through a suite of analytic options, including graph visualizations, and full-text faceted search.

1.8   Business Intelligence

Business intelligence (BI) deals with analyzing the data and presenting the extracted information for business actions to make decisions. It is the process that includes technical infrastructure to collect, store, and analyze the data for different business-related activities.

The overall objective of the process is to make a better decision for the good of the business. Some benefits may include:

- Effective decision-making
- Business process optimization
- Enhanced performance and efficiency
- Increased revenues
- Potential advantages over competitors
- Making effective future policies

These are just a few benefits to mention. However, to do this, effective business intelligence needs to meet four major criteria.

Accuracy   Accuracy is the core of any successful process and product. In the case of business intelligence process, accuracy refers to the accuracy of input data and the produced output. A process with accurate data may not reflect the actual scenario and may result in inaccurate output which may lead to ineffective business decisions. So, we should be especially careful about the accuracy of the input data.

Here the term error is in the general sense. It refers to erroneous data that may contain missing, redundant values and outliers. All these significantly affect the accuracy of the process. For this purpose, we need to apply different cleansing techniques as per the requirement to ensure the accuracy of the input data.

Valuable Insights   The process should generate valuable insight from the data. The insight generated by the business intelligence process should be aligned with the requirements of the business to help it make effective future policies, e.g., for a medical store owner, the information about the customer's medical condition is more valuable than a grocery store.

Timeliness   Timeliness is another important consideration of the business intelligence process. Generating valuable insight is an important component, but the insight should be generated at the right time. For example, for the medical store example discussed above if the system does not generate, the ratio of the people that may be affected by pollen allergy in the upcoming spring. The store may fail to get the full benefit of the process. So, generating the right insight at the right time is essential. It should be noted that here timeliness refers to both the timeliness of the availability of the input data and the timeliness of the insight generated.

Actionable   The insight provided by the business intelligence process should always consider the organizational context to provide effective insight that can be implemented, e.g., although the process may provide a maximum amount of the pollen allergy-related medicines that the medical store should purchase at hand, the budget of the medical store and other constraints like possible sales limit, etc. should also be considered for effective decision-making. Now let's have a brief look at the business intelligence process.

Fig. 1.4 Phases of a BI process

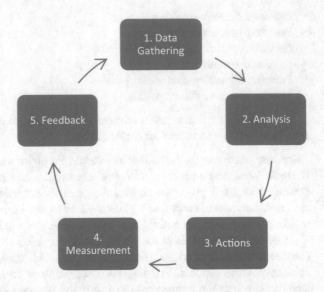

BI Process It has four broad steps that loop over and over. Figure 1.4 shows the process:

1. *Data Gathering:* Refers to collecting the data and cleansing it to convert it into a format suitable for BI processing.
2. *Analysis:* Refers to the processing of data to get insight from it.
3. *Action:* Refers to action taken under the information analyzed.
4. *Measurement:* Results of actions are measured with respect to the required output.
5. *Feedback:* Measurements are then used to improve the BI process.

1.9   Data Analytics vs Data Analysis

Below are the lists of points that describe the key differences between data analytics and data analysis:

- Data analytics is a general term that refers to the process of making decisions from data whereas data analysis is a sub-component of data analytics which tends to analyze the data and get the required insight.
- Data analytics refers to data collection and general analysis, whereas data analysis refers to collecting, cleaning, and transforming the data to get deep insight out of it.
- Tools required for data analytics may include Python, R and Tensorflow, etc., whereas the tools required for data analysis may include RapidMiner, KNIME, Google fusion tables, etc.

- Data analysis deals with examining, transforming, and arranging a given data to extract useful information, whereas data analytics deals with complete management of data including collection, organization, and storage.

Figure 1.5 shows the relationship between data analytics and data analysis:

1.10   Data Analytics vs Data Visualization

In this section, we will discuss the difference between data analytics and data visualization.

- Data analytics deals with tools, techniques, and methods to derive deep insight from the data by finding out the relationships in it, whereas data visualization deals with presenting the data in a format (mostly graphical) that is easy to understand.
- Data visualization helps organization management to visually perceive the analytics and concepts present in the data.
- Data analytics is a process that can help organizations increase operational performance, make policies, and take decisions that may provide advantages over the business competitors.
- Descriptive analytics may, for example, help organizations find out what has happened and find out its root causes.
- Prescriptive analytics may help organizations to find out the available prospects and opportunities and consequently make decisions in favor of the business
- Predictive analytics may help organizations to predict future scenarios by looking into the current data and analyzing it.
- Visualizations can be both static and interactive. Static visualizations normally provide a single view, which the current visualization is intended for. Normally users cannot see beyond the lines and figures.
- Interactive visualizations, as the name suggests, can help the user interact with the visualization and get the visualization as per their specified criteria and requirements.
- Data visualization techniques like charts, graphs, and other figures may help see the trends and relationships in the data much more easily. We can say that it is part of the output of the analytics process. For example, a bar graph can be easily understood by a person to understand the sales per month instead of the numbers and text.

So in general, data analytics performs the analytics-related tasks and derives the information which is then presented to the user in the form of visualizations.

Figure 1.6 shows the relationship between data analytics and data visualizations.

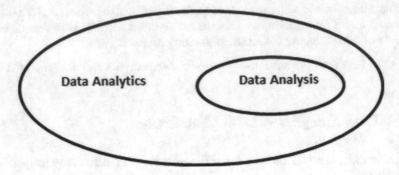

Fig. 1.5 Data analytics vs data analysis

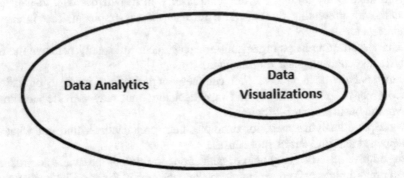

Fig. 1.6 Data analytics vs data visualizations

1.11   Data Analyst vs Data Scientist

Both are prominent jobs in the market these days. A data scientist is someone who can predict the future based on the data and relationships in it, whereas a data analyst is someone who tries to find some meaningful insights from the data. Let's look into both of them.

- Data analyst deals with the analysis of data for report generation, whereas data scientist has a research-oriented job responsible for understanding the data and the relationships in it.
- Data analysts normally look into the known information from a new perspective, whereas a data scientist may involve finding the unknown from the data.
- The skillset required for a data analyst includes the concepts from statistics, mathematics, and various data representation and visualization techniques. The skillset for data scientists includes advance data science programming languages like Python, R, TensorFlow, and various libraries related to data science like NLTK, NumPy, Scipy, etc.

- A data analyst's job includes data analysis and visualization, whereas the job of a data scientist includes the skills to understand the data and find out the relationships in it for deep insight.
- Complexity wise data scientist job is more complex and technical as compared to data analyst.
- Data analyst normally deals with structured data, whereas data scientist may have to deal with both structured, unstructured, and hybrid data.

It should be noted that we cannot prioritize both of the jobs, both are essential and both have their own roles and responsibilities, and both are essential for an organization to help grow the business based on its needs and requirements.

1.12   Data Analytics vs Business Intelligence

Now we will explain the difference between data science and some other domains.
Apparently both the terms seem to be a synonym. However, there are many differences which are discussed below.

- Business intelligence refers to a generic process that is useful for decision-making out of the historical information in any business, whereas data analytics is the process of finding the relationships between data to get deep insight.
- The main focus of business intelligence is to help in decision-making for further growth of the business, whereas data analytics deals with gathering, cleaning, modeling, and using data as per the business needs of the organization.
- The key difference between data analytics and business intelligence is that business intelligence deals with historical data to help organizations make intelligence decisions, whereas the data analytics process tends to find out the relationships between the data.
- Business intelligence tries to look into the past and tends to answer the questions like "What happened?," "When it happened?," "How many times?," etc. Data analytics on the other hand tries to look into the future and tends to answer the questions like "When will it happen again?," "What will be the consequences?," "How much sales will increase if we do this action?," etc.
- Business intelligence deals with the tools and techniques like reporting, dashboards, scorecards and ad hoc queries, etc. Data analytics on the other hand deals with the tools and techniques like text mining, data mining, multivariate analysis, big data analytics, etc.

In short business intelligence is the process that helps organizations make intelligence decisions out of the historical data normally stored in data warehouses and organizational data repositories, whereas data analytics deals with finding the relationships between data for deep insight.

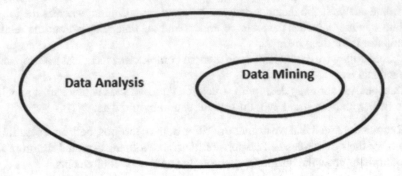

Fig. 1.7 Data analysis vs data mining

1.13 Data Analysis vs Data Mining

Data mining and data analytics are two different processes and terms having their own scope and flow. Here we will present some differences between the two.

- Data mining is the process of finding the existing patterns in data, whereas data analysis tends to analyze the data and get the required insight.
- It may require a skillset like mathematics, statistics, machine learning, etc., whereas the data analysis process involves skillset like statistics mathematics, machine learning, subject knowledge, etc.
- A data mining person is responsible for mining patterns into the data, whereas a data analyst performs data collection, cleaning, and transforming the data to get deep insight out of it.

Figure 1.7 shows the relationship between data mining and data analysis.

1.14 What Is ETL?

In data warehouses, data comes from various sources. These sources may be homogeneous or heterogeneous. Homogeneous sources may contain the same data semantics, whereas heterogeneous sources are the ones where data semantics and schemas are different. Furthermore, the data from different sources may contain anomalies like missing values, redundant values, and outliers. However, the data warehouse should contain homogeneous and accurate data to provide this data for further processing. The main process that enables data to be stored in the data warehouse is called the Extract, Transform, and Load (ETL) process.

ETL is the process of collecting data from various homogeneous and heterogeneous sources and then applying the transformation process to prepare the data for storing in the data warehouse.

It should be noted that the data in different sources may not be able to store in the data warehouse primarily due to different data formats and semantics. Here we will present some examples of anomalies that may be present in source data that required a complex ETL process.

Different Data Formats Consider two databases, both of which store the customer's age. Now there is a possibility that one database may store the date in the format "mm/dd/yyyy," while the other database may have some different format like "yyyy/mm/dd," "d/m/yy," etc. Due to different data formats, it may not be possible for us to integrate their data without involving the ETL process.

Different Data Semantics In previous cases, data formats were different. It may be the case that data formats are the same but data semantics are different. For example, consider two shopping cart databases where the currencies are represented in the form of floating numbers. Note that although the format of the data is the same, semantics may be different, e.g., the floating currency value in one database may represent the currency in dollars, whereas the same value in another database may represent Euros, etc., so, again we need the ETL process for converting the data to a homogeneous format.

Missing Values Databases may have missing values as well. It may be due to certain reasons, e.g., normally people are not willing to show their salaries or personal contact numbers. Similarly, gender and date of birth are among some measures which people may not be willing to provide while filling out different online forms. All this results in missing values which ultimately affect the quality of the output process performed on such data.

Incorrect Values Similarly, we may have incorrect values, e.g., some outliers perhaps resulted due to a theft of a credit card or due to malfunctioning of a weather data collection sensor. Again, such values affect the quality of analysis performed on this data. So, before storing such data for analysis purposes, the ETL process is performed to fix such issues.

Some benefits of the ETL process may include:

- You can compare the data from different sources.
- You can perform complex transformations; however, it should be noted that the ETL process may require separate data stores for intermediate data storage.
- It helps the migration of data into single homogeneous storage suitable for analysis purposes.
- Improves productivity as the transformation process is defined once which is applied to all the data automatically.
- If data in the source repository is changed, the ETL process will automatically be performed to transform data for the target data warehouse.

Figure 1.8 shows the ETL process.

ETL process comprises three steps. We will explain all these steps one by one now.

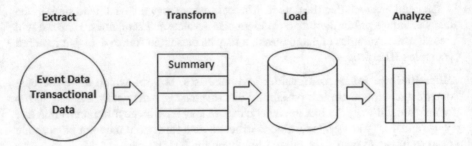

Fig. 1.8 Steps of the ETL process

1.14.1   Extraction

Data extraction is the first step of the ETL process. In this step, data is read from the source database and stored in intermediate storage. The transformation is then performed. Note that the transformation is performed on other systems so that the source database and the system using it are not affected. Once all the transformations are performed, the data becomes ready for the next stage; however, after completing the transformation, it is essential to validate the extracted data before storing it in the data warehouse. The data transformation process is performed when the data is from different DBMSs, hardware, operating systems, and communication protocols. The source of data may be any relational database, conventional file system repositories, spreadsheets, document files, CSVs, etc.

So, it becomes evident that we need schemas of all the source data as well as the schema of the target system before performing the ETL process. This will help us identify the nature of the required transformation.

- *Three data extraction methods*

Following are the three extraction methods. However, it should be noted that irrespective of the transformation method used, the performance and working of source and target databases should not be affected. A critical aspect in this regard is when to perform the extraction as it may require the source database of the company unavailable to the customers or in a simple case and may affect the performance of the system. So, keeping all this in mind, the following are the three types of extractions performed.

1. Full extraction
2. Partial extraction – without update notification
3. Partial extraction – with an update notification.

1.14.2   Transformation

Data extracted from source systems may be in raw format that is not useful for the target BI process. The primary reason is that data schemas are designed according to local organizational systems and due to various anomalies present in the systems. Hence, before moving this data to the target system for the BI process, we need to transform it. Now, this is the core step of the entire ETL process where the value is added to data for BI and analysis purposes. We perform different functions to transform the data; however, sometimes transformation may not be required such data is called direct move or pass through data.

The transformation functions performed in this stage are defined according to requirements, e.g., we may require the monthly gross sale of each store by the end of the month. The source database may only contain the individual timestamped transactions on daily basis. So, here we may have two options, we may simply pass the data to the target data warehouse and the calculation of monthly gross sales can be performed at runtime whenever required. The other option is to perform the aggregation and store the monthly aggregated data in a data warehouse. The latter will provide high performance as compared to calculating the monthly sale at runtime. Now we will provide some examples of why transformation may be required.

1. The same person may have different name spelling, e.g., Jon or John.
2. A company can be represented in different ways, e.g., HBL or HBL Inc.
3. Use of different names like Cleaveland and Cleveland.
4. The same person may have different account numbers generated by different applications.
5. Data may have different semantics.
6. The same name, e.g., "Khursheed" can be of a male or female at the same time.
7. Fields may be missing.
8. We may use derived attributes in the target system, e.g., "Age" which is not present in the source system. So, we can apply the expression "Current date minus DOB" and the resulted value can be stored in a target data warehouse (again to enhance the performance).

There are two types of transformations:

Multistage data transformation – This is a conventional method where data is extracted from the source and stored in intermediate storage, transformation is performed, and data is moved to the data warehouse.

In-warehouse data transformation – This is a slight modification from the conventional method. Data is extracted from the source and moved into the data warehouse, and all the transformations are performed there. Formally it may be called ETL, i.e., Extract, Transform, and Load process.

Each method has its own merits and demerits, and the selection of any of these may depend upon the requirements.

1.14.3   Loading

It is the last step of the ETL process. In this step, the transformed data is loaded into the data warehouse. Normally it requires a huge amount of data loading, so the process should be optimized and performance should not be degraded.

However, if due to some reason, the process results in a failure, then the measures are taken to restart the loading process from the last checkpoint and the failure should not affect the integrity of data. So, the entire loading process should be monitored to ensure its success. Based on its nature, the loading process can be categorized into two types:

Full Load   All the data from the source is loaded into the data warehouse for the first time. However, it takes more time.

Incremental Load   Data is loaded into a data warehouse in increments. The check-points are recorded. A checkpoint represents the timestamp from which onward data will be stored in the data warehouse.

The full load takes more time but the process is relatively less complex as compared to the incremental load which takes less time but is relatively complex.

ETL Challenges
Apparently, the ETL process seems to be an interesting and simple tool-oriented approach where a tool is configured and the process starts automatically. However, certain aspects are challenging and need substantial consideration. Here are a few:

- Quality of data
- Unavailability of accurate schemas
- Complex dependencies between data
- Unavailability of technical persons
- ETL tool cost
- Cost of the storage
- Schedule of the load process
- Maintain the integrity of the data

1.15   Data Science

Data Science is a multidisciplinary field that focuses on the study of all aspects of data right from its generation to processing to convert it into a valuable knowledge source.

As said, it is a multidisciplinary field; it uses concepts from mathematics, statistics, machine learning, data mining and artificial intelligence, etc. having a wide range of applications; data science has become a buzzword now. With the realization of the worth of insight from data, all organizations and businesses are

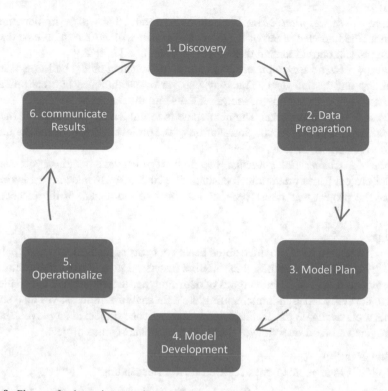

Fig. 1.9 Phases of a data science project

striving for the best data science and analytics techniques for the good of the business.

Bigger companies like Google, Amazon, Yahoo, etc. have shown that carefully storing data and then using it for extraction of the knowledge to make decisions and adopt new policies is always worth it. So, small companies are also striving for such use of data which has ultimately increased the demand for data science techniques and skills in the market. So, efforts are also underway to decrease the cost of providing data science tools and techniques.

Now we will explain different phases of the life cycle of a data science project. Figure 1.9 shows the pictorial form of the process.

- *Phase 1 – Discovery:* The first phase of the data science project is discovery. It is the discovery of the available sources that you have, that you need; your requirements; your output, i.e., what you want out of the project; its feasibility; the required infrastructure; etc.
- *Phase 2 – Data preparation:* In this phase, you explore your data and its worth. You may need to perform some preprocessing including the ETL process.

- *Phase 3 – Model plan:* Now think about the model that will be implemented to extract the knowledge out of the data. The model will work as a base to find out patterns and correlations in data as per your required output.
- *Phase 4 – Model Development:* Next phase is the actual model building based on training and testing data. You may use various techniques like classification, clustering, etc. based on your requirements and the nature of the data.
- *Phase 5 – Operationalize:* The next stage is to implement the project. This may include delivery of the code, installation of the project, delivery of the documents and give demonstrations, etc.
- *Phase 6 – Communicate results:* The final stage is evaluation, i.e., you evaluate your project based on various measures like customer satisfaction, achieving the goal the project was developed far, accuracy of the results of the project, and so on.

Summary

In this chapter, we have provided some basic concepts of data science and analytics. Starting from the basic definition of data, we discussed several of its representations. We provided details of different types of data analytics techniques and the challenges that are normally faced in implementing the data analytics process. We also studied various tools available for developing any data science project at any level. Finally, we provided a broad overview of the phases of a data science project.

Further Reading

The followings are some books referred for further reading:

- "Big Data in Practice: How 45 Successful Companies Used Big Data Analytics to Deliver Extraordinary Results" by Bernard Marr, Wiley, 2016. The book covers the domain of big data with the help of exciting examples. This book explains how the business community is exploiting big data with existing and new technology. This book precisely explains the details and provides detailed insight into how organizations can use big data to refine different areas of their organizations. Big data is used in almost all domains of life these days. Everybody is trying to understand its importance and power with to change the working, thinking, and planning environment in the organizations. Technology giants are taking a great part to develop such general or specific solutions that attract organizations to buy and utilize big data technology as no one has utilized before. Retail businesses, sports franchises, financial institutions, government agencies, and media are developing their in-house departments to use this technology for the growth of their business. Learning outcomes of this book are how predictive analytics helped Amazon, Apple, and John Deere to understand their valuable consumers. It explains the methods to discover patterns that are behind the success of Walmart, Microsoft, and LinkedIn. It describes directions to develop your big data strategy with additional material in each chapter.
- "Big Data at Work: Dispelling the Myths, Uncovering the Opportunities" by Thomas H., Harvard Business Press, 2014. The book will help you understand the importance of big data to you and your business, the kind of technology you need

to manage it, how big data can change your job, and how big data is leading to a new approach. This book explains the meaning of big data and the reasons to use big data in every business these days. The reader gets to know the meaning of big data as a customer, technical person, and business person. The book highlights the cost and opportunities available in big data that impact real-life tasks. The book helps to understand the importance of big data, the need for technology to manage it, and reasons to adopt it as part of the organization. As a scientist, businessman, or academician, the book is a very good resource for understanding the domain of big data and the related concepts.

- "Weapons of Math Destruction: How Big Data Increases Inequality and Threatens Democracy" by Cathy O'Neil, Crown, 2016. The book highlights the facts that how algorithms are increasingly affecting the decision-making process in different fields of life, for example, the selection of schools, car loans, and insurance management. Overall the book is a valuable resource regarding big data.
- "How to Measure Anything: Finding the Value of Intangibles in Business" 3rd Edition by Douglas W. Hubbard, Wily, 2014. This book will show you the measurement methods of different things in business and government organizations. It discussed customer satisfaction, technology risk, and organizational flexibility, and how to measure all of these.
- "Predictive Analytics: The Power to Predict Who Will Click, Buy, Lie, or Die" Revised and Updated Edition by Eric Siegel, Wily, 2013. It is about how predictive analysis affects everyday life. The book fascinates beginner and expert readers by covering trending case studies.
- "How to Measure Anything Workbook: Finding the Value of Intangibles in Business" 1st Edition by Douglas W. Hubbard, Wiley, 2014. This book describes the practical approach to measuring the different intangibles. It explains how to measure customer satisfaction, technology risk, and organizational flexibility.
- "Applied Predictive Analytics: Principles and Techniques for the Professional Data Analyst" 1st Edition by Dean Abbott, Wiley, 2004. This book describes predictive analytics that translates big data into meaningful and usable business information. The book is a good resource for practitioners.

Exercises

Exercise 1.1: Extensive use of mobile phones during driving caused a high death rate in the USA. The government of the county then decided to pass a law against the use of mobile phones while driving. After implementing the law, the government collected data on the death count each year. As a data scientist, a task given to you is to analyze the data and find the trend (of increase or decrease) in death count using a line graph to graphically plot the collected data. Collected data is given below in the table.

Year	Death count
2010	311
2011	276
2012	258
2013	240
2014	199
2015	205
2016	197
2017	194

Exercise 1.2: An experiment has been performed on the soil of the USA to check the growth of different trees in a specific environment. Different types of trees are planted in the new soil to check their growth. The growth of each tree is shown in the table given below. You are required to use the data given in the table and draw a bar chart to show the height comparison of all trees.

Tree name	Height (m)
Coast redwood	116.92
Sitka spruce	101.2
Yellow meranti	99.53
Southern blue gum	91.7
Manna gum	89.9
Brown top stringybark	89.5
Mengaris	86.76
Shoreaargentifolia	85.85
Shining gum	85.3
Western hemlock	84.34

Exercise 1.3: A University uses different graded activities to evaluate their students. They assign different percentages to these activities including assignments, quizzes, practical labs, midterm exams, final term exams, and graded discussion boards. As a data scientist, a task is assigned you to use the information given below in the table and draw a pie chart to show the proportion of each activity.

Activity	Marks (%)
Assignments	10
Quizzes	5
Graded discussion board	5
Practical labs	5
Midterm exams	30
Final term exams	40

Exercise 1.4: Different types of analytics help at certain stages for better decision-making in business and other organizations. One thing common in all types of analytics is the availability of the data for applying the analytics process. The better the quality of data, the better the decisions and results. You are required to mention differences in different types of data analytics and give one practical example of each of them.

Exercise 1.5: List down a few tools (at least 10) commonly used in data analytics, and mention details of any five of them.

Exercise 1.6: Consider the steps of the business intelligence process given below. The sequence of steps is shuffled. You are required to analyze the process and then correct the sequence of steps.

Sequence	Steps
1	Actions
2	Measurement
3	Data gathering
4	Feedback
5	Analysis

Exercise 1.7: What are the key important differences in data analytics and data analysis? You are required to mention only four differences.

Exercise 1.8: Describe the relationship between data analytics and data analysis. You are required to show the relationship with the help of a diagram.

Exercise 1.9: What are the key important differences in data analytics and data visualization? You are required to mention only six differences.

Exercise 1.10: Describe the relationship between data analytics and data visualization. You are required to show the relationship with the help of a diagram.

Exercise 1.11: What are the key important differences in the jobs of data analyst and data scientist? You are required to mention only five differences.

Exercise 1.12: What are the key important differences in data analysis and business intelligence? You are required to mention only five differences.

Exercise 1.13: What are the key important differences between data analysis and data mining? You are required to mention only five differences.

Exercise 1.14: Describe the relationship between data analysis and data mining. You are required to show the relationship with the help of a diagram.

Exercise 1.15: Data is collected from different sources and in different formats. Hence, there is the possibility that collected data can be dirty. So, to clean this dirty data, the Extract, Transform, and Load (ETL) method is used. You are required to provide the details of this method and how it works.

Exercise 1.16: Describe any four issues that may be there in the collected data and that should be removed in the process of data cleaning.

Exercise 1.17: Graphically show the process of ETL.

Exercise 1.18: How many data extraction methods do we have? You are required to provide the detail of every method.

Exercise 1.19: Before analysis, the transformed data is loaded into a data warehouse. We have different types of data loading methods. You have to provide the detail of every data loading method that is used in an ETL process.

Exercise 1.20: Consider the phases of the data science project given below. The sequence of phases is shuffled. You are required to analyze the phases and then correct the sequence.

Sequence	Phases
1	Model development
2	Operationalize
3	Communicate results
4	Discovery
5	Data preparation
6	Model plan

Chapter 2
Applications of Data Science

2.1 Data Science Applications in Healthcare

Data Science has played a vital role in healthcare to predict the patterns that help in curing contagious, long-lasting diseases using fewer resources.

Extracting meaningful information and pattern from the data collected through the patient's history stored in hospital, clinics, and surgeries help the doctors to refine their decision-making and advise the best medical treatment that helps the nations to live longer than before. These patterns provide all the data well before the time enabling the insurance companies to offer packages suitable for patients.

Emerging technologies are providing mature directions in healthcare to dig deep insight and reveal more accurate critical medical schemes. These schemes are taking treatment to next level where a patient has also sufficient knowledge about the existing and upcoming diseases and it makes it easy for doctors to guide patients in a specific way. Nations are getting advantages of data science applications to anticipate and monitor the medical facilities to prevent and solve healthcare issues before it's too late.

Due to the rising costs of medical treatments, the use of data science becomes necessary. Data science applications in healthcare help reduce the cost by various means, for example, providing the information at earlier stages helps the patient avoid expensive medical treatments and medications.

A rise in costs has become a serious problem for healthcare companies over the last 20 years. Healthcare companies are now rethinking optimizing the treatment procedures for patients by using data science applications. Similarly, insurance companies are trying to cut down the cost and provide customized plans as per patient status. All this is guided by data-driven decision-making where insights are taken from the data to make plans and policies.

With this, healthcare companies are getting true benefits by using the power of analytical tools, software as service (SaaS), and business intelligence to extract such patterns and devise schemes that help all stakeholders in reducing the costs and

U. Qamar, M. S. Raza, *Data Science Concepts and Techniques with Applications*, https://doi.org/10.1007/978-3-031-17442-1_2

increasing the benefits. Now doctors not only have their educational knowledge and experience, but they also have access to many verified treatment models of other experts with the same specialization and in the same domain.

1. *Patients Predictions for an Improved Staffing*

Overstaffing and understaffing is a typical problem faced by many hospitals. Overstaffing can overburden the salary and wage costs, while less staffing means that the hospital is compromising on the quality of care that is too dangerous because of sensitive treatments and procedures. Data science applications make it possible to analyze the admission records and patient visiting patterns with respect to weather, days, month, time, and location and provide meaningful insights to place the staff that helps the staffing manager to optimize the staffing placement in line with patient visits.

Forbes report shows how hospitals and clinics, by using the patient's data and history, can predict the patient's future visits and can accordingly place the staff as per the expected number of patient visits. This will not only result in optimized staff placement but will also help patients by reducing the waiting time. Patients will have immediate access to their doctor. The same can be especially helpful in case of emergencies where doctors will already be available.

2. *Electronic Health Records (EHRs)*

Using data science applications in EHR, data can be archived according to patient demographic, lab tests, allergies, and complete medical history in an information system, and doctors can be provided with this data through some security protocols which ultimately can help the doctors in diagnosing and treating their patients with more accuracy. This use of data science applications is called Electronic Health Records (EHRs) where patient data is shared through secured systems with the doctors and physicians.

Most developed nations like the USA have already implemented it; European countries are on the way to implement it while other others are in the process to develop the rules and policies for implementation. Although EHR is an interesting application of data science, it requires special considerations concerning security because EHR involves the patient's data that should be handled with care. This is also one of the reasons that various hospitals and clinics even in developed countries are still reluctant to participate in EHR. However, with secure systems and keeping in mind the benefits of EHR to hospitals, clinics, and patients, it is not far away that EHR will be implemented in a majority of the countries across the globe.

3. *Real-Time Alerting*

Real-time alerting is one of the core benefits of data science applications in healthcare. Using data science for such applications, the patient's data is gathered in real time and analyzed, and the medical staff is informed about the patient's conditions which can ultimately take decisions well before time. Patients can use GPS-guided wearables that report to the patients and doctors about the patient's medical state, for example, blood pressure and heart rate of the patient. So, if the patient's blood pressure changes abruptly or reaches any dangerous level, then the doctor can contact the patient and advise him on the medication accordingly.

Similarly, for patients with asthma, this system records the asthma trends by using GPS-guided inhalers. This data is being used for further research at the clinical level and at the national level to make a general policy for asthma patients.

4. *Enhancing Patient Engagement*

With the availability of the more advance wearables, and convincing the patients about their benefits, we can develop the patient's interest for use of these devices. With these wearables, we can track all the changes to the human body and give feedback with initial treatment to reduce the risk by avoiding critical situations regarding the patient's health.

Insurance companies are also advertising and guiding their clients to use these wearables, and many companies are providing these wearables for free to promote these trackable devices. These wearables are proving a great source in reducing patient visits and lab tests. These devices are gaining huge attention in the health market and giving fruitful results. It is also engaging the researchers to come up with more and more features to be added to these devices with the passage of time.

5. *Prevent Opioid Abuse*

Drug addiction is an extreme problem in many countries even in the developed nations where billions of dollars have already been spent and still programs are underway to devise solutions for it.

The problem is getting worst and worst. Now the research is already in progress to come up with the solutions. Many risk factors have already been identified to predict the patients at risk of opioid abuse with high accuracy.

Although the problem seems to be challenging to identify and reach such persons and convince them to avoid drug issues, we can hope that success can be obtained with little more effort both by doctors and public.

6. *Using Health Data for Informed Strategic Planning*

With the help of data science, health organizations create strategic plans for patients to provide better treatments and reduce costs. These plans help these organizations to examine the latest situation in a particular area or region, for example, the current situation of existing chronic disease in a certain region.

Using advance data science tools and techniques, we can come up with proactive approaches for treating emergencies and critical situations. For example, by analyzing the data of a certain region, we can predict the coming heat waves or dengue virus attacks and can establish clinics and temporary facilities beforehand ultimately avoiding critical situations in the region.

7. *Disease Cure*

Data science applications provide great opportunities to treat diseases like cancer thus giving relief to cancer patients. We can predict the disease and provide directions to cure and treat the different patients at different stages. However, all this requires cooperation at various levels. For example, individuals should be convinced to participate in the information-providing process, so that if any information is required from any cancer victim, he is willing to provide that information. Furthermore, organizations that already have security policies may be reluctant to provide such data. Beyond this, there are many other issues, e.g., technical compatibility of the diagnostic systems as they are developed locally without keeping in mind their integration with other systems. Similarly, there may be legal issues in sharing such information. Furthermore, we also need to change the mindset which hinders an organization to share its success with others perhaps due to business reasons.

All this requires a significant effort to deal with such issues to come up with solutions. However, once all such data is available and interlinked, a researcher can come up with models and cures to help cancer victims with more effectiveness and accuracy.

8. *Reduce Fraud and Enhance Security*

Data breaching is a common hack due to the less security of the data because this data has great value in terms of money and competitive advantages and can be used for unfair means. A data breach can occur due to many reasons, e.g., viruses, cyberattacks, penetration in company's networks through friendly pretending nodes, etc. However, it should be noted that as the systems are becoming more and more victims of possible cyberattacks and data hacks, the security solutions are also getting mature day by day with the use of encryption methodologies, antiviruses, firewalls, and other advance technologies.

By use of data science applications, organizations may predict any possible attacks on their organizational data. Similarly, by analyzing the patient's data, the possible frauds in patients' insurance claims can also be identified.

9. *Telemedicine*

Telemedicine is the process of contacting the doctor and physician using advance technologies without involving personal physical visits. Telemedicine is one of the most important applications of data science. This helps in the delivery of the health services in remote areas and is a special help for developing countries where health facilities are not available in rural areas.

Patients can contact their doctor through videoconferencing or smart mobile devices or any other available services.

The abovementioned use is just a trivial example. Now doctors can perform surgeries through robots even sitting far ways from the real surgery room. People are getting the best and instant treatment that makes them more comfortable and less costly than arranging visits and standing in long lines. Telemedicine is helping hospitals to reduce costs and manage other critical patients with more care and quality and placement of the staff as per requirements. It also allows the healthcare industry to predict which diseases and levels of diseases may treat remotely so that personal visits could be reduced as much as possible.

10. *Medical Imaging*

Medical images are becoming important in diagnosing disease which requires high skills and experience if performed manually. Moreover, hospitals required a huge budget to store these images for a long time as they may be needed anytime in the future for the treatment of the patient whose image is there. Data science tools make it possible to store these images in an optimized way to take less storage as well as these algorithms generate patterns using pixels and convert these pixels into numbers that help the medical assistants and doctors to analysis one particular image and compare it with other images to perform the diagnosis procedure.

Radiologists are the people who are most relevant to generate these images. As their understandability may change due to the human nature of mood swings and many other reasons, the accuracy of the extracted information may be affected. Computers, on the other hand, don't get tired and behave cleanly to predict the true sense of images, alternatively resulting in the extraction of quality information. Furthermore, the process will be more efficient and time-saving.

11. *A Way to Prevent Unnecessary ER Visits*

Better and optimized use of resources like money, staff, and energy is also a main consequence of data science applications and implementation of these systems makes it possible. There was a case where a woman having mental illness visited hospitals 900 times in a year. This is just one case of that type that has been reported, but there can be many similar cases that cause the overburden on healthcare institutions and other taxpayers.

The healthcare system can easily tackle this problem by sharing the patient's information in emergency departments and clinics of hospitals. So, hospital staff can check patient's visits in other hospitals and lab test dates and timings to suggest or not suggest a retest of the patient based on his previous recently conducted test. This way hospitals may save their time and all resources.

ER system may help the staff in the following ways:

- It may help in finding medical history and patients visits to nearby hospitals.
- Similarly, we can find if the patient has already been assigned to a specialist in a nearby hospital.
- What medical advice has been given to the patient previously or currently by another hospital?

This is another benefit of the healthcare system that it helps to utilize all resources in a better way for the care of the patients and taxpayers. Before these systems, a patient could get a medical checkups again and again and causing a great burden on all stakeholders.

2.2   Data Science Applications in Education

1. *To Help Individual Learners*

Learning applications are providing an excellent way to examine a learner's performance and provide them with value able suggestions. Using run-time analysis, these applications provide the students with their strengths and weaknesses, so that students may overcome their deficiencies and may well go through with other classmates. This may be an example of a proactive approach where students are informed about the gray areas where they need to improve. Using the student progress reports and the contents where students were weak, overall, analytic applications can be a great help for students to improve their grades.

 2. *To Help Mentors*

   Not only the students but data science applications can be great assistance to teachers and faculty members to identify the areas where they need to give more focus in the class. So, by identifying the patterns, mentors can help a particular student or group of students, for example, a question was attempted by a few students out of 1000 students which means that there can be any issue with the question so the teacher can modify or change or may re-explain the contents of the questions perhaps in more details.

 3. *To Help Developing Curriculum and Learning Processes*

   Course development and learning process is not a static procedure. Both require continued effort to develop or enhance the courses and learning process. Educational organizations can get insight information to develop the new courses as per competency and levels of students and enhancement in the learning procedures. After examining the students learning patterns, institutions will be in a better position to develop a course that will give great confidence to learners to get their desired skills as per their interests.

 4. *To Help Administrators*

   Institution's administration can devise policies with the help of data science applications that which course should be promoted more, and which areas are need of the time. What marketing or resource optimizing policies they should implement? Which students may get more benefit from their courses and offered programs?

 5. *Prevent Dropouts and Learn from the Results*

   Normally admission to an institution is simple and easy as compared to sustaining competition. This is especially true for high-ranking institutions. With data science applications, institutions can find out the root causes of the increase in student dropouts and alternatively can take measures and devise policies to avoid such cases. They can help students in choosing the right course as per their education level and guide them about the industry situation regarding particular degree programs.

6. *A Better Informed Learning Future*

Data science applications can help the educational institutes in finding out the current and future trends in the market and industry, which alternatively can help them decide the future of their students. They can study the industry requirements in the near future and help out the student in selecting the best study programs. Even industry may share their future trends to help in launching a particular study program. This will ultimately give the students great confidence and risk-free thoughts regarding their job security in the future.

Industries can work on recruitment strategies and share these with the institutes to develop new courses according to the industry needs. Also, internship opportunities can be provided by the industry to the students by assessing their learning skills.

7. *It Helps You Find Answers to Hard Questions*

Sometimes, it seems difficult to identify the causes of the student-related problems, for example, their assessment, admission policy, development of new courses, etc., but analyzing the historical data and mining history can help you find the solutions to these problems that can be hard to fix without applying these technologies or using legacy procedures. For example, with the help of a data science application, you can find the dropout reasons for a particular department or the overall institution. There can be various other issues you can resolve, e.g.:

- Why is there low registration in a certain program?
- Why are students getting poor grades in a certain course?
- How has the grade relaxation policy helped in increasing the student confidence?
- Which deficiency courses are required for a certain group of students?
- Does a particular student need extra class hours to overcome poor grades?

8. *It is Accessible*

Technological infrastructure provides an environment in which all departments are interlinked and can collaborate excellently. Before this advancement, searching any information was a cumbersome task, that perhaps had required verification from different departments, consequently consuming much energy, cost, and time. But now with data science tools, interlinking of department makes it possible to extract required information in seconds and devise a better policy according to the need of the current situation in the market.

As data is available in centralized repositories, creating simple browser base extensions may ease the process of using the tools across the institution even if the campuses are geographically located at different places.

9. *It Can Save Costs*

As there are many departments interlinked and generating lots of data, so, analyzing this data helps the institution at to develop the strategic plans to place the required staff at right place. So, data science applications can help you optimize the resource allocation, hiring process, transportation systems, etc. You can be in better position to develop the infrastructure according to the need to existing and upcoming students with minimum cost.

Similarly, you can develop new admission policy according to the existing available resources; mentor the staff and polishing their skills to market the good will in the community.

So, with the help of data science applications, your institution helps you get deep insight from the information and alternatively perform best utilization of your resource by getting the maximum benefits with minimum costs.

10. *Its Quick*

There is no manual work, all departments generate their own bulk data, and this data is stored at centralized repositories. So, should you need any information, it is already available at hand without and delay. This ultimately means that you have enough time to take actions against any event.

So, with data science tools, you have central repositories that let you take timely decisions and make accurate action plans for the better of all stakeholders of that institution.

2.3   Data Science Applications in Manufacturing and Production

Manufacturing industries are somewhat lacking in adopting technology advancement because these firms gave it less priority and were taking technology as extra burden, consequently were reluctant to apply all this technology. However, with growing population and increase in manufacturing goods resulted in a large amount of data. Added with low lower cost of technological equipment, now this is going to be possible for industries to adopt intelligent tools and apply in manufacturing process with minimum cost.

With this, data science applications are now helping industries to optimize their processes and quality of the products thus resulting in increased satisfaction of customers and the industry itself. This advancement is playing vital roles in expanding the business, and getting maximum benefit with minimum cost is great attraction. Without data science tools, there were various factors adding to the cost like large number of required labors, no prediction mechanism, and resource optimization policies.

1. *Predictive Maintenance*

Use of intelligent sensors in industry provides great benefits by predicting the potential faults and errors in the machinery. Prompt response to fix these minor or major issues saves the industry owners to reinstall or buy new machinery that required huge financial expenditures. However, the prediction analysis of manufacturing process can enhance the production with high accuracy and alternatively giving confidence to buyers to predict their sales and take corrective actions if any is required. With the further advancement in data science tools and techniques, business industry is gaining maximum day by day now.

2. *Performance Analyses*

Performance analysis enables the manufacturing industries to forecast the performance before the production process and enables them to fulfill the market needs. This helps them find out the discrepancies in the production process and fix them so that the output is expected to be same as is required.

Without data science applications, few years back, it was not possible to predict the market trends; however, the availability of data science applications the industrial organizations are now in better position to tune up their performance according to the required levels.

3. *Decrease in Downtime*

Any fault in any part of manufacturing equipment may cause a great loss and change all the delivery dates and production levels. But inclusion of technology in all parts of industry, e.g., biometric attendance recording, handling, and fixing of errors with the help of robotics and smart fixing tools, fault predictions, etc., is giving the excellent benefits in reducing the downtime of manufacturing units, alternatively, giving good confidence to utilize these technologies in manufacturing process.

4. *Improved Strategic Decision-Making*

Data science helps businesses in making the strategic decisions based on ground realities and organizational contexts. Various tools are available including data cleanup tools, profiling tools, data mining tools, data mapping tools, data analysis platforms, data visualization resources, data monitoring solutions, and many more. All these tools make it possible to extract deep insight from the available information and making the best possible decisions.

5. *Asset Optimization*

With the emergence of Internet of Things (IoTs) and data science applications, businesses can now enhance the production process by automating all the tasks thus resulting in optimized use of their assets. With data science tools, it can now be determined about when to use a certain resource and up to what level will be more beneficial.

6. *Product Design*

Launching a new product design is always a risky job. So, without thorough analysis, you cannot purchase a new unit for a new product as it may involve huge financial budget and equal failure risks. So, you can avoid the cost of purchasing equipment for production of a product that you are not sure about its success or failure.

However, with help of data science applications, by looking into the available information, you can definitely find out whether to launch a new design or not. Similarly, you can find out the success or failure ratio of a certain product that has not been launched so far. These provides a potential advantage over competitive businesses as you can perform pre-analysis and launch a much demanding product before some other launches it, thus having more opportunity to capture the certain portion of market.

7. *Product Quality*

Designing the great quality product with the help of end users always produce good financial benefits. End user provides feedback through different channel like social media, online surveys, review platforms, video streaming feedback, and other digital media tools. Finding the feedback of the users in this way always helps in finding out the much-needed products and their features. Even after launching a failed product in the market, it can be enhanced and updated in an excellent way after incorporating the user feedback and investigating customer trends in market using the data science applications.

8. *Demand Forecasting*

Business intelligent tools provide comprehensive platform to forecast the actual demands of certain products in the market. These demands may be based on weather condition, population growth, change in world economic policy, and many other reasons. However, data science applications provide accurate prediction up to maximum level, which was unfortunately not possible in past to rough manual methods.

9. *Customer Experience*

Data science tools provide deep insight into the customer data, their buying trends, customer preferences, their priorities, behaviors, etc. The data can be obtained from various sources and is then normalized to central repositories where data science tools are applied to extract the customer experiences. These experiences then help in policy- and decision-making to enhance the business. For example, the customer experience may help us maintain a minimum stock level of a certain product in a specific season. A very common example of such experience is to place together the products that are normally soled in groups, e.g., milk, butter, bread, etc.

10. *Supply Chain Optimization*

Supply chain is always a complex process. Data science applications help finding out the necessary information, for example, the suppliers with high efficiency, quality of the products they deliver and their production capability, etc.

2.4 Data Science Applications in Sports

Implementation of data science in all fields including healthcare, education or sports, etc. is proving to be the need of the time now days to unhide the patterns in data that were unable to exploit in last few years. Sports is a billion-dollar industry nowadays after starting sports competitions and leagues at multiple national and international levels. These leagues are hiring data scientist to get insight form the numbers that is required to take decision in bidding on best players, team management, and investing in right leagues. Interests have already been developed in finding out the relationships in data for predicting the performance of the players and teams.

1. *Strategic Decisions*

Data science plays a critical role in strategic decision-making when it comes to sports. Player's health monitoring through wearable and his performance prediction using the previous data is nowadays common application of data science. As each game and its complete context can be stored in databases, with this information, data science applications can help team coaches to justify out the validity of their decisions and identify the weak areas and discrepancies which resulted in losing a certain game. This can be a great help in future for overcoming these discrepancies and thus increasing the performance of players and teams.

You can not only find your own team discrepancies but you can also find the gray areas in opponent teams. This will help you come up with better strategy next time which will ultimately increase the success changes in future.

2. *Deep Insight*

Data science applications require a central repository where data from multiple sources is stored for analysis purpose. The more the data is stored, the more accurate are the results. However, the application may have their own cost, e.g., to store and process such large amount of data requires large storage volumes, memory, and processing power. But once all this is available, the insight obtained from the data by using data science tools is always a worth as it helps you predict future events and makes decisions and policies accordingly.

3. *Marketing Edge*

Nowadays, sports have become a great business. The more you are popular, the more you will attract the companies to get their ads. Data science applications provide you better opportunities now to target your ads as compared to earlier days. By finding out the fan clubs, players, and teams ratings and interests of the people, now you are in better position to devise your marketing strategies that will ultimately result in more reach of your product.

4. *On-site Assistance and Extra Services*

By collecting the information from different sources like ticket booths, shopping malls, parking areas, etc., and properly storing and analyzing it can help in providing on-site assistance to people and ultimately increasing the revenue. For example, you can provide better car parking arrangements based on a person's priority of seating in a match. Similarly, in a particular match, providing him with his favorite food at his seating place, thus making him happier by enjoying the favorite food and game at the same time, will ultimately ease him and will increase your revenue as people will also try to attend such matches in future.

2.5   Data Science Applications in Cybersecurity

Cybersecurity has become an important domain nowadays because of widespread use of computers in all walks of life. Now all such devices whether it be your personal laptop, smart device, a wearable, smart fridge, car, etc. are all interlinked. So, once you are linked with other network, you are vulnerable of cyberattacks from all the directions intended to harm you by any means. Data science is playing a wide role in the domain of cybersecurity. Here we will explain few of the contributions of data science applications.

1. *Analyzing Historical Data*

Historical data is backbone of any data science application. Same is the case in domain of cybersecurity. You will have to store large amount of previous data to find out what a normal service request is and what an attack is. Once the data is available, you can analyze it find out the malicious patterns, service requests and resource allocations, etc.

Once a pattern is verified to be an attack, you can monitor the same pattern for future and deny it in case it happens again. This can ultimately save the companies from the huge data and business losses especially in case of sensitive domains like defense-related, finance-related, or healthcare-related data.

2. *Monitoring and Automating Workflows*

All the workflows in a network need to be monitored either in real time or offline. Studies show that data breaches occur mostly due to local employees. One solution to such problem is to implement some authorization mechanism, so that only the authorized persons have access to sensitive information. However, whether the flow is generated from local systems or from some external source, data science tools can help you in automatically identifying the patterns of these flows and consequently helping you avoid the worst consequences.

3. *Deploying an Intrusion Detection System*

As discussed earlier that networked systems, networks, wireless devices, wearables, and others connected computers may get vulnerable to different type of cyberattacks. Furthermore, the type and nature of cyberattacks is increasing day by day. This requires a strong need of intrusion detection systems.

Use of data science plays a very crucial role for development of such systems. With proper use of data science applications you can, beforehand, identify the credibility of a source sending traffic to your system, nature of request, its consequences, etc. and thus can implement corrective actions.

2.6 Data Science Applications in Airlines

Data science applications in airline industry are proving to be a great help to devise policy according to the customer preferences. A simple example may be the analysis of booking system, but analyzing the booking system airlines can provide the customers with personalized deals and thus increasing their revenue. You can find out the most frequent customers and their preferences and can provide them with the best experience as per their demand.

Furthermore, airlines are not just about ticketing, by using the data science applications, you can find the optimized roots and fairs and enhance your customer base. You can come up with deals that most of the travelers will like to take thus ultimately enhancing your revenue.

1. *Smart Maintenance*

Baggage handling is a big problem on airports but data science tools provide better solutions to track baggage on run time using radio frequencies. As air traffic is increasing day by day and new problems are emerging on daily basis, data science tools can be a big helping hand in all these scenarios. For example, with increase air traffic, intelligent routing applications are required. So, in this scenario, intelligent routing applications can make the journey much safer than before. Furthermore, you can predict about future issues that are normally difficult to handle at runtime, once these issues are identified beforehand, you can have contingency plans to fix them. For example, sudden maintenance that is hard to handle at run time but prediction before time enables industry to take reasonable measures. Similarly, you can predict weather conditions and can ultimately inform the customers about the delay in order to avoid any unhappy experience or customer dissatisfaction.

2. *Cost Reduction*

There may be many costs that airlines have to beer with, e.g., one aspect is that of baggage lost. With the use of real-time bag tracking, such costs can significantly be avoided and thus avoiding the customer dissatisfaction.

One of the major costs of airline companies is that of fuel. Using more fuel may expensive and less fuel may be dangerous. So, maintaining a specific level is much necessary. Data science applications can dig out the relevant data like jet engine information, weather conditions, altitude, rout information, distance, etc. and can come up with optimal fuel consumption, ultimately helping companies to decrease fuel consumption cost.

3. *Customer Satisfaction*

Airline companies take all measures just to satisfy their customers by enhancing their experience with the company. User satisfaction depends on number of factors as everybody has individual level of satisfaction, but creating an environment that can satisfy large number of customers is difficult, but data science tools through analysis of customers previous data can provide maximum ease like their favorite food, preferred boarding, preferred movies, etc. which will ultimately make customers to choose same airline again.

4. *Digital Transformation*

Transformation of existing processes into digital model is giving high edge to airline industries whether it is related to customer or other monitoring factors. Attractive dashboards and smart technological gadgets are making it possible to provide greater level of services on time and airlines companies are able to receive and analyze instant feedback in order to provide better experience and enhanced service quality.

5. *Performance Measurements*

Airline companies normally operate at international level and thus face tough competition. So, in order to remain in the business, they not only have to take performance measures but have to make sure that they are ahead of their competitors.

With the help of data science applications, airlines can automatically generate their performance reports and analyze them, e.g., the total number of passengers that travelled last week preferred the same airline again or the ratio of the fuel consumption one the same route as compared to the previous flight with respect to the number of customers, etc.

6. *Risk Management*

Risk management is an important area where data science applications can help the airline industry. Whenever a plan takes off, there are multiple risks that are adhered to the flight including the changing weather conditions, sudden unavailability of a rout, malfunctioning issues, and most importantly pilots fatigue due to constant flights.

Data science applications can help airlines to overcome these issues and come up with contingencies plan to manage all these risks, e.g., using dynamic routing mechanisms to rout the flight to a different path at run time in order to avoid any disaster. Similarly, in order to avoid pilots from flying fatigues resulting from long hour flying, optimal staff scheduling can be done by analyzing the pilot's medical data.

7. *Control and Verification*

In order to reduce the cost and make the business successful, airlines need in depth analysis of the historical data. Here data science applications can help by using a central repository containing data from various flights. One such example may be the verification of expected number of customers as compared to actual customers that travelled airline.

8. *Load Forecasting*

Airlines regularly need load forecasting models in order to arrange availability of seats and plan other activities like amount of assigned staff, quantity of available food, etc. Data science applications can be special help in this scenario. By using data science applications, we can predict the load of a certain flight with maximum accuracy.

Summary

In this chapter, we discussed few of the applications of data science. As the size of data is increasing every second with immense rate, the manual and other conventional automation mechanisms are not sufficient, so the concepts of data science is very much relevant now. We have seen that how organizations are benefiting from the entire process in all walks of life. The overall intention was to emphasize the importance of the data science applications in daily life.

Further Reading

Following are some valuable resources for further reading:

- Geospatial Data Science Techniques and Applications (Editors: Hassan A. Karimi, Bobak Karimi), CRC Press, 2020. The book deals with different geospatial data and its different characteristics. It discusses the different methods and techniques for obtaining the information from the geospatial data. It focuses on various state of the art techniques used for processing geospatial data and provides details about how these techniques work to get the information from this data.
- Big Data Science in Finance: Mathematics and Applications (Authors: Irene Aldridge, M. Avellaneda), Wiley, 2021. This book provides a complete account of big data that includes proofs, step-by-step applications, and code samples. It explains the difference between principal component analysis (PCA) and singular value decomposition (SVD) and covers vital topics in the field in a clear, straightforward manner
- Data Science for Beginners (Author: Prof John Smith), Amazon Digital Services LLC, 2018. This book explains the topics like what is data science, need of the data science, data science life cycle along with some important applications.
- Data Science: Theory, Analysis, and Applications (Authors: Qurban A Memon, Shakeel Ahmed Khoja), CRC Press, 2021. This book provides collection of scientific research methods, technologies, and applications in the area of data science. It discusses the topics like theory, concepts, and algorithms of data science, data design, analysis, applications, and new trends in data science.
- "Data Science for Healthcare" 1st ed. 2019 Edition by Consoli, Springer & Co, 2019. This book exploits the use of data science in healthcare systems. Author focuses on different methods related to data analytics and how these methods can be used to extract new knowledge from the healthcare systems.

- "Healthcare Analytics Made Simple: Techniques in healthcare computing using machine learning and Python" by Vikas (Vik) Kumar, Packt Publishing, 2019. This book discusses about health-related analytics and discussed the concepts with the help of examples. This book is intended for those who have strong programming skills.
- Visualizing Health and Healthcare Data: Creating Clear and Compelling Visualizations to "See How You're Doing" 1st Edition by Lindsay Betzendahl, Cambria Brown, and Katherine Rowell, Wiley, 2020. This book is worthy addition that covers several general and best practices for data visualization. The book seeks the different ways to present data according to the background of audience.
- "Healthcare Analytics for Quality and Performance Improvement" 1st Edition by Trevor L. Strome, Wiley, 2013. This book is an excellent resource on health informatics. You will find it easy to read and follow the instructions given to enhance understandings about the concepts covered in this book.
- "Healthcare Analytics 101" by Sanket Shah, 2020. A short book that provides an overview of American healthcare trends and challenges according to the viewpoint of healthcare analytics.
- "Healthcare Data Analytics: Primary Methods and Related Insights" by Bob Kelley. This book provides excellent analyses used with healthcare data.
- "The U.S. Healthcare Ecosystem: Payers, Providers, Producers" 1st Edition by Lawton Robert Burns, McGraw Hill/Medical, 2021. The book is a very relevant resource that provides knowledge about US healthcare system and various of its stakeholders.
- "Healthcare Data Analytics (Chapman & Hall/CRC Data Mining and Knowledge Discovery Series)" 1st Edition by Chandan K. Reddy and Charu C. Aggarwal, Chapman and Hall/CRC, 2015. This is a comprehensive book about healthcare data analytics that presents different aspects of EMR data formats with along with their classifications.
- "A Practical Approach to Analyzing Healthcare Data" 4th Edition by Susan White, AHIMA Press, 2021. The book presents a comprehensive approach for the analysis the healthcare data. Overall, the book is a very good resource for professionals that are engaged with healthcare systems and organizations.

Exercises

Exercise 2.1: How can data can be useful in healthcare services? Explain the concept with the help of real-life applications.

Exercise 2.2: Overstaffing and understaffing is a big problem in healthcare services. How can data science be used to overcome this problem?

Exercise 2.3: A healthcare services organization builds a center in a remote location where a large number of doctors is not a possibility due to heavy snow and blockage of roads. They set up different healthcare devices which can be operated by a few trained persons. How can data science be helpful to increase the efficiency of such remote healthcare service centers? Explain with the help of real-life scenarios and examples.

Exercise 2.4: How can data science help an individual learner to examine and improve his performance?

Exercise 2.5: How can data science help educational institutes to design and improve educational content and offer new programs according to the demand of the market?

Exercise 2.6: What in your opinion is the most important usage of data science in healthcare? Discuss.

Exercise 2.7: How can data science be used for predicting education trends?

Exercise 2.8: How can data science be utilized for Industry 4.0?

Exercise 2.9: Discuss a case study where data science may be used for cybersecurity.

Exercise 2.10: How can airlines use data science for improving their profits?

Exercise 2.11: How can data science be utilized for climate change?

Chapter 3
Widely Used Techniques in Data Science Applications

3.1 Supervised Learning

The majority of machine learning algorithms these days are based on supervised machine learning techniques. Although it is a complete domain that requires separate in-depth discussion, here we will only provide a brief overview of the topic.

In supervised machine learning, the program already knows the output. Note that this is opposite to conventional programming where we feed input to the program and the program gives output. Here, in this case, we give input and output at the same time, to make the program learn that in case of any of this or related input what program has to output.

This learning process is called model building. It means that through provided input and output, the system will have to build a model that maps the input to output, so that the next time whenever the input is given to the system, the system provides the output using this model. Mathematically speaking, the task of a machine learning algorithm is to find the value of the dependent variable using the model with provided independent variables. The more accurate the model is, the more efficient the algorithm, and the more efficient the decisions based on this model.

The dependent and independent variables are provided to the algorithm using the training dataset. We will discuss the training data set in the upcoming sections. Two important techniques that use supervised machine learning are as follows:

- *Classification:* Classification is one of the core machine learning techniques that use supervised learning. We classify the unknown data using supervised learning, e.g., we may classify the students of a class into male and female, we can classify the email into two classes like spam and no-spam, etc.
- *Regression:* In regression, we try to find the value of dependent variables from independent variables using the already provided data; however, a basic difference is that the data provided here is in the form of real values.

© The Author(s), under exclusive license to Springer Nature Switzerland AG 2023
U. Qamar, M. S. Raza, *Data Science Concepts and Techniques with Applications*,
https://doi.org/10.1007/978-3-031-17442-1_3

3.2 Unsupervised Learning

Just like supervised learning, there is another technique called unsupervised learning. Although not as common as supervised learning, we still have a number of applications that use unsupervised learning.

As discussed above that in supervised learning, we have training data set comprising both the dependent and independent variables. The dependent variable is called the class, whereas independent variables are called the features. However, this is not the case every time. We may have applications where the class labels are not known. The scenario is called unsupervised learning.

In unsupervised learning, the machine learns from the training dataset and groups the objects based on similar features, e.g., we may group the fruits based on colors, weight, size, etc.

This makes it challenging as we do not have any heuristics to guide the algorithm. However, it also opens new opportunities to work with scenarios where the outcomes are not known beforehand. The only thing available is the set of operations available to predict the group of the unknown data.

Let's discuss it with the help of an example. Suppose we are given a few shapes including rectangles, triangles, circles, lines, etc., and the problem is to make the system learn about the shapes so that it may recognize the shapes in the future.

For a simple supervised learning scenario, the problem will be simple because the system will already be fed with the labeled data, i.e., the system will be informed that if the shape has four sides it will be a rectangle, and a three sides shape will be a triangle, and a round closed end shape will be a circle, etc. So, the next time whenever a shape with four sides is provided to the system, the system will recognize it as a rectangle. Similarly, a shape will three sides will be recognized as a triangle, and so on.

However, things will be a little messier in the case of unsupervised learning as we will not provide any prior labels to the system. The system will have to recognize the figures and will have to make groups of similar shapes by recognizing the properties of the shapes. The shapes will be grouped and will be given system-generated labels. Due to this reason, it is a bit more challenging than the supervised classification.

This also makes it more error-prone as compared to supervised machine learning techniques. The more accurately the algorithm groups the input, the more accurate the output and thus the decisions based on it. One of the most common techniques using unsupervised learning is clustering. There is a number of clustering algorithms that cluster data using the features. We will provide a complete chapter on clustering and related technologies later. Table 3.1 shows the difference between supervised and unsupervised learning.

Table 3.1 Supervised vs unsupervised learning

Supervised learning	Unsupervised learning
Uses labeled data	Uses unlabeled data
Tries to predict something, e.g., a disease	Tries to group things based on their properties
We can directly measure the accuracy	Evaluation is normally indirect or qualitative
Underlying techniques: Classification, regression	Underlying techniques: Clustering

Table 3.2 Common algorithms in each type of learning

Learning type	Common algorithm
Supervised learning	Support vector machines Linear regression Logistic regression Naïve Bayes K nearest neighbors
Unsupervised learning	K-Means Hierarchical clustering Principal component analysis t-distributed stochastic neighbor embedding (t-SNE)
Reinforcement learning	Q-learning Temporal difference (TD) Deep adversarial networks

3.3   Reinforcement Learning

Reinforcement learning is another important technique of machine learning that is gaining widespread use. It is the process of training a model so that it could make a series of decisions. In reinforcement learning, a machine agent interacts with its environment in uncertain conditions to perform some actions. Now the agent is guided to achieve the intended output with the help of rewards and penalties.

The overall goal is to increase the total number of rewards. The designer sets the rewards policy. Now it is up to the model to perform the actions to maximize the rewards. In this way, the overall training of the model is guided.

Table 3.2 shows all the three types of learning and corresponding commonly used algorithms:

3.4   AB Testing

AB testing is one of the strategies used to test your online promotions and advertising campaigns, website designs, application interfaces, etc. Basically the test is to analyze user experience. We present two different versions of the same thing to the user and try to analyze the user experience of both. The item which performs better is considered to be the best.

So, this testing strategy helps you prioritize your policies and you can find out what is more effective as compared to others alternatively giving you the chance to improve your advertisements and business policies. Now we will discuss some important concepts related to AB testing.

3.4.1   AB Test Planning

Conducting an AB test requires proper planning including what to test, how to test, and when to test. Giving thorough consideration to these aspects will make you run a successful test with more effective results because it will help you narrow down your experiments to the exact details you want out of your customers.

For example, either you want to test your sale promotion or your email template. Either this test will be conducted online or offline. If it is to be conducted online, then either you will present your actual site to the user or will adopt some other way. Suppose you want to test your sale promotion through your online website, you can better focus on those parts of the website that provide the users with sales-related data. You can provide different copies and find out which design is realized in the business. This will not online provide you with more business but also the justification to spend more on effective sale promotions.

Once you have decided on what to test, then you will try to find the variables that you will include in your test. For example, if you want to test your sales ad, your variables may include:

- Color scheme
- The text of the advertisement
- The celebrities to be hired for the campaign

It is pertinent to carefully find out all these variables and include them in the test which may require finding out these variables that may have a strong impact on your campaign.

Similarly, you should know the possible outcomes you are testing for. You should consider all the possibilities for each option should be tested so that the option which provides better results will be considered.

You need to simultaneously conduct your tests so that the effectiveness of whatever you are testing could be determined continually with time and you keep on making dynamic decisions concerning the current scenario and situation.

3.4.2   AB Testing Can Make a Huge Difference

AB testing can have a huge impact on your business by finding out user experience related to something you are testing. It should be noted that starting something without knowing user experience may always be expensive. AB testing provides you with the opportunity to find out customer interests before launching something. By knowing the user preferences and the outcomes of the ads that performed better, you can have justification for spending more on such campaigns that realize into a profitable business.

This will also help you avoid the strategies that do not worth to customers and thus have a little contribution. You can find out customer priorities and alternatively provide them with what they want and how they want.

3.4.3   What You Can Test

In the context of AB testing, an important question is what you can test. You can test anything from the format of your sales letter to a single image on the website of your business. However, it should be noted that it does not mean that you should spend time and resources on testing everything related to your business. Carefully find out the things that have a strong impact on your business and it will be worth testing them only.

Similarly, once you have decided to test, e.g., two newsletters A and B, make sure to test all possible options, e.g., header of newsletter A with B, header of newsletter B with A, and so on. This may require some careful workout before conducting the tests.

Some examples of what you can test:

- News Letter

  - Header
  - Body text
  - Body format
  - Bottom image

- Website

  - Website header
  - Website body
  - Sales advertisement
  - Product information
  - Website footers

- Sales advertisement

  - Advertisement text
  - Product images
  - Celebrities images
  - Position of the advertisement in the newspaper

3.4.4 For How Long You Should Test

You should carefully select the time period for which the test will be conducted and it depends on the frequency of response you get, e.g., if you are conducting a test on your website and you have a lot of traffic per day, you can conduct your test for few days, and vice versa.

The insufficient time period you allocate for your test, the insufficient results you will get, and thus the skewed will be your results, so, before selecting the time make sure to open the test for time period that you get enough responses to make accurate decisions.

Similarly, giving more time to a test may also give you skewed results. So, note that there are no accurate guidelines and heuristics about the time period of a test, but as discussed above, you should select the time period very carefully keeping in mind the frequency of responses you get. Experience in this regard can be helpful.

3.5 Association Rules

Association rules are another machine learning technique that tries to find out the relationship between two items. An important application of association rules is market basket analysis. Before going into details of association rules, let's discuss an example.

In market stores, it is always important to place similar items (i.e., the items that sell together) close to each other. This not only saves the time of the customers but also promotes cross-item sales. However, finding out this item relationship requires some prior processing to find out the relationship between the items.

It should be noted that association rules do not reflect the individual's interests and preferences but only the relationship between the products. We find out these relationships by using our data from previous transactions.

Now let's discuss the details of association rules.

It consists of two parts, i.e., an antecedent and a consequent as shown below. Both of which are a list of items.

$$\text{Antecedent} \rightarrow \text{Consequent}$$

It simply shows that if there is antecedent, then there is consequent as well which means that implication represents the co-occurrence. For a given rule, itemset is the list of all the items in the antecedent and the consequent for example:

$$\{\text{Bread, Egg}\} \rightarrow \{\text{Milk}\}$$

Here itemset comprises Bread, Egg, and Milk. It may simply mean that the customers who purchased Bread and Eggs also purchased Milk most of the time which may simply result in placing these products close to each other in the store.

There is a number of metrics that can be used to measure the accuracy and other aspects of association rules. Here we will discuss a few.

3.5.1 Support

Support defines how frequently an itemset appears in a transaction. For example, consider that there are two itemsets, i.e., itemset1 and itemset2. Itemset1 contains {Bread}, whereas itemset2 contains {Cake}. As there is more likely that a large number of transactions will contain Bread as compared to those containing Cake, the support of itemset1 will be higher than itemset2. Mathematically:

$$\text{Support}\left(\{X\} \rightarrow \{Y\}\right) = \frac{\text{Transactions containing both } X \text{ and } Y}{\text{Total Number of Transactions}} \quad (3.1)$$

The "Support" metric helps identify the items that are sold more than the others and thus the rules that are worth more consideration. For example, a store owner may be interested in finding out the items that are sold 50 times out of all 100 transactions. So:

$$\text{Support} = 50/100 = 0.5$$

3.5.2 Confidence

Confidence determines the likeliness of occurrence of consequent in a transaction with a given antecedent. For example, we can answer the questions like, in all the transactions containing Bread and Butter, how many were containing milk as well. So, we can say that confidence is the probability of occurrence of consequent given the probability of antecedent. Mathematically:

$$\text{Confidence}(\{x\} \rightarrow \{y\}) = \frac{\text{Transactions containing both } X \text{ and } Y}{\text{Transactions containing } X} \quad (3.2)$$

Suppose we want to find the confidence of the {Butter} → {Milk}. If there are 100 transactions in total, out of which 6 have both Milk and Butter. 60 have Milk without Butter and 10 have Butter but no Milk, the confidence will be:

$$\text{Confidence} = 6/16 = 0.375$$

3.5.3 Lift

Lift is the ratio of the probability of consequent being present with the knowledge of antecedent over the probability of consequent being present without the knowledge of antecedent. Mathematically:

$$\text{Lift}(\{x\} \rightarrow \{y\})$$
$$= \frac{(\text{Transactions containing both } X \text{ and } Y)/(\text{Transactions containing } X)}{\text{Transactions containing } Y} \quad (3.3)$$

Again, we take the previous example. Suppose there are 100 transactions in total, out of which 6 have both Milk and Butter. 60 have Milk without Butter and 10 have Butter but no Milk. Now: the probability of having Milk with the knowledge that Butter is also present = 6/(10+6)= 0.375.

Similarly:

Probability of having Milk without knowledge that Butter is also present = 66/100= 0.66

Now Lift = 0.375/0.66=0.56

3.6 Decision Tree

A decision tree is used to show the possible outcomes of different choices that are made based on some conditions. Normally we have to make decisions based on different options, e.g., cost, benefits, availability of resources, etc. This can be done by using decision trees where following a complete path lets us reach a specific decision.

Figure 3.1 shows a sample decision tree. It is clear that we start with a single node that splits into different branches. Each branch refers to an outcome; once an outcome is reached, we may have further branches thus all the nodes are connected. We have to follow a path based on different values of interest.

There are three different types of nodes:

Chance nodes: Chance nodes show the probability of a certain decision. They are represented by circles.

Decision nodes: As the name implies, the decision node represents a decision that is made. It is represented by square.

End nodes: End nodes represent the outcome of the decision. It is represented by a triangle.

Figure 3.2 shows some decision tree notations.

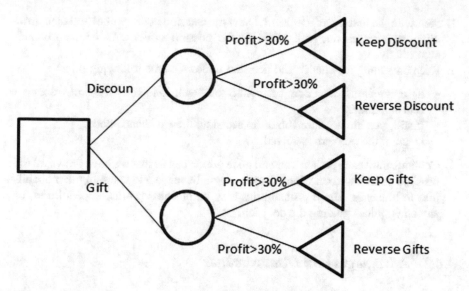

Fig. 3.1 Decision tree

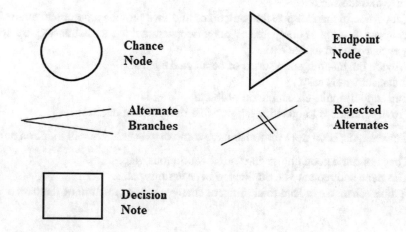

Fig. 3.2 Decision tree

3.6.1 How to Draw a Decision Tree

To draw a decision tree, you will have to find all the input variables on which your decisions are based and their possible values. Once you have all this, you are ready to work on decision trees. Following are the steps to draw a decision tree:

1. Start with the first (main) decision. Draw the decision node symbol and then draw different branches out of it based on the possible options. Label each branch accordingly.
2. Keep on adding the chance and decision nodes with the following rules:

   - If a previous decision is connected with another decision, draw a decision node.
   - If still, you are not sure about the decision, draw a chance node.
   - Leave if the problem is solved.

3. Continue until each path has an end node which means that we have drawn all the possible paths based on which decisions will be made. Now assign the probabilities to branches. These probabilities may be business profits, expenditures, or any other values guiding the decision.

3.6.2   Advantages and Disadvantages

A decision tree may have the following advantages:

- Easy to understand.
- They force to map all possible outcomes and thus provide a complete analysis.
- They are flexible as compared to other decision-making tools that require comprehensive quantitative data.
- Provide the mapping of a series of connected events.
- Enhancement is easy.
- Can support multi-classification problem.
- Provides chances to quantify the possible outcomes of decisions.

  However, decision trees may also have some drawbacks. We will mention a few:

- They are not a good choice in case of continuous values.
- The same subtree may be duplicated over various paths.
- A small change can lead to significant changes in the structure of the tree.

3.6.3   Decision Trees in Machine Learning and Data Mining

Decision trees have also their application in machine learning, data mining, and statistics. We can build various prediction models using decision trees. These models take input the values related to different items and try to predict the output based on the item's value.

In the context of machine learning, these types of trees are normally named classification trees. We will discuss classification trees in detail in upcoming chapters. The nodes in these trees represent the attributes based on which classification is performed. The branches are made on the basis of the values of the attributes. They can be represented in the form of if then else conditions, e.g., if $X > 75$ then TaxRate $= 0.34$. The end nodes are called the leaf nodes, which represent the final classification.

As we may have a series of events in decision trees, their counterparts, i.e., classification trees in machine learning may also have several nodes connected thus forming a hierarchy. The deeper the tree structure, the more accurate classification it may provide.

However, note that we may not have the discrete values all the time and there may be scenarios where continuous numbers such as price or height, person, etc. need to be modeled for such situations we may have another version of decision trees called regression trees.

It should be noted that an optimal classification tree is the one that models most of the data with a minimum number of levels. There are multiple algorithms for creating classification trees. Some common are CART, ASSISTANT, CLS, ID3/4/5, etc.

3.6.4 Advantages and Disadvantages

The use of decision trees in machine learning has the following advantages:

- Simple and easy to build.
- Fast as following a branch let's avoid traversal of others.
- Works with multi-classification problems.
- For simple problems, accuracy can be compared with other classification techniques.
- Can handle both numerical and categorical data.

Disadvantages
- Too small data may result in underfitting.
- More training data may result in overfitting for real-world data.
- It may not be good for continuous data values.
- The same subtree may be repeated over multiple branches.

3.7 Cluster Analysis

Just like classification, cluster analysis is another important technique, commonly known as clustering. Cluster analysis is the process of examining the properties of objects and then grouping them in such a way that similar objects are always arranged in similar groups.

For example, we can group similar customers based on their properties into different groups and can offer each group different sales promotions, e.g., discounts, promotional tokens, etc.

3.7.1   Different Approaches of Cluster Analysis

There are several methods that can be used to perform cluster analysis. Here we will discuss a few of them.

Hierarchical Methods
These methods can be further divided into two clustering techniques as follows:

- *Agglomerative methods:* In this method, each object forms its cluster. Then similar clusters are merged. The process goes on until we get the one bigger or K bigger clusters.
- *Divisive methods:* This scheme is the inverse of the abovementioned method. We start with one single cluster and assign all the objects to it. Then we split it into two least similar clusters. The process continues until each object belongs to its cluster.

Non-hierarchical Methods   Non-hierarchical clustering technique is the most common one. Also known as K-Means clustering, we group similar objects into similar clusters by maximizing or minimizing some evaluation criteria.

3.7.2   Types of Data and Measures of Distance

The objects in cluster analysis can have different types of data. To find out the similarity, you need to have some measure that could determine the distance between them in some coordinate system. The objects that are close to each other are considered to be similar. There are several different measures that can be used to find out the distance between objects. The use of a specific measure depends upon the type of data. One of the most common measures is Euclidean distance.

3.7.2.1   Euclidean Distance

Euclidean distance between two points is simply the length of the straight line between the points in Euclidean space. So, if we have points $P_1, P_2, P_3, \ldots, P_n$ and a point i is represented by $y_{i1}, y_{i2}, \ldots, y_{in}$ and point j are represented by $y_{j1}, y_{j2}, \ldots, y_{jn}$, then, the Euclidean distance d_{ij} between points i and j will be calculated as:

$$d_{ij} = \sqrt{\left(y_{i1} - y_{j1}\right)^2 + \left(y_{i2} - y_{j2}\right)^2 + \ldots + \left(y_{in} - y_{jn}\right)} \tag{3.4}$$

Here Euclidean space is a two-, three-, or n-dimensional space where each point or vector is represented by n real numbers $(x_1, x_2, x_3, \ldots, x_n)$.

3.7.2.2 Hierarchical Agglomerative Methods

There are different clustering techniques used in the hierarchical clustering method. Each technique has its mechanism that determines how to join or split different clusters. Here we will provide a brief overview of a few.

3.7.2.3 Nearest Neighbor Method (Single Linkage Method)

This is one of the simplest methods to measure the distance between two clusters. The method considers the points (objects) that are in different clusters but are closest to each other. The two clusters are then considered to be close or similar. The distance between the closest points can be measured using the Euclidean distance. In this method, the distance between two clusters is defined to be the distance between the two closest members, or neighbors. This method is often criticized because it doesn't take into account cluster structure.

3.7.2.4 Furthest Neighbor Method (Complete Linkage Method)

It is just an inverse of the single linkage method. Here we consider the two points in different clusters that are at the farthest distance from each other. The distance between these points is considered to be the distance between two clusters. Again a simple method but just like single linkage, it also does not take cluster structure into account.

3.7.2.5 Average (Between Groups) Linkage Method (Sometimes Referred to as UPGMA)

The distance between two clusters is considered to be the average distance between each pair of objects from both of the clusters. This is normally considered to be the robust approach for clustering.

3.7.2.6   Centroid Method

In this method, we calculate the centroid of the points in each cluster. Then the distance between the centroids of two clusters is considered to be the distance between the clusters. The clusters having a minimum distance between centroids are considered to be close to each other. Again, this method is also considered to be better than the single linkage method or furthest neighborhood method.

3.7.3   Selecting the Optimum Number of Clusters

Unfortunately, there are no proper heuristics to determine the optimal number of clusters in a dataset. K-Means, one of the most common clustering algorithms, takes the number of clusters K as user input. The optimal number of clusters in a dataset may depend on the nature of the data and the technique used for clustering.

In the case of hierarchical clustering, one simple method may be to use a dendrogram to see if it suggests a particular number of clusters.

3.8   Advantages and Disadvantages of Clustering

Clustering may have a number of advantages:

- Has vast application in business, e.g., grouping the customers
- Works for unsupervised data which we often come around in the real world
- Helpful for identifying the patterns in time or space

  Clustering may have a number of disadvantages as well:

- It may be complex as the number of dimensions increase.
- No heuristics are available to find out the optimal number of clusters.
- Using different measures may result in different clusters.

3.9   Pattern Recognition

A pattern is a structure, sequence, or event that repeats in some manner. We come across many patterns in our daily life, e.g., a clock ringing a bell every hour, a pattern of round stairs in a home, a pattern of white line markings on road, etc.

In terms of computer science and machine learning, pattern recognition is the process of reading the raw data and matching it with some already available data to see if the raw data has the same patterns as we have applied. Figure 3.3 shows the process:

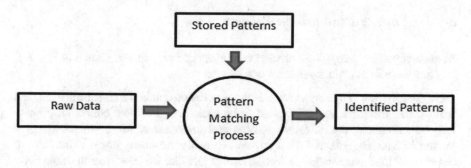

Fig. 3.3 Components of the pattern recognition process

Table 3.3 A sample dataset for pattern recognition

Salary	Age	Homeowner
15,000	45	Yes
5000	50	No
16,000	55	Yes
20,000	48	Yes
12,000	36	No
20,000	42	Yes
10,000	35	No
18,000	50	Yes

For example, consider the following dataset given in Table 3.3.

Now our target is to find the patterns in this data as apparently, it is explicit from the dataset that two patterns exist in it. There are two categories, those with a salary greater than or equal to 15,000 and age greater than 42 years and the other having a salary less than 15,000 and age less than 40 years. In the first case, the people seem to have their own home, while in the other case, people do not seem to own their own home.

This is what pattern recognition is all about. We examine the data and try to find the regularities in it.

As input, we can provide any type of data for pattern recognition including:

- Structured data, e.g., Table 3.3
- Images
- Textual data
- Signals, sounds, etc.
- Emotions.

3.9.1 Pattern Recognition Process

In general pattern recognition process comprises of four steps as shown in Fig. 3.4. Now we will explain these steps one by one.

Data Acquisition Data acquisition is the process of obtaining and storing the data from which the patterns are intended to be identified. The data collection process may be automated, e.g., a sensor sensing real-time data or maybe manual, e.g., a person entering the data of the employees. Normally the data is stored in the form of objects and their corresponding feature values just like we have data in Table 3.4.

Data Preprocessing Data collected in the data store for pattern recognition may not be in ready-to-process form. It may contain many anomalies, noise, missing values, etc. All of these deficiencies in data may affect the accuracy of the pattern recognition process and thus may result in inaccurate decisions. This is where preprocessing step helps to fix the problem; in this step, all of such anomalies are fixed and data is made ready for processing.

Feature Extraction/Selection Once the data has been preprocessed, the next step is to extract or select the features from the data that will be used to build the model. A feature is the characteristic, property, or attribute of an object of interest. Normally the entire set of data is not used as datasets are large in number with hundreds of thousands of features. Furthermore, datasets may have redundant and irrelevant features as well. So, feature extraction/selection is the process to get the relevant features for the problem. It should be noted that the accuracy of the model for pattern extraction depends upon the accuracy of the values of the features and their relevance to the problem.

Fig. 3.4 A generic pattern recognition process

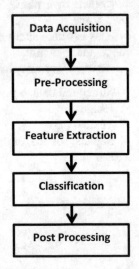

Classification Once we have found the feature vector from the entire dataset, the next step is to develop a classification model for the identification of the patterns. The step of classification is discussed in detail in upcoming chapters. Here it should be noted that a classifier just reads the values of the data and based on the values tries to assign a label to the data as per identified pattern. We have various classifiers available, e.g., decision trees, artificial neural networks, Naïve Bayes, etc. The step of building a classifier requires two types of data, i.e., training data and testing data. We will discuss both of these types of data in the next section.

Post-processing Once, the patterns are identified and decisions are made, we try to validate and justify the decisions made based on the patterns. Post-processing normally includes the steps to evaluate the confidence in these decisions.

3.9.2 Training and Test Datasets

The process of building the classification model requires training and testing the model on available data. So, the data may be classified into two categories.

Training dataset: Training dataset, as the name implies, is required to train our model, so that it could predict with maximum accuracy for some unknown data in the future. We provide the training data to the model, check its output, and compare the output with the actual output to find the amount of error. Based on comparison, we adjust (train) different parameters of the model. The process continues until we get the minimum difference (error) between the provided input and the produced output.

Once the model is trained, the next step is to evaluate or test the model. For this purpose, we use a test dataset. It should be noted that both training and test data are already available and we know the values of the features and classes of the objects. The test dataset is then provided to the model and the produced output is compared with the actual output in the test model. We then take different measures like precision, accuracy, recall, etc.

3.9.3 Applications of Pattern Recognition

There are various applications of pattern recognition. Here we will mention just a few:

Optical Character Readers Optical character reader models read the textual image and try to recognize the actual text, e.g., the postal codes written on the letters. In this way, the model can sort out the letters according to their postal codes.

Biometrics Biometrics may include face recognition, fingerprint identification, retina scans, etc. It is one of the most common and widely used applications of pattern recognition. Such models are helpful in automatic attendance marking, personal identification, criminal investigations, etc.

Diagnostics Systems Diagnostic Systems are advance forms of applications that scan medical images (e.g., X-ray sheets, internal organ images, etc.) for medical diagnostics. This facilitates in diagnosing diseases like brain tumors, cancer, bone fractures, etc.

Speech Recognition Speech recognition is another interesting application of pattern recognition. This helps in creating virtual assistants, speech-to-text converters, etc.

Military Applications Military applications of pattern recognition may include automated target recognition to identify hidden targets, differentiating between friendly and enemy entities, radar signal analysis, classification, etc.

3.10 Summary

In this chapter, we provided a broad overview of the various techniques used in data science applications. We discussed both classification and clustering methods and techniques used for this purpose. It should be noted that each technique has its advantages and disadvantages. So, the selection of a particular technique depends on the requirements of the organization including the expected results and the type of analysis the organization requires along with the nature of the data available.

Further Reading
Following are some valuable resources for further reading:

- Intelligent Techniques for Data Science (Authors: Akerkar, Rajendra, Sajja, Priti Srinivas), Springer & Co, 2018. The book focuses on methods significantly beneficial in data science and clearly describes them at an introductory level, with extensions to selected intermediate and advanced techniques. The authors discuss different approaches and their advantage and disadvantages. The book provides the base for the professionals to help them develop real-life data analytics solutions. The book provides real-world scenarios about how we can extract the values from the existing real-world data in different domains, e.g., health, aviation, tourism, retails, etc.
- Data Science with Matlab. Classification Techniques (Author: A Vidales), Amazon Digital Services, 2019. This book discusses parametric classification supervised techniques such as decision trees and discriminant analysis along with non-supervised analysis techniques such as cluster analysis.
- Information-Theoretic Methods in Data Science (Coordinators: Rodrigues Miguel R. D., EldarYonina C.), Cambridge University Press, 2020. The book provides details about how we can use the information-theoretic methods for the acquisition of data, its presentation, and analysis. Book provides in-depth details

about the signal acquisition, compression of data, compressive sensing, data communication, etc.

- Supervised learning techniques. Time series forecasting. Examples with neural networks and matlab Kindle Edition by César Pérez López, Lulu Press, 2020. This book elaborates the machine learning using supervised learning; predicting categorical responses, classification techniques, and regression techniques. It also covers the time series forecasting techniques using neural networks.
- "The Supervised Learning Workshop: A New, Interactive Approach to Understanding Supervised Learning Algorithms," 2nd Edition by Blaine Bateman, Ashish Ranjan Jha, Benjamin Johnston, and Ishita Mathur, Packt Publishing, 2020. This book describes the understanding of supervised machine learning. The reader will get a grip on the fundamentals of supervised machine learning and discover python libraries for implementing the tasks related to it.
- "Supervised and Unsupervised Learning for Data Science," by Michael W. Berry, Azlinah Mohamed, et al, Springer & Co, 2019. This book covers supervised and unsupervised methods and their applications in data science. Overall the book is a good resource for anyone who wants to step into machine learning and data science.
- "Machine learning with R. Supervised learning techniques: Regression" Kindle Edition by César Pérez López, Lulu Press, 2022. In this book, regression-related supervised learning techniques are discussed. The contents of this book provide details of PLS regression, robust regression, linear models, generalized linear models, probabilistic regression, logistic regression, negative binomial regression, Poisson regression, normal regression, exponential regression, count models, and different supervised techniques based on regression.
- "Supervised Learning with Linear Regression: An Executive Review of Hot Technology" by Stephen Donald Huff, Independently published, 2018. This book describes supervised machine learning theory with a detailed expert-level review. It is a very good book for technology professionals that have a basic knowledge of machine learning concepts.
- "Sampling Techniques for Supervised or Unsupervised Tasks" by Frédéric Ros and Serge Guillaume, Springer & Co, 2020. This book gives a detailed review of sampling techniques that can be used in supervised and unsupervised case studies. It covers the machine learning algorithms with different sampling methods to manage scalability.

Exercises

Exercise 3.1: What are the three types of machine learning? Provide a brief introduction to each type.

Exercise 3.2: What are the two cores techniques in supervised learning? Provide a brief introduction of each type.

Exercise 3.3: What are the similarities and differences between supervised and unsupervised learning? Provide three differences and three similarities.

Exercise 3.4: Consider the data given below, from supervised and unsupervised learning techniques, which one will you use for the given data and why?

Health	Gender	Age
Healthy	Male	42
Healthy	Female	49
Healthy	Male	66
Healthy	Female	55
Healthy	Female	49
Sick	Male	31
Sick	Female	53
Sick	Female	79
Sick	Male	53
Sick	Male	31

Exercise 3.5: Consider the data given below, from supervised and unsupervised learning techniques, which one will you use for the given data and why?

Gender	Age
Male	70
Female	66
Male	38
Female	40
Female	64
Male	42
Female	67
Female	72
Male	62
Male	66

Exercise 3.6: Explain the terms "labeled" and "unlabeled" data. How can you identify whether the given data is labeled or unlabeled?

Exercise 3.7: Consider the names of machine learning algorithms given below, you are required to categorize the given algorithms into supervised, unsupervised, and reinforcement learning categories.

- Support vector machines
- K nearest neighbors
- K-Means
- Q-learning
- Temporal difference (TD)
- Deep adversarial networks
- Hierarchical clustering
- Principal component analysis

- Linear regression
- Logistic regression
- Naïve Bayes
- t-distributed stochastic neighbor embedding (t-SNE)

Exercise 3.8: What is meant by the term AB Testing? How is this type of testing planned? What type of applications can be tested using this technique?

Exercise 3.9: Use the data given below in the table and find the association rules for the given data.

TID	Items
1	Milk, Butter, Eggs, Cheese, Yoghurt
2	Butter, Eggs, Chicken, Beef, Rice
3	Chicken, Beef, Rice, Milk
4	Butter, Chicken, Jam, Honey, Ketchup
5	Oil, Olives, Milk, Eggs, Honey
6	Ketchup, Oil, Olives, Eggs, Rice
7	Yogurt, Chicken, Oil, Olives, Eggs
8	Cheese, Yoghurt, Chicken, Milk, Rice
9	Cheese, Yoghurt, Beef, Rice, Butter
10	Milk, Cheese, Eggs, Chicken, Butter

Exercise 3.10: Consider the data given in Exercise 3.9 and find support, confidence, and lift.

Exercise 3.11: A company has a promotion policy where gifts and discounts are awarded to increase the profit. Consider the following rules and construct a designing tree.

If profit is greater than or equal to 30%, then keep the discount.
If profit is less than 30%, then reverse the discount.
If profit is greater than or equal to 30%, then keep the gifts.
If profit is less than 30%, then reverse the gifts.

Exercise 3.12: What are the advantages and disadvantages of a decision tree?

Exercise 3.13: How many approaches do we have for cluster analysis? Give a brief introduction of each.

Exercise 3.14: How many clustering techniques do we have in hierarchical clustering analysis? Give a brief introduction of each.

Exercise 3.15: Consider the data given below and find out two objects that have minimum Euclidean distance (in case more than two objects have the same distance, mention all).

	X1	X2
A	8	2
B	1	8
C	3	3
D	6	4
E	3	7
F	8	8
G	3	1
H	7	3
I	6	9
J	8	1

Exercise 3.16: Consider the data given below and find out two objects that have maximum Euclidean distance (in case more than two objects have the same distance, mention all).

	X1	X2
A	1	1
B	6	5
C	3	3
D	3	3
E	3	1
F	5	2
G	1	8
H	1	3
I	1	9
J	4	3

Exercise 3.17: The steps of the pattern reorganization process are given below. You are required to provide the detail of each step.

- Data acquisition
- Data preprocessing
- Feature extraction/selection
- Classification
- Post-processing

Exercise 3.18: What is the difference between training data and test data?

Chapter 4
Data Preprocessing

4.1 Feature

A feature is an individual characteristic or property of anything existing in this world. It represents the heuristics of different objects and entities existing in the real world. For example, features of a book may include "Total Pages," "Author Name," "Publication Date," etc. Features help us get certain ideas about the objects they are representing, e.g., from the "Total Pages" feature, we can perceive the size of the book, and from "Publication Date," we can find how old the contents are. It should be noted that these features form the basis of the models developed at later stages, so, the better quality of features results in better models.

The following dataset shows five objects and their corresponding features as shown in Table 4.1:

Based on the nature of the domain, the attribute it is representing, we can classify the attributes into two types, i.e., Numerical features and Categorical features. Now we will discuss each of them.

4.1.1 Numerical

Numerical attributes represent numbers, e.g., "Legs" in Table 4.1 is a numerical attribute. The values of the numerical attributes can be discrete or continuous.

Numerical features are of two types:

Interval-Scaled In this type of feature, the difference is meaningful between two values, e.g., the same difference exists between 10 and 20 as can be found between 20 and 30 degrees for the "Temperature" feature.

U. Qamar, M. S. Raza, *Data Science Concepts and Techniques with Applications*,
https://doi.org/10.1007/978-3-031-17442-1_4

Table 4.1 Sample dataset

Specie	Legs	Feathers
Human	2	No
Dog	4	No
Cat	4	No
Hen	2	Yes
Peacock	2	Yes

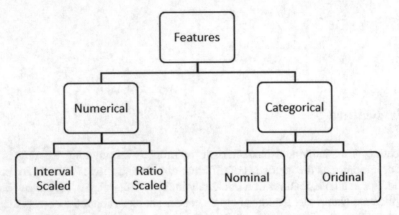

Fig. 4.1 Feature types

Ratio-Scaled This type of feature has the same properties as that of interval-scaled features. Additionally, ratios also hold, e.g., we can say that a location that is 20 km away is twice away as compared to the location which is 10 km away.

4.1.2   Categorical Features

Categorical features consist of symbols to represent domain values. For example, to represent "Gender," we can use "M" or "F." Similarly, to represent Employee type, we can use "H" for hourly, "P" for permanent, etc.

Categorical attributes are of two types further:

Nominal In the Nominal category of attributes, the order does not make sense. For example, in "Gender" attribute, there is no order, i.e., the operators equal to, less than, or greater than do not make any sense.

Ordinal In contrast with Nominal attributes, in ordinal features, comparison operators make full sense, for example, the grade "A" is greater than grade "B" for the "Grade" feature.

Figure 4.1 shows the types of features and their categorization.

4.2 Feature Selection

We are living in a world where we are bombarded with tons of data every second. With digitization in every field of life, the pace with which data is originating is staggering. It is common to have datasets with hundreds of thousands of records just for experimental purposes. This increase in data results in the phenomenon called the curse of dimensionality. However, the increase is two dimensional, i.e., we are not only storing the attributes of the objects in the real world but the number of objects and entities which we are storing is also increasing. The ultimate drawback of these huge volumes of data is that it becomes very tough to process these huge volumes for knowledge extraction and analytics which requires a lot of resources. So, we have to find alternate ways to reduce the size ideally without losing the amount of information. One of the solutions is feature selection.

In this process, only those features are selected that provide most of the useful information. Ideally, we should get the same amount of information that is otherwise provided by the entire set of features in the dataset. Once such features have been found; we can use them on behalf of the entire dataset. Thus, the process helps identify and eliminate irrelevant and redundant features. Based on the facts mentioned above, efforts are always made to select the best quality features. Overall, we can categorize dimensionality reduction techniques into two categories:

(a) Feature selection: A process that selects features from the given feature subset without transforming or losing the information. So, we can preserve data semantics in the transformation process.
(b) Feature extraction: Feature extraction techniques, on the other hand, project current feature space to a new feature subset space. This can be achieved either by combining or applying some other mechanism. The process however has a major drawback that we may lose information in the transformation process. This means the reverse process may not provide the same information present in the original dataset.

The simplest approach to feature selection is to evaluate all subsets of features. However, this is feasible only when the dataset has a small number of attributes or features. This approach terms an exhaustive search. In another approach, subsets of features are randomly selected and evaluated. This process continues for a certain number of iterations. The most widely used approach is the heuristic approach, which is based on a heuristic function to find a reduced feature set.

The process of feature selection reduces features by removing irrelevant and redundant features. Features that do not affect the selection of the target class are considered irrelevant features, and redundant features are those which do not define some new aspects of an instance. However, they are causing performance degradation due to high computational costs. A feature that plays a decisive role in the selection of decision class but is highly unrelated to other features is the most relevant. Such relevancy can be categorized as strong relevant features and weak relevant features.

The rough set-based approach is one such technique and uses to not only reduce feature sets but also can be used as a data mining tool to probe hidden knowledge from the large data.

Rough set theory (RST) is a set theory-based tool for data reduction and classification tasks. It is a very popular dimensionality reduction technique that reduces the data size without destroying the original meaning of the data. RST is an excellent tool to handle inconsistent information. It is one of the first non-statistical approaches. Due to such features of RST, it has been applied in many fields, i.e., image segmentation, classification, and diagnosis of faults.

The dominance-based rough set approach (DRSA) focuses on the preference order of features and categorizes the classes into upward and downward classes to consider the preference relation. The basic notions of DRSA include dominance relations, equivalence classes, approximations, and dependency. In RST and DRSA, approximations are considered the core concept of all algorithms. However, for large datasets, the computation process defined for approximations proves computationally expensive which ultimately affects the performance of algorithms based on DRSA approximations.

Now, the data in real-world applications can either be supervised or unsupervised. In a supervised dataset, the class labels are already given for each object or entity in the dataset, e.g., for a student dataset, a class label may be given in the form of the attribute named "Status" having values "Passed" or "Not Passed." However, the majority of the time we may come across datasets where class labels may not be given which is called unsupervised data, so feature selection algorithms have to proceed without class labels.

4.2.1   Supervised Feature Selection

Most real-world applications require supervised data for operations. As discussed above data may contain huge volumes and hundreds of thousands of attributes. Noisy, irrelevant, and redundant attributes are other problems. To overcome these issues, we perform supervised feature selection. In this process, only those features are selected that provide most of the information relevant to the concept. For example, if our concept is to identify the students who have passed a grade, the relevant features are the only ones that provide exact information and do not mislead. A classification model can be used to help select these features. So, we will select the features that provide the same classification accuracy that is otherwise achieved by using the entire feature set. So, a feature subset with maximum accuracy is always desirable. This alternatively means getting the same results but now using a smaller number of features which ultimately increases the efficiency and requires a less number of resources. Therefore, in most cases, feature selection is preprocessing step that reduces the amount of data that is fed to the classification model. A generic supervised feature selection process is shown in Fig. 4.2:

Fig. 4.2 A generic
supervised feature selection
approach

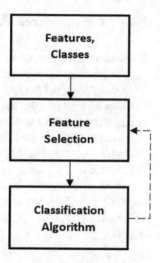

Table 4.2 A sample supervised dataset

U	C1	C2	C3	C4	D
X1	0	3	1	2	0
X2	1	1	2	1	0
X3	1	1	1	1	0
X4	2	3	1	1	1
X5	1	2	1	2	1
X6	0	2	2	0	1
X7	0	3	0	2	2
X8	0	3	0	0	2
X9	1	3	0	1	2
X10	0	2	2	2	1

A complete feature set is provided to the feature selection algorithm which selects features according to specified selection criteria. The selected features are then given as input to the classification algorithm which performs classification based on provided features. It should be noted that the feature subset selected by the feature selection algorithm can be modified using various iterations by considering the feedback provided by the classification algorithm which is shown by dotted lines in the diagram above.

A sample supervised dataset along with four features {C1, C2, C3, C4} and Class {D} is given in Table 4.2:

Here {C1, C2, C3, C4} are normal (Conditional) features and "D" is (Decision) class label. We can use all four features for different knowledge extraction tasks. However, an alternate option is to select some features having the same significance. We can use many criteria for selecting those features, e.g., Information gain, Dependency, Entropy, etc. Here we can see that the attributes C1, C2, and C3 also have the same significance as that of the entire feature set. So, we can use only three instead of four.

4.2.2   Unsupervised Feature Selection

As mentioned earlier, class labels are not given all the time; we have to proceed without labeling information which makes unsupervised classification a little bit tough task. To select the features here, we can use clustering as selection criteria, i.e., we can select the features that give the same clustering structure that can be obtained by the entire feature set. Note that a cluster is just a chunk of similar objects.

A simple unsupervised dataset is shown in Table 4.3:

Dataset contains four objects {X1, X2, X3, X4} and three features {C1, C2, C3}. Objects {X1, X4} form one cluster and {X2, X3} form other one. Note that clustering is explained in detail in upcoming chapters. By applying the nearest neighbor algorithm, we can see that the same clusters are obtained if we use only the features {C1, C2}, {C2, C3}, or {C1, C3}. So, we can use any of these feature subsets.

4.3   Feature Selection Methods

Over the years, many strategies have been developed for dimensionality reduction to evaluate large datasets, but a majority of such techniques perform DR at the cost of structural destruction of data, i.e., principal component analysis (PCA), multidimensional scaling (MDS), etc. The presented DR techniques are categorized as transformation-based techniques and selection-based techniques. Figure 4.3 presents the hierarchy of dimensionality reduction techniques.

Table 4.3 A sample dataset in unsupervised learning

	C1	C2	C3
X1	X	Y	I
X2	Y	Z	I
X3	Z	Y	K
X4	X	X	J

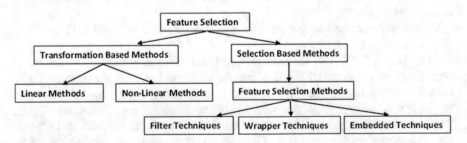

Fig. 4.3 Dimensionality reduction techniques

4.3.1   Transformation-Based Reduction

Transformation-based techniques include those methods which perform dimensionality reduction at the cost of destruction of the semantics of the dataset. These are further categorized into two categories: linear and nonlinear techniques.

Most of the conventional and deep-rooted linear DR methods use factor analysis. Principal component analysis and singular value decomposition belong to this category. Multidimensional scaling is another mathematical-based linear DR technique.

4.3.1.1   Principal Component Analysis

Principal component analysis (PCA) is a well-known multivariate statistical dimensionality reduction process used for data analysis; it identifies those variables that cause a large portion of the variance in large datasets. PCA is a mathematical tool that has been proven as a popular multicriteria decision analysis tool in many fields, i.e., decision-making, chemometrics, commercial airline industry, and life cycle assessment. The basic goals of PCA can be defined as:

1. Identifying the important information from the dataset
2. Data reduction by filtering the important information
3. Converts the complex data into a simple description
4. Structural analysis of the variables

To achieve these goals, PCA converts a large number of related variables to a smaller, more manageable number of linearly uncorrelated variables, called principal components. As a result of this conversion, the first variable or principal component accounts for most information about the dataset. Similarly, subsequent principal components represent maximum variability of the data. These components are computed by decomposing original data into eigenvectors and eigenvalues. The eigenvector term is a principal component that holds variance in the data and the eigenvalue of each eigenvector represents the amount of variability in the data. So, the eigenvector with the maximum eigenvalue is the first principal component and the eigenvector with the second highest eigenvalue will be the second principal component. All computed components or eigenvectors are projecting the multivariate dataset into a new dimension. Thus, DR performs by selecting an appropriate number of first k components and discarding the rest. Although PCA is the most used method for data analysis, it suffers from the following drawbacks:

1. It reduces dimensions but at the cost of damaging data underlying semantics.
2. It can be applied to numerical datasets only.
3. It studies only the linear structure of data.
4. The number of principal components decides by human intervention, to perform DR and maintain information loss as per application requirement.

4.3.1.2   Classical Multidimensional Scaling

Classical multidimensional scaling (MDS), also called principal coordinate analysis (PCoA), is a transformation-based technique that has been widely used in many applications, e.g., analysis of the stock market, medical field-molecular analysis, and pattern recognition. MDS is another linear-based dimensionality reduction approach that reduces dimensions by measuring the distance or dissimilarity between two data points. These dissimilarities are interpreted with the distance of data points on a graph in MDS, i.e., the objects that are similar are closely located on the graph than those objects that are less similar. In the process of reducing high dimensional data to low dimension, data instances that are closer in higher dimension remain closer after conversion to lower space, e.g., data related to shoe brands displaying two brands near to each other in terms of price, in a high dimensional space, MDS converts the data into lower space and these two brands still project closer to each other. So, these methods are named distance methods. If we have p-dimensional space, we can define the MDS algorithm as:

1. Place features of each instance in p-dimensional space.
2. Compute the distance of all points using the Euclidean distance formula (different distance approaches can be used, especially for large datasets Euclidian distance may not suitable):

$$d(y(i), y(j)) = \sqrt{(y(i) - y(j) * y(i) - y(j))} \tag{4.1}$$

3. Compare the matrix with the input matrix by considering the criterion function called stress, i.e., the smaller the value, the greater the similarity between two points. The general formula to measure stress is:

$$\sqrt{\frac{\sum\sum (f(x_{ij}) - d_{ij})^2}{\text{scale}}} \tag{4.2}$$

where d_{ij} refers to Euclidean distance between two data points for all dimensions and scale denotes a constant scaling factor, to keep the value of stress in a range of 0–1. When the stress value is smaller, the representation of data seems good.

Recent versions of MDS, called non-metric MDB, can also handle nonlinear complex data. Classical or metric MDS is also an old approach like PCA which maps data from higher to lower dimensions (normally two to three dimensions) by keeping the distance between data instances at maximum and is suitable for the linear data structure. It requires large computational resources to compute the dissimilarity matrix in each iteration. Therefore, it is hard to include new data in MDS. Moreover, it also destroys the underlying structure of the dataset while mapping from high dimension to low dimensional space.

PCA and MDS are no doubt old and popular approaches used for dimensionality reduction, but they are handling only linear datasets. The advent of nonlinear complex data structure in modern times, e.g., graphs, trees, etc., account for the rapid growth in approaches for nonlinear dimensionality reduction. Few of such approaches are locally linear embedding (LLE), Isomap, etc.

4.3.1.3   Locally Linear Embedding

Locally linear embedding (LLE) is one of the unsupervised approaches which reduces dimensions while conserving geometric features of the nonlinear dataset. In the LLE algorithm, the eigenvector optimization technique has been used. At the start, we find the nearest neighbor of each data point using Euclidean distance for k nearest neighbors. Then, the weight of each data point is computed which helps in the reconstruction of data over the linear plane. Error while reconstructing the data points, also called cost function, can be computed by:

$$E\left(W\right) = \sum_i \left| X_i - \sum_j W_{ij} X_j \right|^2 \tag{4.3}$$

where X_i is data point with neighbor X_j and W_{ij} is a weight matrix used to reconstruct data points on linear dimensions. The error can be minimized by the following constraints:

1. To reconstruct each data point, a neighboring point is used and if there is no neighboring point, the algorithm assigns zero to W_{ij}.
2. Summation of each row in the weight matrix W_{ij} is "1."

But LLE algorithm does not define the procedure to map reconstructed data points to the original manifold space.

4.3.1.4   Isomap

Isomap is a dimensionality reduction for handling nonlinear data structures. It is a generalization of the multidimensional scaling method which uses geodesic interpoint distances instead of Euclidean distances. Isomap is a very effective method based on a simple approach of a rough estimation of the nearest neighbors of each data point on manifold space. If we have r data points on manifold space with distance matrix d, Isomap performs the eigen decomposition of matrix d. The procedure of the Isomap algorithm is:

1. Find nearest neighbors of each data point on manifold data space d using the distance between points.

2. Compute the geodesic distance between points by measuring the shortest path distance in the graph and construct a distance matrix.
3. Perform eigenvalue decomposition on distance matrix to find low dimensional space.

The efficiency of Isomap decides the size of the neighborhood, i.e., selection of the large size of the neighborhood may cause difficulty to find the shortest path on the graph and too small size of the neighborhood may cause sparsity of graph.

4.3.2   Selection-Based Reduction

In the above section, we have discussed a few dimensionality reduction techniques. Although these techniques are sufficiently successful in reducing dimensions, it is required to map data in lower space with the least destruction in the data structure. Selection-based DR techniques are referred to as semantic preserving techniques. Now we will focus on the selection-based dimensionality reduction technique, for processing large datasets. These techniques can be categorized as:

1. Filter techniques
2. Wrapper techniques
3. Embedded techniques

4.3.2.1   Filter-Based Methods

The filter-based approach is the simplest method for feature subset selection. In this approach, features are selected without considering the learning algorithm. So, the selection remains independent of the algorithm. Now each feature can be evaluated either as an individual or as a complete subset. Various feature selection criteria can be used for this purpose. In individual feature selection, each feature is assigned some rank according to a specified criterion, and the features with the highest ranks are selected. In other cases, entire feature subset is evaluated. It should be noted that the selected feature set should always be optimized, that is, the minimum number of features in the feature subset is better.

A generic filter-based approach is shown in Fig. 4.4:

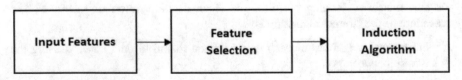

Fig. 4.4 A generic filter-based approach

It should be noted that only a unidirectional link exists between the feature selection process and the induction algorithm.

Pseudo code of a generic feature selection algorithm is shown in Listing 4.1:

Listing 4.1 A generic filter approach

```
Input
S - Data sample with C number of features
E - Evaluation measure
SGO - Successor generation operator
Output
S' - Output solution
I := Start I;
Si := { Best of I with respect to E };
repeat
I := Search (I, SGOI,C);
C' := {best I according to E };
if E(C')≥E(S') or (E(C')=E(S') && |C'| < |S'|) then S' :=C';
until Stop(E,I).
```

X is the input feature set and "E" is the measure to be optimized (selection criteria). S' is the final result that will be output. GSO is the operator to generate the next feature subset. We may start with an empty feature set and keep on adding features according to criteria or we may start with full features set and keep on removing unnecessary features. The process continues until we get the optimized feature subset.

There are many feature selection methods based on the filter technique. FOCUS is a filter-based approach to select good feature subsets, using the breadth-first search technique. It starts evaluating all subsets until a consistent subset has been achieved. However, the algorithm is not suitable for datasets having large feature sets. SCRAP is another feature selection approach, and it selects relevant feature sets by doing a sequential search. It identifies the importance of features by analyzing the effect on each object of the dataset in terms of change in the value of the decision attribute. SCRAP starts evaluation with any random object, called PoC (first point of the class of change), and then evaluates against the nearest neighboring object; the process continues until all objects have been assessed. Although filter methods are faster, a major drawback is these methods select a large number of features as a relevant subset.

4.3.2.2 Wrapper Methods

In contrast with filter approaches, where a selection of features is independent of the learning model, in wrapper-based approaches, the selection is done by keeping in mind the learning model (the algorithm that will use the features) and the features that enhance the accuracy and performance of the learning model that are selected.

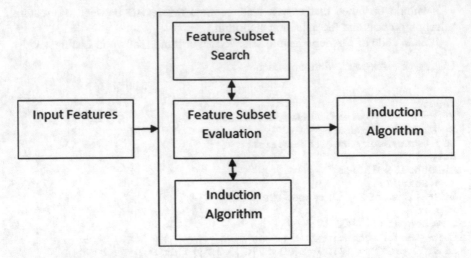

Fig. 4.5 A generic wrapper-based approach

So, the selected features should be aligned with the learning model. A generic wrapper-based approach is shown in Fig. 4.5:

As you can see, the feature selection process comprising of feature search and evaluation is interconnected with the induction algorithm; the overall process comprises three steps:

1. Feature subset search.
2. Feature evaluation considering the induction algorithm.
3. Continue the process till optimization.

It should be noted that the feature subset is sent to the induction algorithm and the quality of the subset is determined based on its feedback. The feedback may be in the form of any measure, e.g., error rate.

A few popular wrapper-based feature selection strategies are:

1. *Forward selection:* In this method, we start with an empty feature set, and then iteratively we add a feature that gives the best performance. This process continues until the addition of a feature does not increase performance.
2. *Backward selection*: In backward selection, we have a complete set of features and then we eliminate features one by one which enhances performance. We repeat this process till no further improvement is seen upon removal of features.
3. *Recursive feature elimination*: This is an exhaustive search method, which creates a subset of features, iteratively, and then identifies the best or worst features. It is termed a greedy search strategy.

Wrapper methods are comparatively expensive as compared to filter methods, but wrapper methods always provide the best subset. However, wrapper models have more chances of overfitting.

Table 4.4 Comparison of feature selection approaches

	Filter	Wrapper	Embedded
Advantage	Simple approach More efficient	Feature dependencies are considered Feedback from the classification algorithm is considered, so high-quality features are selected	Combines advantages of both filter-based and wrapper-based approaches
Disadvantage	Feedback from the classification algorithm is not considered, so the selected features may affect the accuracy of the algorithm	Computationally expensive as compared to the filter-based approach May result in high overfitting	Specific to machine learning

4.3.2.3 Embedded Methods

Both of the previous approaches discussed have their advantages and disadvantages. Filter-based approaches are efficient, but the features selected may not be quality features because feedback from the learning algorithm is not considered. In the case of wrapper-based approaches, although quality features are selected, getting feedback and updating features, again and again, is computationally inefficient.

Embedded methods take advantage of the plus points of both of these approaches. In the embedded method, features are selected as part of the classification algorithm, without taking feedback from the algorithm. A generic embedded approach works as follows:

1. Feature subset initialization.
2. Feature subset evaluation using evaluation measure.
3. It becomes a new current subset if better than the current subset.
4. Evaluation of subset using evaluation criteria of classifier.
5. It becomes a new current subset if better than the current subset.
6. Repeat Step 2 to Step 5 until specified criteria are met.

The generic comparison of each of the abovementioned approaches is given in Table 4.4:

4.4 Objective of Feature Selection

The core objective of the feature selection process is to select those features that could represent the entire dataset in terms of the information provided by them. So, the lesser the number of features, the more the performance. Various objectives of feature selection are provided in the literature. Some of these are:

1. Efficiency: Feature selection results in a lesser number of features that provide the same information but enhance the quality. Here it is shown using an example given in Table 4.5:

Note that if we classify the objects in the above dataset using all three features, we get the classification given on the left side of the diagram. However, if we only use features C1 and C2, we also get the same classification as shown in the Fig. 4.6. So, we can use the feature subset {C1, C2} instead of the entire dataset.

The abovementioned is a very simple example. Consider the case when we have hundreds and thousands of features.

2. Avoid overfitting: Since feature selection helps us remove the noisy and irrel-evant features, the accuracy of the classification model also increases. In the example given above, we can safely remove the feature C3 as removing this feature does not increase or decrease the accuracy of classification.

3. Identifying the relationship between data and process: feature selection helps us the relationship between features to understand the process that generated those features. For example, in the table, we can see that "Class" is fully dependent on {C1, C2}, so to fully predict the class we should have values of these two features.

Table 4.5 Symptoms table

Object	C1	C2	C3	Class
X1	T	T	T	T
X2	T	T	F	T
X3	T	F	T	T
X4	F	T	T	T
X5	F	F	F	F
X6	F	T	F	T
X7	F	F	T	F

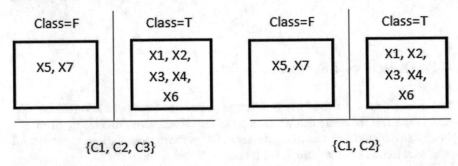

Fig. 4.6 Classification: using selected features vs the entire feature set

4.5 Feature Selection Criteria

Selection of the criteria according to which features are selected is an important step in the entire selection process. In Table 4.5, our criterion is comprised of selecting the features that provided the same classification structure. It should be noted that multiple feature subsets may exist in the same dataset that may full the same evaluation criteria. Here we present a few measures that may be used as underlying criteria for feature set selection.

4.5.1 Information Gain

Information gain relates to the uncertainty of the feature. For two features X and Y, feature X will be preferable if $IG(X) > IG(Y)$. Mathematically:

$$IG(X) = \sum_i U(P(C_i)) - E\left[\sum_i U(P(C_i|X))\right] \qquad (4.4)$$

Here:

U = Uncertainty function
$P(C_i)$ = Probability of C_i before considering feature "X."
$P(C_i|X)$ = Probability if C_i after considering feature "X."

4.5.2 Distance

Distance measure represents how effectively a feature can discriminate the class. The feature with high discrimination power is preferable. Discrimination, here, can be defined in terms of the difference in the probabilities $P(X|C_i)$ and $P(X|C_j)$. Here C_i and C_j are two classes and $P(X|C_i)$ means the probability of X when the class is C_i. Similarly, $P(X|C_j)$ shows the probability of class C_j X when class is C_j. So, for two features X and Y, X will be preferable if $D(X) > D(Y)$.

4.5.3 Dependency

Dependency is another measure that can be used as a selection criterion. Dependency measure specifies how uniquely the value of a class can be determined from a feature or a set of features. The more the dependency of a class on the feature, the lesser the uncertainty in the feature, hence the more preferable that feature is.

The majority of feature selection algorithms use this measure. Starting with an empty feature set, they keep on adding features until we get a feature subset having maximum dependency, i.e., one (1).

Note that a feature R will be considered redundant if D(X,D)=D(XU{R},D).

4.5.4   Consistency

Consistency is another measure that can be used for feature subset selection. A consistent feature is the one that provides the same class structure otherwise provided by the entire feature subset, so:

$$P(C|Subset) = P(C|Entire\ Feature\ Set). \tag{4.5}$$

So, the more the feature is consistent, the more it is preferable.

4.5.5   Classification Accuracy

This measure is commonly used in wrapper-based approaches. The features that enhance the accuracy of the model are always preferable. The quality of features is enhanced gradually using the feedback from the model. The measure provides better features in terms of quality of feature subset; however, it may be expensive as compared to other criteria.

4.6   Feature Generation Schemes

One of the important steps in feature selection algorithms may be to generate the next subset of features. It should be noted that the next feature subset should contain more quality features as compared to the current one. Normally three feature subset selection schemes are used for this purpose.

Firstly, we have a forward feature generation scheme, where the algorithm starts with an empty set and keeps on adding new features one by one. The process continues until the desired criteria are met up to the maximum.

A generic forward feature generation algorithm is given in Listing 4.2:

Listing 4.2 A generic forward feature generation algorithm

```
Input
S – Dataset containing X features
E – Evaluation criteria
Initialization
    c) S' ← {ϕ}
do
b) ∀x ∈ X
c) S' ← S' ∪ {x_i}    i = 1..n
    d) until Stop(E, S')
Output
d) Return S'.
```

S' here is the final feature subset that the algorithm will output. Initially, it is initialized with an empty set, and one by one features are added until the solution meets the evaluation criteria "E."

Secondly, we can use the backward feature generation scheme. This is opposite to the forward feature generation mechanism. We keep on eliminating the feature subset until we come to a subset from which no more features can be eliminated without affecting the criterion.

Listing 4.3 shows a generic backward feature selection algorithm.

Listing 4.3 A sample backward feature generation algorithm

```
Input
S – Data sample with features X
E – Evaluation measure
Initialization
a) S' ← X
do
b) ∀x ∈ S'
c) S' ← S' - {x_i}    i = n..1
d) until Stop(E, S')
Output
d) Return S'.
```

Note that we have assigned the entire feature set to the S' and are removing features one by one. A feature can safely be removed if, after removing the feature, the criteria remain intact. We can also use the combination of both approaches where an algorithm may start with an empty set and a full set. On one side we keep on adding features while on the other side we keep on removing features.

Thirdly, we have a random approach. As the name employees, in a random feature generation scheme, we randomly include and skip features. The process continues until the mentioned criterion is met. Features can be selected or skipped by using any scheme. Listing 4.4 shows a simple one where a random value is generated between 0 and 1. If the value is less than one, the feature will be included else excluded. This gives each feature an equal opportunity to be part of the solution.

A generic random feature generation algorithm is given in Listing 4.4:

Listing 4.4 A random feature generation algorithm using hit and trial approach

```
Input
S – Dataset containing X features
E – Evaluation measure
S'←{φ}
For j=1 to n
If (Random(0,1) <= 0.5) then
S'←S' ∪ {Xᵢ}
Until Stop(E, S')
Output
d) Return S'.
```

Details given above provide the basis for three types of feature selection algorithms. Initially, we have exhaustive search algorithms where the entire feature space is searched for selecting an appropriate feature subset. Exhaustive search provides an optimum solution, but due to the limitation of resources, it becomes infeasible for datasets beyond smaller size.

Random search, on the other hand, randomly searches the feature space and keeps on continuing the process until we get some solution before the mentioned time. The process is fast but an alternate drawback is that we may not have an optimal solution.

The third and most common strategy is to use heuristics-based search, where the search mechanism is guided by some heuristic function. A common example of such types of algorithms is the genetic algorithm. The process continues until we find a solution or a specified time threshold has been met.

So, overall feature selection algorithms can be characterized by three parts:

1. Search organization
2. Generation strategy for next feature subset selection
3. Selection criteria

Figure 4.7 shows different parts of a typical feature selection algorithm.

As discussed before, rough set theory and dominance-based rough set theory can be used for feature selection; we will provide a detailed discussion on both of these theories and how they can be used for performing feature selection.

4.7   Rough Set Theory

The feature selection process involves the selection of a minimal subset of features while sustaining the maximum information about the problem domain and accuracy must also not be compromised. One of the widely used data analysis methods is

Fig. 4.7 A generic feature
selection process

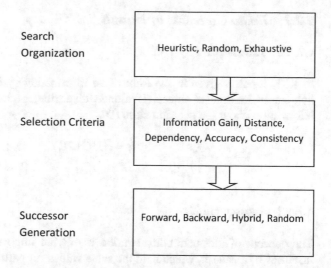

Search
Organization — Heuristic, Random, Exhaustive

Selection Criteria — Information Gain, Distance,
Dependency, Accuracy, Consistency

Successor
Generation — Forward, Backward, Hybrid, Random

rough set theory-based techniques which have been successfully used to perform data mining and knowledge discovery, mainly feature selection. Rough set theory (RST) can be used to successfully identify data dependencies and perform data reduction without destroying original data semantics. RST is a mathematical-based data analysis tool that was proposed by Pawlak in 1982. Feature selection algorithms based on rough set theory consider filter techniques. Filter techniques are most widely used for feature selection as compared to wrapper techniques due to the computational complexity of wrapper techniques where a selection of relevant feature sets depends on learning algorithms. Filter techniques are independent of learning algorithms.

Data reduction can be made using RST by using two approaches: (1) considering dependency measure and (2) discernibility matrix. RST generates a reduct set, a subset of relevant features, which does not affect the accuracy while performing classification, prediction, decision-making, and other subsequent data mining tasks.

The most prominent features of RST include:

(a) Ability to analyze hidden data patterns.
(b) Does not require any additional information about the data, i.e., the probability distribution for Bayesian decision theory.
(c) Results are easy to interpret.
(d) Reduce the size of knowledge representation.

RST is based on the conventional set theory of mathematics that can extract minimal but relevant knowledge from the underlying data. In the underlying section, we discuss a few basic concepts of RST.

4.7.1   Basic Concepts of Rough Set Theory

4.7.1.1   Information System

In RST, input is given in the form of an information system which consists of a non-empty, finite set of objects (U) along with attributes (conditional attribute C) and information about the decision class (D).

$$\alpha = (U, C \cup D) \tag{4.6}$$

4.7.1.2   Indiscernibility

The concept of indiscernibility relation is of vital importance in RST. All those instances of a dataset, which hold the same values for particular attributes, consider indiscernible objects for those attributes or criteria. Dataset can be reduced by finding indiscernible objects. Indiscernibility can be named as the equivalence relation between objects. Two objects A and B considered indiscernible in terms of attributes Q ($Q \subseteq C$):

$$IND\ (P) = \{(A,\ B) \in U \times U \mid \forall = q \in Q, f(A,\ q) = f(B,\ q)\} \tag{4.7}$$

4.7.1.3   Lower and Upper Approximations

RST defines two approximation sets to approximate classes that are not distinguishable in terms of conditional attributes:

The lower approximation set of decision class X includes those instances that with certainty belong to class X.

$$X : \underline{P}X = \{x[x]_B \subseteq X\} \tag{4.8}$$

The upper approximation set of decision class X contains all those instances which may belong to class X.

$$X : \overline{P}X = \{x[x]_B \cap X \neq 0\} \tag{4.9}$$

4.7.1.4   Boundary Region

The difference between the two approximation sets of decision class X defines the boundary region:

$$X : BN_B(X) = \overline{P}X - \underline{P}X \tag{4.10}$$

If the resultant set is empty, it is called a crisp set; alternatively, it is named rough. If the boundary region is empty, the set is called "crisp" otherwise set is termed "rough."

4.7.1.5   Dependency

Dependency measure is used to construct a reduct set, a minimal subset of features. It measures the relationship between attributes. If the dependency value between two attributes is "1," it shows full dependency, and if the value is in the range of 0–1, it shows partial dependency.

$$k = \gamma(C, M) = \frac{|POS_C(M)|}{|U|} \tag{4.11}$$

POS is the positive region or lower approximation.

4.7.1.6   Reduct Set

RST generates reduct sets using dependency measure:

$$\gamma(C, D) = \gamma(\acute{C}, D) \quad \text{for } \acute{C} \subseteq C \tag{4.12}$$

An attribute set $\acute{C}$ can be called reduct, if dependency of C and $\acute{C}$ is same.

4.7.1.7   Discernibility Matrix

Two objects are considered discernable if their values are different in at least one attribute. For example, consider the information system given in Table 4.6:

In this table, conditional attributes are {a, b, c} and decision attribute is {d}. x1 and x2 are discernible in terms of conditional attributes "a" and "c" as shown in Table 4.7. Objects which belong to the same decision class, represented by "λ" in the matrix, e.g., objects x2 and x3, belong to different decision classes.

$$\text{Reduct sets} = (a \cup c) \cap (b) \cap (c) \cap (a \cup b)$$
$$= (b) \cap (c)$$

Therefore, the reduct set for the information system given in Table 4.6 is {b, c}.

Table 4.6 Information system

U	A	b	c	d
x1	a0	b1	c1	Yes
x2	a1	b1	c0	No
x3	a0	b2	c1	No
x4	a1	b1	c1	Yes

Table 4.7 Discernibility matrix

	x1	x2	x3
x2	a, c		
x3	B	λ	
x4	Λ	c	a, b

4.7.2 Rough Set-Based Feature Selection Techniques

Many algorithms used rough set theory for feature selection. In this section, we reviewed a few feature selection techniques based on rough set theory.

4.7.2.1 Quick Reduct

Quick Reduct (QR) is a rough set-based feature selection that used the forward feature selection technique. In this technique, the algorithm started with an empty attribute set and then attributed selected one by one based on the dependency measure. The algorithm continued till the dependency of the feature subset was equal to the dependency value of the whole dataset. Although QR is one of the most commonly used RST-based feature selection algorithms, the algorithm used a POS (positive origin)-based approach to compute dependency which caused high computational cost.

There are many hybrid feature selection techniques based on rough set theory. This technique performed feature selection using particle swarm optimization and rough set theory. It also started with an empty feature set and selected attributes one by one by computing the fitness function using RST-based dependency measure. The feature which has the highest dependency measure would be selected. The algorithm continued until the stopping criteria reached a predefined maximum number of iterations.

There is another technique that is based on Quick Reduct and Improved Harmony Search Algorithm. It used a rough set-based dependency measure to check the fitness of the subset, and the algorithm stopped when maximum iterations were completed.

4.7.2.2 Rough Set-Based Genetic Algorithm

It is a rough set-based genetic algorithm for feature selection. The selected features are then provided to a neural network for classification. This algorithm also used POS-based dependency measure to select feature sets.

4.7.2.3   Incremental Feature Selection Algorithm

This algorithm starts with a feature subset and computes dependency value using rough set theory. It then incrementally selects a new feature set and measures dependency value and compares the dependency value of the selected feature subset with the dependency value of the whole dataset. If both values are equal to each other, the subset is selected as a reduct; otherwise new feature subset is selected. Finally, the redundant and irrelevant features are removed from the dataset.

4.7.2.4   Fish Swarm Algorithm

Fish Swarm Algorithm is another technique based on rough set theory. Initially, the algorithm has a feature subset, called fish. With time, fish change to find the best fit. A fish that achieves maximum fitness becomes part of reduct. The algorithm continues till a defined number of iterations.

4.7.2.5   Feature Selection Approach Using Random Feature Vector

This technique is based on the random feature vector selection method. In the first step, the algorithm forms a feature vector from the dataset by randomly selecting features using the dependency measure of rough set theory. Then it removes the irrelevant features from the subset to obtain the optimal subset.

It is evident from the above discussion that most feature selection techniques are using rough set-based dependency measure. To calculate dependency value, a lower approximation needs to be calculated. Therefore, it can be stated that the calculation of approximations is of central importance in RST.

4.7.3   Dominance-Based Rough Set Approach

Dominance-based rough set approach (DRSA) is an extension of rough set theory to consider the preference relation of attributes. It proves a useful approach to reducing the data size in the case of the multicriteria classification domain. The main concept of DRSA is based on the consideration of dominance relation by defining upward and downward union of classes. Using these classes, lower and upper approximation sets are calculated separately. DRSA has been widely used to deduce decision rules based on knowledge provided by lower approximation sets instead of a complete dataset. Therefore, it is an effective technique that successfully reduces data size while maintaining critical information about the dataset. Based on that, classification, prediction, analysis, etc. can be performed with more accuracy. A few features of dominance-based rough set theory are as follows:

1. DRSA handles preference order attributes.
2. DRSA deals with data inconsistencies.
3. It successfully handles the monotonic relationship between values of conditional attributes.
4. It can deduce critical information from the data in the form of approximation sets.
5. DRSA can be used to generate decision rules.

There are many DRSA-based data analysis algorithms that can be applied in medical classifications, customer analysis, behavior analysis, fraud detection, etc. However, the computational cost of applying DRSA may turn high, particularly for large datasets. In the following section, we will discuss the basic concepts of DRSA and the computational cost of generating approximation sets.

Now we will discuss basic notations of the DRSA.

4.7.3.1 Decision System

The decision system consists of a finite number of objects (U) and conditional attributes (C) against each object along with information about the targeted class (D). In DRSA, the decision system must also include a value set for conditional attributes (V) and defined dominance relation of attributes (f).

$$IS = (U, (CUD), V, f) \tag{4.13}$$

4.7.3.2 Dominance Relation

The main concept of DRSA is a dominance relation. In terms of conditional attributes, the dominance relation of objects is defined as positive and negative dominance relation:

1. If an object u2 is highly preferred over object u1 for the mentioned feature set, it is named as a dominance positive set of objects u1.

$$D_p^+(x) = \{y \in U; \mathbf{y}D_p x\} \tag{4.14}$$

2. If an object u1 is more preferred than object u1 for mentioned criteria, it is called a dominance negative set of objects u1.

$$D_p^-(x) = \{y \in U; \mathbf{x}D_p y\} \tag{4.15}$$

4.7.3.3   Upward and Downward Union of Classes

To handle the preference order of decision classes, DRSA categorize decision classes as the upward and downward union of classes:

1. Upward union of classes of decision class X includes objects which belong to X and more preferred decision class.

$$Cl_X^{\geq} = U_{s \geq x} Cl_s \tag{4.16}$$

2. Downward union of classes of decision class X contains all objects which belong to X and less preferred decision class.

$$Cl_X^{\leq} = U_{s \leq x} Cl_s \tag{4.17}$$

4.7.3.4   Lower and Upper Approximations

In DRSA, lower and upper approximations are computed for the upward and downward union of classes, individually, to address the preference ordered domain. Lower approximation sets include all those objects which certainly belong to the upward/downward union of classes. The definition is as under:

Lower approximation for upward union of classes:

$$\underline{P}(Cl_t^{\geq}) = \{x \in U : D_P^+(x) \subseteq Cl_t^{\geq}\} \tag{4.18}$$

Lower approximation for downward union of classes:

$$\underline{P}(Cl_t^{\leq}) = \{x \in U : D_P^-(x) \subseteq Cl_t^{\leq}\} \tag{4.19}$$

Similarly, the upper approximation consists of objects which may belong to an upward/downward union of classes:

Upper approximation for upward union of classes:

$$\bar{P}(Cl_t^{\geq}) = \{x \in U : D_P^-(x) \cap Cl_t^{\geq} \neq \emptyset\} \tag{4.20}$$

Upper approximation for downward union of classes:

$$\bar{P}(Cl_t^{\leq}) = \{x \in U : D_P^+(x) \cap Cl_t^{\leq} \neq \emptyset\} \tag{4.21}$$

4.7.3.5   Applications of Dominance-Based Rough Set Theory

Since its inception, dominance-based rough set theory has been widely applied in many complex domains, i.e., artificial intelligence, expert systems, etc. It has gained attention as a successful tool for the removal of irrelevant and redundant information from large datasets for accurately and efficiently performing subsequent tasks, i.e., classification, behavior analysis, decision-making, and predictions. It has been applied in many real-life applications, i.e., medical classification systems, spare parts management systems, airline customer analysis, bank fraud detection, risk analysis, etc. Dominance-based rough set theory approximates the complete dataset in terms of approximation sets which can then be used to perform other data mining tasks. DOMLEM algorithm is used for induction of decision rules based on approximation sets of dominance-based rough set theory. Instead of analyzing the complete dataset, the DOMLEM algorithm generates decision rules using lower approximation sets, a concise representation of the whole data. Therefore, DRSA successfully deduces critical information from the dataset which consequently improves the efficiency and accuracy of data mining tasks. It has been used in many applications for data extraction, rule generation, and pattern identification. It has been used to predict which students may drop out from an online course in the coming week by analyzing data from the previous week. Similarly, DRSA proves more beneficial for the prediction of bank frauds.

In this section, we will describe the implementation of DRSA in a real-life medical classification system. A digital system used to classify the patient by analyzing the symptoms termed a medical classification system. The flow graph of the medical classification is given in Fig. 4.8.

To diagnose breast cancer patients using a medical classification system based on DRSA, the following steps are performed:

1. Input dataset to system.
2. Apply DRSA algorithm to compute lower and upper approximations.
3. Input approximation sets to DOMLEM algorithm to generate decision rules.
4. Based on decision rules, perform diagnosis of the patient, either benign or malignant.

For experimental purposes, the UCI dataset named Breast Cancer Wisconsin dataset was used. The characteristics of the dataset are given in Table 4.8:

Applying DRSA conventional algorithm, we computed lower and upper approximations of the upward and downward union of classes for "benign" and "malignant" classes $\left\{ Cl_b^{\leq}, Cl_m^{\leq}, Cl_b^{\geq}, Cl_m^{\geq} \right\}$. DOMLEM algorithm deduces decision rules

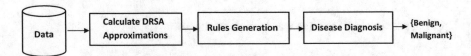

Fig. 4.8 Flow graph: Medical classification system for cancer patient

Table 4.8 Dataset characteristics

Characteristics of dataset	Values
Total number of instances	699
Total number of attributes	9 {q1, q2, q3, q4, q5, q6, q7, q8, q9}
Total number of decision classes	2 {benign, malignant}
Characteristics of attributes	Integer

based on DRSA approximations. After generating DRSA approximation sets, DOMLEM algorithm was applied to generate decision rules for classifying the patient as benign or malignant.

DRSA transforms data to a lower dimension by generating approximation sets. Using lower approximation sets, the DOMLEM algorithm designed decision rules. The medical classification system diagnoses the patient based on these rules. It has been evident that the DOMLEM algorithm that is based on DRSA uses lower approximation sets. Therefore, it can be noted that approximation is the central idea of DRSA. However, the computation of approximations using a conventional DRSA algorithm may cause a performance bottleneck while knowledge discovery from the large dataset.

Computational Cost of Calculation of Approximations of DRSA

Although DRSA-based data analysis algorithms are widely used for data reduction, particularly in the multicriteria classification domain, these algorithms incur high computational costs due to the lengthy computations of approximations. A conventional DRSA algorithm needs to navigate the whole dataset multiple times to compute lower and upper approximation sets. For example, the decision system given in Table 4.9 has three decision classes {Bad, Medium, Good}.

To compute a lower approximation for the upward and downward union of classes, the traditional DRSA algorithm performs heavy computations. In the first step, compute the upward/downward union of classes of all decision classes.

Table 4.9 Decision system

Object	q1	q2	q3	D
x_1	Medium	Medium	Bad	Bad
x_2	Good	Medium	Bad	Medium
x_3	Medium	Good	Bad	Medium
x_4	Bad	Medium	Good	Bad
x_5	Bad	Bad	Medium	Bad
x_6	Bad	Medium	Medium	Medium
x_7	Good	Good	Bad	Good
x_8	Good	Medium	Medium	Medium
x_9	Medium	Medium	Good	Good
x_{10}	Good	Medium	Good	Good

$$Cl_1^\geq = \{x_1, x_2, x_3, x_4, x_5, x_6, x_7, x_8, x_9, x_{10}\}$$
$$Cl_1^\leq = \{x_1, x_4, x_5\}$$
$$Cl_2^\geq = \{x_2, x_3, x_6, x_7, x_8, x_9, x_{10}\}$$
$$Cl_2^\leq = \{x_1, x_2, x_3, x_4, x_5, x_6, x_8\}$$
$$Cl_3^\geq = \{x_7, x_9, x_{10}\}$$
$$Cl_3^\leq = \{x_1, x_2, x_3, x_4, x_5, x_6, x_7, x_8, x_9, x_{10}\}$$

In the second step, compute the dominance positive and negative relation of all objects identified in the first step. To compute the dominance-positive and dominance-negative relation of a single object, it requires passing through a complete dataset multiple times and comparing each object in terms of all conditional attributes. For example, to compute $\underline{P}(Cl_1^\leq)$, compute dominance-negative relations of all three objects present in $Cl_1^\leq$. For this computation, the algorithm traverses the whole dataset three times and compares each object with the remaining objects in terms of all conditional attributes. Therefore, it is evident from the above discussion that, in the case of a large dataset, this step adversely affects the efficiency of the algorithm.

In the third step, the algorithm computes a lower approximation for the upward or downward union of classes by finding proper subsets from dominance sets in comparison with sets identified in the first step.

Similarly, to compute an upper approximation for the upward and downward union of classes of decision class, conventional DRSA algorithm needs to pass through all three computationally expensive steps. The algorithm repeats this whole process for computation of lower and upper approximation for the upward and downward union of classes for each decision class. Therefore, a solution requires which can efficiently compute approximations.

4.7.4 Comparison of Feature Selection Techniques

Table 4.10 shows the comparison of different feature selection techniques.

4.8 Miscellaneous Concepts

Now we will provide some concepts related to features and feature selection.

Table 4.10 Comparison of feature selection techniques

No	Techniques	Advantages	Disadvantages
1	Principal component analysis (PCA)	Removes correlated features Improves algorithm performance Reduces overfitting Improves visualization	Data standardization required Destroy original semantics of data Human input about the number of principal components is required
2	Classical multidimensional scaling (MDS)	Can handle linear complex data Reduce dimensions by measuring the distance between two objects only	Destroy underlying semantics of data Required large computational resources
3	Locally linear embedding	Handle nonlinear data Reduce data to low dimensions Better computational time	Inability to deal with the new data Destroy the original meaning of the data
4	Isomap	Handle nonlinear complex data Excellently map relationships between data points Reduce features	Destroy the original meaning of the data Need many observations
5	Rough set theory-based feature selection techniques	Reduce data dimensions Preserve the original meaning of the data Handle vague concepts	Inability to handle preference order of values of features The high computational time of traditional RST for large datasets
6	Dominance-based rough set theory	Handle preference order data Reduce data size Handle inconsistent data Preserves data semantics	The high computational cost for large datasets

4.8.1   Feature Relevancy

The relevancy of a feature is always related to the selection criteria. A feature will be said irrelevant if:

$$S(C - X_i) = S(X)$$

Here:

S is the selection criteria (e.g., dependency).
C is the current feature subset.
X is the entire feature set.

So, even if after removing the feature, the dependency on the current feature subset is equal to that of the entire feature subset, the feature can be termed as irrelevant and thus can be safely removed. Depending on the selection criteria a feature can either be strongly relevant, relevant, or irrelevant.

4.8.2 Feature Redundancy

Along with irrelevant features, a dataset may have redundant features as well. A redundant feature is one that does not add any information to the feature subset. So, a feature will be declared redundant if:

$$D(X \cup \{x_j\}) = D(X) \tag{4.22}$$

As it can be seen that the selection criterion remains the same even after removing the feature, the feature X_j will be declared redundant, and like an irrelevant feature, it can also be removed.

4.8.3 Applications of Feature Selection

Today we are living in a world where the curse of dimensionality is a common problem for applications to deal with. Within the limited resources, it becomes infeasible to process such huge volumes of data. So, feature selection remains an effective approach to deal with this issue. It has become a common preprocessing task for the majority of domain applications. Here we will discuss only a few domains using feature selection.

4.8.3.1 Text Mining

Text mining is an important application of feature selection. On daily basis, tons of emails, text messages, blogs, etc. are generated. This requires their proper processing (e.g., their classification, finding their conceptual depth, developing vocabularies, etc.). The feature selection approach has become very handy in such cases by enhancing the performance of the entire process. Literature is full of feature selection algorithms meant only for text mining.

4.8.3.2 Intrusion Detection

With the widespread use of the internet, the vulnerability of systems has increased for being attacked by hackers and intruders. Here the static security mechanisms become insufficient where the hackers and intruders are finding new mechanisms on daily basis, so we need to have dynamic intrusion detection systems. These systems may need to monitor the huge volume of traffic passed through them every minute and second. So feature selection may help them identify the important and more relevant features to select for inspection and detection thus alternatively enhancing their performance.

4.8.3.3 Information Systems

Information systems are one of the main application areas of feature selection. It is common to have information systems processing hundred and thousands of features for different tasks. So feature selection becomes a handy tool for such systems. To get the idea, you can simply consider the classification task using 1000 features versus the classification performed using 50 features by getting the same/sufficient classification accuracy.

4.9 Feature Selection: Issues

Feature selection is a great help in the era of the curse of dimensionality; however, the process has some issues which still need to be handled. Here we will discuss a few.

4.9.1 Scalability

Although feature selection provides a solution to handle the huge volumes of data but an inherent issue a feature selection algorithm faces is the required number of resources. For a feature selection algorithm to rank the features, the entire dataset is required to be kept in memory until the algorithm completes. Now keeping in mind the large volume of data, it requires large memory. The issue is unavoidable as to produce quality features entire data has to be considered. So, scalability of feature selection algorithms keeping in mind the dataset size is still a challenging task.

4.9.2 Stability

A feature selection algorithm should be stable that is it should produce the same results even with perturbations. Various factors such as the amount of data available, the number of features, distribution of data, etc. may affect the stability of a particular feature selection algorithm.

4.9.3   Linked Data

Feature selection algorithms make an important assumption that the available features are independent and ideally distributed. An important factor that is ignored is that features may be linked with other features and perhaps sometimes in other datasets, e.g., a user may be linked with its posts and posts may be liked by other users. Such a scenario especially arises in the domain of social media applications. Although research is underway to deal with this issue, handling linked data is still a challenge that should be considered.

4.10   Different Types of Feature Selection Algorithms

Based on how a feature selection algorithm searches the solution space, feature selection algorithms can be divided into three categories:

- Exhaustive algorithms
- Random search-based algorithms
- Heuristics-based algorithms

Exhaustive algorithms search the entire feature space for selecting a feature subset. These algorithms provide optimal results, i.e., the resultant feature subset contains the minimum number of features. This is one of the greatest benefits of feature selection methods; however, unfortunately, an exhaustive search is not possible as it requires a lot of processing resources. The issue becomes more serious when the size of datasets increases. For a dataset with n number of attributes, an exhaustive algorithm will have to explore 2^n solutions. Beyond requiring the computational power, such algorithms also require a lot of memory, and it should be noted that now it is common to have datasets with tens of thousands and millions of features. Exhaustive algorithms almost become impossible to apply to such datasets.

For example, consider the following dataset given in Table 4.11:

An exhaustive algorithm will have to check the following subsets:

Subset:{}{a}{b}{c}{d}{e}{a,b}{a,c}{a,d}{a,e}{b,c}{b,d}{b,e}{c,d}{c,e}{d,e}
{a,b,c}{a,b,d}   {a,b,e}{a,c,d}{a,c,e}{a,d,e}{b,c,d}{b,c,e}{b,d,e}{c,d,e}{a,b,c,d}{a,b,c,e}{a,b,d,e}{a,c,d,e}{b,c,d,e}{a,b,c,d,e}

Table 4.11 Sample dataset

U	a	b	c	d	E	Z
1	L	J	F	X	D	2
2	M	K	G	Y	E	2
3	N	J	F	X	E	1
4	M	K	H	Y	D	1
5	L	J	G	Y	D	2
6	L	K	G	X	E	2

Now consider a dataset the 1000 features or objects; the computational time will substantially increase. As discussed earlier that feature selection is used as preprocessing step for various data analytics-related algorithms, so, if this step is computationally expensive, it will result in serious performance bottlenecks for the underlying algorithm using it. However, we may have semi-exhaustive algorithms, which do not have to explore the entire solution space. Such algorithms search solution space until the required feature subset is found. These algorithms can use both forward elimination and backward elimination strategy.

The algorithm keeps on adding or removing features until a specified criterion is met. Although such algorithms are more efficient than fully exhaustive algorithms, even then, it is not possible to use such algorithms for medium or larger dataset sizes. Furthermore, such algorithms suffer from a serious dilemma, i.e., distribution of the features. For example, if a high-quality feature is found in the beginning, the algorithm is expected to stop earlier as compared to the one where the high-quality attributes are indexed in a later position in the dataset.

One optimization for such algorithms provided in filter-based approaches is that instead of starting a subset, we first rank all the features. Note that here we will use a rough set-based dependency measure for ranking and feature selection criteria. We check the dependency of the decision class ("Z" in the above given dataset) on each attribute one by one. Once all the attributes are ranked, we start combining the features. However, instead of combining the features in sequence as given above, we combine the features in decreasing order of their rank. Here we have made two assumptions.

- Features having high ranks are high-quality features, i.e., having minimum noise and high classification accuracy.
- Combining the features with high ranks is assumed to generate high-quality feature subsets more quickly as compared to combining the low-ranked features. Features having the same rank may be considered redundant.

This approach results in the following benefits:

- The algorithm will not depend on the distribution of features as the order of the features becomes insignificant while combing the features for subset generation.
- The algorithm is expected to generate results quickly, alternatively increasing the performance of the corresponding algorithm using it.

However, these algorithms provide more efficiency as compared to the previously discussed semi-exhaustive algorithms, even these algorithms can be computationally expensive when the number of features and objects in datasets increases beyond a smaller size.

Then we have random feature selection algorithms. These algorithms use the hit and trial method for feature subset selection. A generic random feature selection algorithm is given in Listing 4.4.

The advantage of these algorithms is that they can generate feature subsets more quickly as compared to other types of algorithms, but the downside is that the generated feature subset may not be optimized, i.e., it may contain redundant or irrelevant features that do not participate in feature subset. Finally, we have a heuristics-based algorithm where the solution space is searched based on some heuristics; we start with random solutions and keep on improving the solutions with the passage of time. Examples of such algorithms may include genetic algorithms and swarm-based approaches. Here we will discuss the genetic algorithm and particle swarm algorithm in detail.

4.11   Genetic Algorithm

A genetic algorithm is one of the most common heuristics-based approaches inspired by the nature. The algorithm is based on the concept of survival of the fittest. The species that adopt themselves with the passage of time and strengthen their capacities to survive are supposed to survive as compared to the ones that do not adopt and vanish with the passage of time.

Applying the concept of an optimization problem, we select an initial solution and keep on updating it so that finally we get our required solution.

The Listing 4.5 shows the pseudo code of the genetic algorithm:

Listing 4.5 Pseudo code of the genetic algorithm

```
Input
C: Feature Subset X: Objects in the dataset

1) Initialize the initial population.
2) Check the fitness of each individual.
Repeat
3) Select the best Individuals.
4) Perform their crossover and generate offsprings.
5) Perform mutation.
6) Evaluate the fitness of each individual.
7) Replace the bad individuals with new best individuals.
8) Until (Stopping Criteria is met).
```

Now we will explain each step one by one.

Initializing the Population

In a genetic algorithm, we randomly initialize the population. The population in the genetic algorithm is a collection of chromosomes. A chromosome represents a potential solution. For the feature selection problem, a chromosome represents the features that are part of the selected subset. The chromosome is generated using a binary encoding scheme where the value "1" represents the presence of a feature and the value "0" represents its absence. These features are selected randomly. For the dataset, given in Table 4.6, a random chromosome may be:

1	0	0	1	1

Here the first "1" represents the presence of the feature "a," and the second and third "0" values show that features "b" and "c" will not be part of the solution. Similarly, "d" and "e" will be included as shown by the values "1" in fourth and fifth place. Each value is called a "Gene." So, a gene represents a feature. The resulting chromosome shown above represents the feature subset {a,d,e}.

A population may contain many chromosomes depending on the requirements of the algorithm.

Fitness Function

Once the population is initialized, the next task is to check the fitness of each chromosome. The chromosomes having higher fitness are considered and those with low one are discarded. The fitness function is the one that specifies our selection criteria. It may be Information Gain, Gini Index, or dependency in the case of rough set theory. Here we will consider the dependency measure as selection criteria. The chromosome with higher dependency is preferable. An ideal chromosome is one having the fitness value of "1," i.e., equal to the maximum dependency of decision class on the complete set of features present in the dataset.

After this, we start our iterations and keep on checking until the required chromosome is found.

It should be noted that when chromosomes are initialized, different chromosomes may have different fitness. We will select the best ones having higher dependency and will use them for further crossover. This is in line with the concept of survival of the fittest inspired by the nature.

Crossover

Applying crossover is one of the important steps of the genetic algorithm by using which we generate offsprings. The offsprings form the next population. It should be noted that only the solutions having high fitness participate in the crossover, so that the resulting offsprings are also healthy solutions. There are various types of crossover operators. Following is the example of a simple one-point crossover operator in which two chromosomes simply replace one part of each other to generate new offspring.

1	0	1	0	1

1	1	0	1	0

We replace the parts of chromosomes after the third gene. So, the resulting offsprings become:

1	0	1	1	0

1	1	0	0	1

Similarly, in a two-point crossover, we select two random points, and the chromosomes replace the genes between these points as shown below:

1	0	0	1	1

1	1	1	0	0

There may be other types of crossover operators, e.g., uniform crossover, order-based crossover, etc. The choice of selecting a specific crossover operator may depend on the requirements.

Mutation

Once crossover is performed, the next step is to perform mutation. The mutation is the process of randomly changing a few bits. The reason behind this is to introduce diversity in genetic algorithms and introduce new solutions so that previous solutions are not repeated. In its simplest form, a single bit is flipped in the offspring. For example, consider the following offspring resulting after the crossover of parent chromosomes:

1	0	1	1	0

After flipping the second gene (bit) the resultant chromosome becomes:

1	1	1	1	0

Just like the crossover, there are many types of mutation operators. Few may include flip bit mutation, uniform, nonuniform, Gaussian mutation, etc.

Now we evaluate the fitness of each chromosome in the population, and the best chromosomes replace the bad ones. The process continues until the stopping criteria are met.

Stopping Criteria

There can be three stopping criteria in a genetic algorithm. The first and the ideal one is we find the ideal chromosomes, i.e., the chromosomes that have the fitness required. In the case of dependency measure, the chromosomes having the dependency value of "1" will be the ideal ones. The second possibility is that we do not get the ideal chromosome, in such cases the algorithm is executed till the chromosomes keep on repeating. So, in this case, we can select the chromosomes with maximum fitness. However, in some cases, we do not get this scenario even. So, in all such cases, we can use generation threshold, i.e., if the algorithm does not produce a solution after n generations, we will stop, and the chromosomes with maximum fitness may be considered as a potential solution.

Table 4.12 Comparison of heuristics and exhaustive approaches

	Heuristics-based approaches	Exhaustive approaches
Execution time	Take less time	Take less time
Result optimality	Do not ensure optimal results	Ensure optimal results
Dataset size	Can be applied to a dataset of any size	Can only be applied to small datasets
Resources	Take less resource	Take more resources

Heuristics-based algorithms are most commonly used. However, they may have certain drawbacks, e.g.:

- They may not produce the optimal solution. In the case of a genetic algorithm, the resulting chromosomes may contain irrelevant features having no contribution to the feature subset.
- As heuristics-based algorithms are random in nature, different executions may take different execution times, and the same solutions may not be produced again.

Table 4.12 shows the comparison of both exhaustive and heuristics-based approaches.

4.12 Feature Engineering

One of the important tasks that data scientists deal with is to come up with the features that make the data science applications work. Feature engineering is the process of creating the features relevant to the underlying problem using the knowledge about the available data. Once such features are prepared, the data is fed to data science applications for performing their tasks.

Feature engineering deals with different types of problems, e.g., feature encoding, handling missing values, feature transformation, etc. We will discuss a few problems here.

4.12.1 Feature Encoding

Feature encoding is the process of converting features into a form that is easy for data science applications to process. There are different data encoding techniques depending upon the types of the data. Now we will discuss a few of these techniques.

4.12.1.1 One-Hot Encoding

One-hot encoding is the process of converting the categorical features to the corresponding binary values for the data science application to process conveniently.

This technique deals with encoding the feature values based on their presence or absence for a certain object.

For example, consider the following dataset as shown in Table 4.13:

Now as you can see, the "Color" feature is of the nominal type which means that there is no particular order between the values of this feature. Normally processing such types of variables is a tough job for data science applications as processing these string values takes a lot of time. Now the one-hot encoding helps us replace these string values with simple binary values which are relatively easy to process. The process will be carried out as follows.

First, we will see the number of distinct categories in the feature. Here in the "Color" feature, we have three categories, so we will define three extra columns as shown in Table 4.14:

Note that we have named these new features based on the values in the "Color" features. Secondly, assign a binary "1" in each feature value, where that particular value is present, otherwise, assign a binary "0." Binary "1" means that the value is present and binary "0" means that the value is not present. For example, in the feature "Color_Red," we have assigned a value "1" for ObjectID 1 and 3 because these objects have a value of "Red" in the "Color" feature. Similarly, for the feature "Color_Blue," we have assigned the value "1" for ObjectID "2" and "4."

You can see from the three newly generated features that the value "100" can be used to represent the "Red" value and "010" can be used to represent the "Blue" value. Similarly, "001" can be used to represent the "Yellow" value. It should, however, be noted that the third feature, i.e., "Color_Yellow" is a redundant one, i.e., we can still recognize each value uniquely even if the third feature is not present. It means that the value "10" can represent the "Red" color, "01" can represent the "Blue" color, and the value "00" can represent the "Yellow" color. So, if there are "K" values in a nominal feature, we can generate the new "K-1" features to represent

Table 4.13 Sample dataset for one-hot encoding

ObjectID	Color	Size	Grade
1	Red	Small	0
2	Blue	Small	1
3	Red	Large	1
4	Blue	Large	1
5	Yellow	Extra Large	1

Table 4.14 Dataset with three new features for "Color"

ObjectID	Color	Size	Grade	Color_Red	Color_Blue	Color_Yellow
1	Red	Small	0	1	0	0
2	Blue	Small	1	0	1	0
3	Red	Large	1	1	0	0
4	Blue	Large	1	0	1	0
5	Yellow	Extra Large	1	0	0	1

Table 4.15   Dataset with two new features for "Size"

ObjectID	Color	Size	Grade	Size_Small	Size_Large
1	Red	Small	0	1	0
2	Blue	Small	1	1	0
3	Red	Large	1	0	1
4	Blue	Large	1	0	1
5	Yellow	Extra Large	1	0	0

all the unique values in that feature. Similarly, for the Size feature, we will have to create two new features as shown in the Table 4.15.

So, in this way, we have converted the nominal feature into binary (integer) features which are relatively easy to process. One of the biggest benefits of one-hot encoding is that it is a simple encoding scheme that we can easily use for any type of data science application. However, one of the major issues with this technique is that it cannot be used for features having a large number of unique feature values. For example, consider a feature named "ProductCategory" in a shopping store database. If there are 100 product categories in that store, we will have to generate at least 99 new features to handle just this "ProductCategory" feature which ultimately may result in the form of a dilemma called the "urse of dimensionality" which is another problem the data science applications have to deal with.

4.12.1.2   Label Encoding and Ordinal Encoding

After one-hot encoding, the label encoding and the ordinal encoding is the simplest one. In one-hot encoding, we have to generate new features which may increase the data volume. In label encoding and ordinal encoding, the data size does not change. The process is very simple; we simply assign integer values to each category value in the categorical feature. The assigned value is then used to refer to that particular value of the feature. For example, if there are three category values "Red," "Blue," and "Yellow," we can simply assign an integer value to each of them. For example, we may assign the color "Red" with the integer value of "0," "Blue" color with the integer value of "1," and the "Yellow" color with the value of "2." Note that we can assign any value as per our convenience. For example, values may be 1, 2, and 3 as well in our case.

So, the "Color" feature in our dataset shown in Table 4.13 will be encoded as shown in Table 4.16.

As you can see, the entire feature is now converted to integer values which are easy to deal with as compared to original categorical values.

The values to categories in categorical features are assigned arbitrarily and depend on the user. There is no particular relationship between these values as each value is independent of the others. However, in the ordinal types of features, some values may be more significant than others, e.g., according to user perception

Table 4.16 Sample dataset for one-hot encoding

ObjectID	Color	Size	Grade
1	1	Small	0
2	2	Small	1
3	1	Large	1
4	2	Large	1
5	3	Extra Large	1

Table 4.17 Sample dataset for one-hot encoding

ObjectID	Color	Size	Grade
1	1	1	0
2	2	1	1
3	1	2	1
4	2	2	1
5	3	3	1

the value "Large" represents something that has more volume or size as compared to something that is "Small." So, in simple words, we can say that the order should be considered while encoding such types of values. This is what "ordinal encoding" deals with. In ordinal encoding, we assign the values based on the relative relationship between the values of the features. The assigned values should represent the same type of relationship. For example, in our case, we can assign the value "1" to the feature value "Small," "2" to the feature value "Large," and "3" to the feature value "Extra Large" in our "Size" feature. As the value "2" is bigger than the value "1," it shows that "Large" is bigger than "Small."

Note that the new values are assigned to the categories based on the knowledge of the domain. So, the "Size" feature in our dataset shown in Table 4.13 will be encoded as shown in Table 4.17.

As you can see, the size of the data remains the same and we do not need to deal with the extra data volumes.

4.12.1.3   Frequency Encoding

Frequency encoding is perhaps the simplest of all feature encoding schemes in the context of categorical features. In this technique, the feature values are encoded based on their frequencies in the feature. The value that appears more is assigned a higher value. For example, in our dataset, the feature value "Red" appears twice and there are "5" values in total, so the frequency of feature value "Red" is 2/5, i.e., 0.4. Similarly the feature value "Blue" also appears twice, so it will have the label "0.4" as well. The feature value "Yellow" will have the label "0.2." So, based on the frequency encoding, the features "Color" and "Size" will be labeled as shown in Table 4.18.

As you can see, the feature values are labeled between the values 0 and 1.

Table 4.18 Sample dataset for one-hot encoding

ObjectID	Color	Size	Grade
1	0.4	0.4	0
2	0.4	0.4	1
3	0.4	0.4	1
4	0.4	0.4	1
5	0.2	0.2	1

Table 4.19 Sum of the target feature values

Feature value	Sum of the target
Red	1
Blue	2
Yellow	1

Table 4.20 Count of the target feature values

Feature value	Count
Red	2
Blue	2
Yellow	1

Table 4.21 Mean of every feature value

Feature value	Count
Red	$1/2 = 0.5$
Blue	$2/2 = 1$
Yellow	$1/1 = 1$

4.12.1.4   Target Encoding

Target encoding or mean encoding is also one of the commonly used categorical feature encoding techniques. In target encoding, we use the mean value to assign the labels, but the mean value is calculated with respect to the target feature. Note that this is different from the mean encoding technique where we calculate the mean of the feature values only. The technique is performed in three steps.

First, we calculate the sum of the target values for the current categorical feature values. For example, for the feature value "Red," the sum of the target values in the "Grade" feature is "1." Similarly, for the feature value "Blue," the sum of the target values in the "Grade" feature is "2." We will calculate the sum of the target values in the target feature for each current categorical value. This is represented in Table 4.19.

Secondly, we will calculate the sum of each feature value in our current categorical feature. In our case, both the "Red" and the "Blue" values appear twice, whereas the "Yellow" value appears only once. This is shown in the Table 4.20.

Finally, in the third step, we will calculate the mean as shown in the Table 4.21. So, values in the feature "Color" will be labeled as shown in the Table 4.22.

ObjectID	Color	Size	Grade
1	0.5	Small	0
2	1	Small	1
3	0.5	Large	1
4	1	Large	1
5	1	Extra Large	1

Table 4.22 Sample dataset for one-hot encoding

4.13 Binning

Binning, also called "Discretization," can be called the inverse process of encoding. In encoding, we convert the categorical features into numerical features. However, in binning or discretization, we convert the numerical features especially the continuous features into categorical features.

A bin specifies a range of values. For example, a bin [1, 10] specifies a range of values that fall between "1" and "10."

The binning process works in two steps:

1. Define the bins based on feature values.
2. Smooth the feature values within each bin.

As mentioned above, the first step is to define the bins based on the feature values. Bins can be defined in two ways as follows:

4.13.1 Equal Width Bins

In the equal width binning method, the bins are defined such that each bin has the same width. The value of the "width" is decided based on the number of bins required by the user. In the equal width binning method, the total number of required bins is entered by the user. The width is calculated by using the following formula:

$$w = \frac{\text{max\_feature\_value} - \text{min\_feature\_value}}{\text{number of bins}} \tag{4.23}$$

For example, if the minimum feature value is "1" and the maximum feature value is "13" and the user requires three bins, the width of each bin will be calculated as follows:

$$w = \frac{13 - 1}{4} = 3$$

Now once the bin width is defined, the next step is to find the range (i.e., the lower and the upper bound) of each bin. This is done by using the following formula:

Range of bin-1: [min,min+w−1]
Range of bin-2: [min+w,min+2*w−1]
Range of bin-3: [min+2*w, min+3*w−1]
. . .
Range of bin-n: [min+(n−1)*w, max]

Here:

W = width of the bins
Min = the minimum feature value
Max = the maximum feature value

Let's now consider the example of equal width bins. Suppose we have a feature X that contains the following values:

$$X = \{1, 3, 6, 8, 9, 10, 12, 17, 19, 21, 34, 38, 40, 41, 47, 53\}$$

Suppose that we want four bins of equal width. So first we define the bin width:

$$w = \frac{53 - 1}{4} = 13$$

So in our case:

W = 13
Min = 1
Max = 53

The range of each bin will be as follows:

Range of bin-1: [1,1+13−1]
Range of bin-2: [1+13,1+2*13−1]
Range of bin-3: [1+2*13, 1+3*13−1]
Range of bin-4: [1+3*13, max]

So the range of the bins will be as follows:

Range of bin-1: [1, 13]
Range of bin-2: [14, 26]
Range of bin-3: [27, 39]
Range of bin-4: [40, 53]

Now our feature values will fall in these ranges as shown in Table 4.23.

Table 4.23 Equal width bins

Feature value	Bin
1	[1, 13]
3	[1, 13]
6	[1, 13]
8	[1, 13]
9	[1, 13]
10	[1, 13]
12	[1, 13]
17	[14, 26]
19	[14, 26]
21	[14, 26]
34	[27, 39]
38	[27, 39]
40	[40, 53]
41	[40, 53]
47	[40, 53]
53	[40, 53]

4.13.2   Equal Frequency Bins

In contrast to equal width bins, in equal frequency bins, we create the bins in such a way that each bin contains the same number of values from the feature. The number of values in each bin (which we call the frequency of the bin) depends upon the number of bins we require and the number of values in the feature.

The formula to calculate the bins is as follows:

$$\text{frequency} = \frac{\text{number of values in the feature}}{\text{required number of bins}} \qquad (4.24)$$

Now after sorting the feature values (if not already sorted), we will pick the first "frequency" number of values, and the first value will be considered "Min" and the last value will be considered as "Max" of that bin.

Similarly, we will then pick the next "frequency" number of values, and again the first value will become the "Min" and the last value will become the "Max" of the bin. The process will be repeated for the required number of bins.

For example, consider the same feature "X" as shown above. The feature has 16 features and we want 4 bins. The frequency will be calculated as follows:

$$\text{frequency} = \frac{16}{4} = 4$$

Now we will pick the first four values i.e., {1, 3, 6, 8}. The first feature value "1" will become the "Min" of the bin and the feature value "8" will become the "Max" of the range. Similarly, we will pick the second four values, i.e., {9, 10, 12, 17}. The

Table 4.24 Equal
frequency bins

Feature values	Bin
1	[1, 8]
3	[1, 8]
6	[1, 8]
8	[1, 8]
9	[9, 17]
10	[9, 17]
12	[9, 17]
17	[9, 17]
19	[19, 38]
21	[19, 38]
34	[19, 38]
38	[19, 38]
40	[40, 53]
41	[40, 53]
47	[40, 53]
53	[40, 53]

first feature value "9" will become the "Min" of the second bin, and the value "17" will become the "Max" of the second bin. So, finally, the feature values will be assigned to the bins as shown in Table 4.24.

Note that each bin has an equal number of feature values.

Once we have defined the bins, the next task is to smooth the feature values according to each bin. There are two common methods that can be used to smooth the feature values. We will discuss both of them.

4.13.3 Smoothing by Bin Means

In this method, each feature value that falls in a certain bin is replaced by the "Mean" value of that bin. So, the method works in the following three steps:

1. Take all the values that fall in a certain bin.
2. Calculate the mean of the bin.
3. Replace all the feature values with the mean value.

Let's consider the feature X that we have shown above. Suppose we have already calculated the bins using the "equal frequency binning" method. We will calculate the means of the first bin, i.e., [1, 8]. The feature values that are assigned to this bin are {1, 3, 6, 8}. The mean of these values will be:

$$\text{Mean} = \frac{1 + 3 + 6 + 8}{4} = \frac{18}{4} = 5 \text{ (Rounded)}$$

Table 4.25 Smoothing by bin means

Feature value	Bin
5	[1, 8]
5	[1, 8]
5	[1, 8]
5	[1, 8]
12	[9, 17]
12	[9, 17]
12	[9, 17]
12	[9, 17]
28	[19, 38]
28	[19, 38]
28	[19, 38]
28	[19, 38]
45	[40, 53]
45	[40, 53]
45	[40, 53]
45	[40, 53]

Now we will replace all the bin values with "5." Similarly, the mean of the second bin, i.e., [9, 17] is "12" so each feature value in this bin will be replaced with "12." The feature "X" after using the equal frequency binning method and then smoothing the feature values by using bin means will become as shown in Table 4.25.

4.13.4 Smoothing by Bin Boundaries

In this method, each feature value in a bin is assigned to its nearest boundary. Note that in a bin [1, 8], "1" is the lower boundary whereas "8" is the upper boundary of the bin. Each bin value is replaced with the boundary value that is close to it. For example, in the bin [1, 8], the feature value "3" will be replaced with the lower boundary value i.e., "1" because it is closer to "1" as compared to "8." Similarly, the value "5" will be replaced with the boundary value "8" as it is closer to the upper boundary than the lower boundary. So, in this way, each feature value is replaced with the closest boundary value.

If we consider our feature "X" and the equal frequency binning method, then in the first bin, the value "1" will remain the same, the value "3" will also be replaced with "1," the value "6" will be replaced with "8" and the value "8" will remain the same.

Similarly, in the second bin, the feature value "9" will remain the same, "10" will be replaced with "9," "12" will be replaced with "9," and "17" will remain the same. Table 4.26 shows the values of feature "X" after using the smoothing by bin boundaries method.

Table 4.26 Equal
frequency bins

Feature values	Bin
1	[1, 8]
1	[1, 8]
8	[1, 8]
8	[1, 8]
9	[9, 17]
9	[9, 17]
9	[9, 17]
17	[9, 17]
19	[19, 38]
21	[19, 38]
38	[19, 38]
38	[19, 38]
40	[40, 53]
40	[40, 53]
53	[40, 53]
53	[40, 53]

4.14   Remove Missing Values

It is common to have datasets with missing values, especially in applications that contain real-life data. For example, consider an application that requires the personal information of a person including the salary, age, etc. Handling the missing values is very important for data science applications especially when the decisions need to be made based on the data. Missing values affect the accuracy of the algorithms. Therefore, removing the missing values is one of the important tasks of data scientists. There are different techniques that have been adopted to remove the missing values. Here we will discuss only four.

1. Remove the objects containing the missing values.
2. Remove the missing value by using the default values.
3. Remove the missing values by mean/mode.
4. Remove by using the close distance.

Let's discuss all of them one by one.

4.14.1   Remove the Objects Containing Missing Values

This is the simplest method to remove the missing values that does not require any logic. All the objects that contain a missing value in any of their features are simply deleted. This results in a decreased number of objects in the final dataset. For example, consider the dataset given in the Table 4.27.

Table 4.27 Sample dataset containing the missing values

ObjectID	C1	C2	C3	C4	Class
X1	1	1	2	No	2
X2	1	2	1	No	1
X3	2	2	2	No	2
X5	3	1	2	?	1
X6	3	2	2	Yes	1
X7	1	2	2	No	2
X8	3	2	1	Yes	1
X9	1	1	?	No	1

Table 4.28 Dataset after deleting the objects containing missing values

ObjectID	C1	C2	C3	C4	Class
X1	1	1	2	No	2
X2	1	2	1	No	1
X3	2	2	2	No	2
X4	3	2	2	Yes	1
X5	1	2	2	No	2
X6	3	2	1	Yes	1

Here the objects X5 and X9 have missing values in features "C4" and "C3," respectively. The missing values are represented by the symbol "?". We will simply navigate through the entire dataset and will remove the objects that have a missing value in any of the features. So, here the object X5 and X9 will be removed. The resulting dataset is shown in Table 4.28:

As you can see that the original dataset comprising eight objects has been reduced to six objects. The advantage of using this technique is that it is the simplest technique to work with. However, the major disadvantage is that it works only for the datasets that have large number of objects, because in all such datasets, deleting the few objects does not affect the accuracy of the algorithm. This is because the deletion of a few objects does not result in the loss of a large amount of information. However, for smaller datasets where there are already a small number of objects, deleting even a few objects results in the loss of information which ultimately affects the accuracy of the algorithm using this technique.

4.14.2   Remove Using the Default Values

This is the next simplest method to remove the missing values. Here, instead of removing the entire objects, some default values are assigned to each missing value in a feature. The default values may be different for different features depending on the information stored by the features. The simplest example may be the age feature. Suppose that the age feature that stores the age of the graduate level applicants has missing values. We can assume that normally a student completes his graduation at

Table 4.29 Dataset with missing values removed using default values

ObjectID	C1	C2	C3	C4	Class
X1	1	1	2	No	2
X2	1	2	1	No	1
X3	2	2	2	No	2
X5	3	1	2	Yes	1
X6	3	2	2	Yes	1
X7	1	2	2	No	2
X8	3	2	1	Yes	1
X9	1	1	1	No	1

the age of almost 21–25 years. So, here we can assume the default value as 22 years. Now each missing value will be assigned the new value after adding the duration of "22" years to the current date.

In the case of our sample dataset shown above, we can assume the default value of "1" for the feature "C3" and the default value of "Yes" for the feature "C4." Table 4.29 shows our resulting dataset after removing missing values using the selected default value.

4.14.3 Remove Missing Values Using Mean/Mode

As the name implies, the missing values are removed by generating a new value. The new value is generated by using the other values in the same feature. For this purpose, we calculate the mean of the entire feature in case the feature contains the numeric values and the Mode in case the feature contains the categorical value. All the missing values in that feature are replaced with this newly generated value.

Now let's consider the feature C3. As it is a numeric feature, so we will use the mean operation to find the alternative to the missing values. The mean value becomes "1.7" which becomes "2" after rounding it. So, all the missing values will be replaced with "2" in feature "C3."

Now let's talk about the feature "C4." As you can see that it contains the categorical values, so we cannot apply the mean operation. For this purpose, we will use the mode operation to find the missing values. The mode operation results in a value that appears more than other values in the dataset. In the feature "C4," the mode operation generates the feature value "No." So, all the missing values will be replaced with this value. Table 4.30 shows the resulting table after applying the mean/mode operation.

The advantage of using mean/mode operation is that the resulting dataset does not lose or add any new information. Rather the information is added based on the already available information in the dataset. So, the net effect of removing the missing values is almost minimum.

Table 4.30 Dataset with missing values removed using default values

ObjectID	C1	C2	C3	C4	Class
X1	1	1	2	No	2
X2	1	2	1	No	1
X3	2	2	2	No	2
X5	3	1	2	No	1
X6	3	2	2	Yes	1
X7	1	2	2	No	2
X8	3	2	1	Yes	1
X9	1	1	2	No	1

4.14.4   Remove Missing Values by Using the Closest Distance

This is perhaps one of the most technical methods for removing the missing values in a dataset. We measure the distance of the object (containing missing values) with all other objects and the missing values are replaced with the corresponding feature values of the object having a minimum distance from the current object (containing missing values). To calculate the distance, we can use any method. However, here we will explain the simplest one. In this method, the feature values of one object are subtracted from the feature values of the other object (in the case of numerical features) and the results are added. For the categorical features, we add "1" in case the values are different and "0" in case the values are the same. For example, let us calculate the distance between objects X1 and X2. The features "C1," "C2," "C3," and "Class" are nominal features, so we sum the difference of their values:

$$X1 - X2 = |1 - 1| + |1 - 2| + |2 - 1| + |2 - 1|$$
$$X1 - X2 = 0 + 1 + 1 + 1 = 3$$

Since "C4" is a categorical feature, so we will compare its values for objects X1 and X2. Since the values are the same so we will add "0" to the already calculated distance. So, the distance between objects "X1" and "X2" is that of the factor "2." In this way, we can calculate the distance between any two objects.

In our sample dataset, the object X5 is closest to object "X6" (the distance between X5 and X6 is minimum as compared to the distance of object X5 with other objects). So, we will replace the value of feature "C4" of object X5 with that of "X6."

For the object X9, it has the closest distance with objects "X1" and "X2," i.e., its distance with both of these objects is the same and minimum than its distance with other objects, so in this case, we can use any object to remove the missing values of object "X9." As object X1 comes first, so we can use this object. Table 4.31 shows the resulting dataset after removing the missing values using the closest distance.

Table 4.31 Dataset with missing values removed using closest distance

ObjectID	C1	C2	C3	C4	Class
X1	1	1	2	No	2
X2	1	2	1	No	1
X3	2	2	2	No	2
X5	3	1	2	Yes	1
X6	3	2	2	Yes	1
X7	1	2	2	No	2
X8	3	2	1	Yes	1
X9	1	1	2	No	1

Table 4.32 Dissimilarity matrix

	1	2	3	4	...	...	N
1	0						
2	$d(1,2)$	0					
3	$d(1,3)$	$d(2,3)$	0				
4	$d(1,4)$	$d(2,4)$	$d(i,j)$	0			
...					0		
...						0	
N	$d(1,N)$	$d(2,N)$	$d(3,N)$	$d(4,N)$			0

4.15 Proximity Measures

The proximity measures refer to the similarity or dissimilarity between data. As each object is described in terms of its features, the proximity measures specify how much similar or dissimilar two objects are in terms of the values of their features. Both of the terms are relative which means that if the similarity is high, dissimilarity will be low and vice versa.

4.16 Dissimilarity Matrix

The dissimilarity matrix specifies the proximities between every two objects of the dataset. As the name implies, it is a two-dimensional matrix where each element refers to a proximity measure. A dissimilarity matrix can be represented as shown in Table 4.32.

Here the term $d(i,j)$ represents the dissimilarity between the objects "i" and "j." For example, consider the dataset given in Table 4.33.

The dissimilarity matrix of the above dataset is shown in the Table 4.34.

Table 4.33 Sample dataset

Objects	Feature-1	Feature-2
X1	0	2
X2	1	3
X3	0	2
X4	1	3

Table 4.34 Dissimilarity matrix

	X1	X2	X3	X4
X1	0			
X2	1	0		
X3	0	1	0	
X4	1	0	1	0

Table 4.35 Sample dataset

	X-axis	Y-axis	Z-axis
X1	0	2	4
X2	2	8	7
X3	3	9	3
X4	2	4	1
X5	8	4	7

4.16.1 *Manhattan Distance*

Manhattan distance specifies the difference between the vectors. It can be obtained by summing the absolute difference of attribute values of two vectors. The formula for Manhattan distance can be written as:

Suppose there are two vectors V1 and V2 as follows:

V1 = (V1X, V1Y)
V2 = (V2X, V2Y)

The Manhattan distance will be:

$$d = |V1_x - V2_x| + |V1_x - V2_y|$$

For example, consider the following dataset given in Table 4.35.

Now we will calculate the Manhattan distance and draw the dissimilarity matrix

D(x1,x2) = |0-2|+|2-8|+|4-7| = 2+6+3 = 11
D(x1,x3) = |0-3|+|2-9|+|4-3| = 3+7+1 = 11
D(x1,x4) = |0-2|+|2-4|+|4-1| = 2+2+3 = 7
D(x1,x5) = |0-8|+|4-2|+|7-4| = 8+2+3 = 13

D(x2,x3) = |2-3|+|8-9|+|7-3| = 1+1+4 = 6
D(x2,x4) = |2-2|+|8-4|+|7-1| = 0+4+6 = 10
D(x2,x5) = |2-8|+|8-4|+|7-7| = 6+4+0 = 10

$D(x3,x4) = |3-2|+|9-4|+|3-1| = 1+5+2 = 8$
$D(x3,x5) = |3-8|+|9-4|+|3-7| = 5+5+4 = 14$

$D(x4,x5) = |2-8|+|4-4|+|1-7| = 6+0+6 = 12$

Now the dissimilarity matrix based on Manhattan distance is given in the Table 4.36.

4.16.2 Euclidian Distance

Euclidian distance between two points specifies the distance of the straight line between two points. It can be calculated just like the Manhattan distance; however, we take the square of the absolute difference and then the square root of the summation specifies the distance.

Suppose there are two vectors V1 and V2 as follows:

$V1 = (V1_x, V1_y)$
$V2 = (V2_x, V2_y)$

The Manhattan distance will be:

$$d = \sqrt{|V1_x - V2_x|^2 + |V1_x - V2_y|^2}$$

For the dataset given in Table 4.35, the Euclidian distance and then the dissimilarity matrix will be calculated as follows:

$D(x1,x2) = |0-2|+|2-8|+|4-7| = 2+6+3=11 = \text{Sqrt} (2{\sim}2+6{\sim}2+3{\sim}2) = 7$
$D(x1,x3) = |0-3|+|2-9|+|4-3| = 3+7+1=11 = \text{Sqrt} (3{\sim}2+7{\sim}2+1{\sim}2) = 7.68$
$D(x1,x4) = |0-2|+|2-4|+|4-1| = 2+2+3=7 = \text{Sqrt} (2{\sim}2+2{\sim}2+3{\sim}2) = 4.12$
$D(x1,x5) = |0-8|+|4-2|+|7-4| = 8+2+3=13 = \text{Sqrt} (8{\sim}2+2{\sim}2+3{\sim}2) = 8.77$

$D(x2,x3) = |2-3|+|8-9|+|7-3| = 1+1+4=6 = \text{Sqrt} (1{\sim}2+1{\sim}2+4{\sim}2) = 4.24$
$D(x2,x4) = |2-2|+|8-4|+|7-1| = 0+4+6=10 = \text{Sqrt} (0{\sim}2+4{\sim}2+6{\sim}2) = 7.21$
$D(x2,x5) = |2-8|+|8-4|+|7-7| = 6+4+0=10 = \text{Sqrt} (6{\sim}2+4{\sim}2+0{\sim}2) = 7.21$

$D(x3,x4) = |3-2|+|9-4|+|3-1| = 1+5+2=8 = \text{Sqrt} (1{\sim}2+5{\sim}2+2{\sim}2) = 5.47$
$D(x3,x5) = |3-8|+|9-4|+|3-7| = 5+5+4=14 = \text{Sqrt} (5{\sim}2+5{\sim}2+4{\sim}2) = 8.12$

$D(x4,x5) = |2-8|+|4-4|+|1-7| = 6+0+6=12 = \text{Sqrt} (6{\sim}2+0{\sim}2+6{\sim}2) = 8.48$

Table 4.36 Dissimilarity matrix based on Manhattan distance		X1	X2	X3	X4	X5
	X1	0				
	X2	11	0			
	X3	11	6	0		
	X4	7	10	8	0	
	X5	13	10	14	12	0

Table 4.37 specifies the dissimilarity matrix based on Euclidean distance.

4.16.3   Supremum Distance

Supremum distance between two vectors is equal to the maximum difference that exists between any two axes of the points. For example, if we consider X1 and X2, the maximum difference between the Y-axis, the supremum distance between X1 and X2 will be equal to "6."

$D(x1,x2) = \max(|0-2|, |2-8|, |4-7|) = \max(2, 6, 3) = 6$
$D(x1,x3) = \max(|0-3|, |2-9|, |4-3|) = \max(3, 7, 1) = 7$
$D(x1,x4) = \max(|0-2|, |2-4|, |4-1|) = \max(2, 2, 3) = 3$
$D(x1,x5) = \max(|0-8|, |4-2|, |7-4|) = \max(8, 2, 3) = 8$

$D(x2,x3) = \max(|2-3|, |8-9|, |7-3|) = \max(1, 1, 4) = 4$
$D(x2,x4) = \max(|2-2|+|8-4|+|7-1|) = \max(0, 4, 6) = 6$
$D(x2,x5) = \max(|2-8|, |8-4|, |7-7|) = \max(6, 4, 0) = 6$

$D(x3,x4) = \max(|3-2|, |9-4|, |3-1| = \max(1, 5, 2) = 5$
$D(x3,x5) = \max(|3-8|, |9-4|, |3-7| = \max(5, 5, 4) = 5$

$D(x4,x5) = \max(|2-8|, |4-4|, |1-7| = \max(6, 0, 6) = 6$

Table 4.38 shows the dissimilarity matrix based on the supremum distance.

Table 4.37 Dissimilarity matrix based on Euclidian distance

	X1	X2	X3	X4	X5
X1	0				
X2	7	0			
X3	7.68	4.24	0		
X4	4.12	7.21	5.47	0	
X5	8.77	7.21	8.12	8.48	0

Table 4.38 Dissimilarity matrix based on Supremum distance

	X1	X2	X3	X4	X5
X1					
X2	6				
X3	7	4			
X4	3	6	5		
X5	8	6	5	6	

4.17 Summary

Feature selection is an important data preprocessing step in data analytic applications, so, it should be given equal importance. In this chapter, we have provided a strong base for this topic. Starting from basic concepts toward different types and feature selection mechanisms, we have provided deep insight into the entire types of feature selection algorithms along with their issues and the challenges that still need to be fixed.

Further Reading
Following are some valuable resources for further reading:

- Advances in Feature Selection for Data and Pattern Recognition (Editors: Urszula Stanczyk, Beata Zielosko, Lakhmi C. Jain), Springer & Co, 2018. Book provides state-of-the-art developments and trends in feature selection and pattern recognition. The book is divided into four parts. The first part provides details about the data and the different in its representation. The second part deals with different methods that can be used to explore and rank features. The third part is about the detection and recognition of images, shapes, motion, audio, etc. The fourth part deals with the decision support systems.
- Hierarchical Feature Selection for Knowledge Discovery (Authors: Cen Wan), Springer & Co, 2019. Book systematically describes the procedure of data mining and knowledge discovery on Bioinformatics databases by using state-of-the-art hierarchical feature selection algorithms. Moreover, the book also provides the details of the biological patterns minded by the hierarchical feature selection methods that are relevant to aging-association genes. The patterns show different aging-associated factors that provide future guidelines for research related to Ageing Biology.
- Data pre-processing for machine learning in Python Kindle Edition by Gianluca Malato, independently published, 2022. This book presents the python programming language with a practical approach using preprocessing activities of machine learning projects.
- Feature Extraction, Construction, and Selection: A Data Mining Perspective (The Springer International Series in Engineering and Computer Science Book 453) by Huan Liu, Springer & Co, 2018. This book explains the methods of extracting and identifying features for data mining and pattern recognition and machine learning. The main objective of this book is to provide details of the feature extraction, selection, and generation techniques from different data processing perspectives.
- Modern Data Mining Algorithms in C++ and CUDA C: Recent Developments in Feature Extraction and Selection Algorithms for Data Science 1st ed. Edition, by Timothy Masters, Apress, 2020. Feature selection is one of the important tasks in the domain of machine learning. This book provides details of local-level feature selection techniques. It provides in-depth details about the feature selection and extraction algorithms and the related concepts.

- Computational Methods of Feature Selection (Chapman & Hall/CRC Data Mining and Knowledge Discovery Series) Part of: Chapman & Hall/CRC Data Mining and Knowledge Discovery (53 Books) | by Huan Liu and Hiroshi Motoda, Chapman and Hall/CRC, 2007. Book explores different methods for feature selection. A detailed discussion is provided on these methods along with how they work for the selection of the relevant features.
- Practical Data Science with Jupyter: Explore Data Cleaning, Pre-processing, Data Wrangling, Feature Engineering and Machine Learning using Python and Jupyter (English Edition) 1st Edition, by Prateek Gupta, BPB Publications, 2021. The book provides details about data science and its practical applications. Real-life use cases are presented in the form of case studies for data visualization, data preprocessing, unsupervised machine learning, supervised machine learning, and handling time-series data, and time series methods.
- Practitioner's Guide to Data Science: Streamlining Data Science Solutions using Python, Scikit-Learn, and Azure ML Service Platform (English Edition) by Nasir Ali Mirza, BPB Publications, 2022. The book provides a strong base for a data science learner. It provides good details about the data science-related concepts. Different data science-related platforms are also discussed. Details about the real-life applications are provided as well. Furthermore, the book provides in-depth details about the Azure ML service architecture and the different of its capabilities.
- Graph Machine Learning: Take graph data to the next level by applying machine learning techniques and algorithms by Claudio Stamile, Aldo Marzullo, et al, Packt Publishing, 2021. This book explains graph machine learning, sets of tools used for network processing, analytics task, modeling, graph theory, graph representation learning, model training, data processing, text analytics, and natural language processing. The book also explored the latest trends on a graph.
- Feature Engineering for Machine Learning and Data Analytics (Chapman & Hall/CRC Data Mining and Knowledge Discovery Series) Part of: Chapman & Hall/CRC Data Mining and Knowledge Discovery by Guozhu Dong and Huan Liu, CRC Press, 2021. This book presents concepts, examples, applications, and feature engineering methods for machine learning and data analytics.

Exercises

Exercise 4.1: What is the difference in numerical and categorical features?

Exercise 4.2: Numerical features are divided into how many categories? Elaborate each category with an example.

Exercise 4.3: Categorical features are divided into how many categories? Elaborate each category with an example.

Exercise 4.4: Draw a tree to show that if the "age" of a person is greater than 18, he can go to the cinema, or else he cannot go to the cinema.

Exercise 4.5: What is the difference between feature selection and feature extraction?

Exercise 4.6: What are the main categories of feature selection? You are required to give a brief introduction to each category.

Exercise 4.7: What are the major advantages and disadvantages of filter-based, wrapper-based, and embedded features selection approaches?

Exercise 4.8: What are the main benefits of feature selection?

Exercise 4.9: Consider the following pseudo code of a generic filter-based feature selection approach. Implement it on any dataset using any programming language.

```
Input
S - Data sample with C number of features
E - Evaluation measure
SGO - successor generation operator
Output
S' - Output solution
I := Start I;
Si := { Best of I with respect to E };
repeat
I := Search (I, SGOI,C);
C' := {Best I according to E };
if E(C')≥E(S') or (E(C')=E(S') && |C'| < |S'|) then S' :=C';
until Stop(E,I).
```

Exercise 4.10: Following are a few methods that can be used for the selection of the features. You are required to give a brief introduction to each method.

- Information gain
- Distance
- Dependency
- Consistency
- Classification accuracy

Exercise 4.11: Consider the following feature selection algorithms. Categorize them into filter-based, wrapper-based, and embedded feature selection techniques.

- ANOVA
- Backward elimination
- Chi-square
- Forward selection
- Lasso regression
- LDA
- Memetic algorithm
- Pearson's correlation
- Recursive feature elimination
- Regularized trees
- Ridge regression

Exercise 4.12: Consider the following pseudo code of a generic forward feature selection algorithm and implement it on any dataset using any programming language. You can assume any measure (discussed in this chapter) as evaluation criteria.

```
Input
S - Dataset containing X features
E - Evaluation criteria
Initialization
    c) S' ← {φ}
do
b) ∀x ∈ X
c) S' ← S' ∪ {x_i}    where i = 1...n
    d) until Stop(E, S')
Output
d) Return S'.
```

Exercise 4.13: Consider the following given pseudo code of backward feature selection algorithm and implement it on any dataset using any programming language. You can assume any measure (discussed in this chapter) as evaluation criteria.

```
Input
S - Data sample with features X
E - Evaluation measure
Initialization
a) S' ← X
do
b) ∀x ∈ S'
c) S' ← S' - {x_i}   i = n. ...1
d) until Stop(E, S')
Output
d) Return S'.
```

Exercise 4.14: Consider the following pseudo code of feature selection algorithm using the hit and trial approach and implement it on any dataset using any programming language.

```
Input
S - Dataset containing X features
E - Evaluation measure
S' ← {φ}
For j=1 to n
If (Random(0,1) <= 0.5) then
S' ← S' ∪ {X_j}
Until Stop(E, S')
Output
d) Return S'.
```

Exercise 4.15: What is the difference between feature relevancy and feature redundancy?

Exercise 4.16: What are the issues of feature selection, give a brief introduction of every feature selection issue.

Exercise 4.17: Consider the following categories of feature selection algorithms and provide a list of algorithms that may fall in these categories.

- Exhaustive algorithms
- Random search-based algorithms
- Heuristics-based algorithms.

Exercise 4.18: Consider the following pseudo code of genetic algorithm and implement it on any dataset using any programming language.

```
Input
C - Feature Subset X: Objects in the dataset

1) Initialize the initial population.
2) Check the fitness of each individual.
Repeat
3) Select the best Individuals.
4) Perform their crossover and generate offsprings.
5) Perform mutation.
6) Evaluate the fitness of each individual.
7) Replace the bad individuals with new best individuals.
8) Until (Stopping Criteria is met).
```

Exercise 4.19: Consider the following dataset and calculate the Manhattan distance between each object.

	X-axis	Y-axis	Z-axis
X1	0	1	3
X2	5	5	9
X3	2	7	4
X4	6	9	0
X5	7	3	6

Exercise 4.20: Consider the following dataset and calculate the Euclidean Distance between each object.

	X-axis	Y-axis
X1	5	5
X2	4	3
X3	8	1
X4	0	7
X5	6	9

Exercise 4.21: Consider the following dataset and calculate the supremum distance between each object.

	X-axis	Y-axis
X1	9	6
X2	6	8
X3	1	5
X4	7	0
X5	4	3

Chapter 5
Classification

5.1 Classification

Classification is the problem of assigning objects to predefined groups based on their characteristics, for example, classification of pictures into two classes, e.g., rectangles and circles based on their shape as shown in Fig. 5.1, assigning the email messages to two classes, e.g., spam and non-spam based on their source, etc.

We input the dataset to a process called the classification process and the process assigns the labels to data based on the features of the dataset. For example, we may enter the source of the emails and subject, based on which the classifier may label a message either as spam or non-spam. Figure 5.2 shows the clustering process at a high level.

The input of the classification process is a dataset with n rows (tuples) and m attributes (also called features, vectors, properties, etc.). Table 5.1 shows a sample dataset.

As shown, the dataset defines what classification of different people based on their property type. "C1," "C2," "C3," etc. are class labels, whereas "has Garden," "Occupation," "Home type," etc. are all the attributes, based on which the class of the people is decided.

Classification can be used for predictive modeling, i.e., we can predict based on the classification process, the class of a record whose class is not already known.

For example, consider the record given in Table 5.2.

We can use the dataset given in Table 5.1 to train the classification model and then use the same model to predict the class of the abovementioned record. It should be noted that classification techniques are effective for prediction for binary classes, i.e., where we have attributes with nominal or ordinal category and less effective for other types of attributes.

A classification process is a systematic approach to build a classifier model based on training records. The classifier model is built in such a way that it predicts the class of the unknown data with a certain accuracy. The model identifies the class of

© The Author(s), under exclusive license to Springer Nature Switzerland AG 2023
U. Qamar, M. S. Raza, *Data Science Concepts and Techniques with Applications*,
https://doi.org/10.1007/978-3-031-17442-1_5

Fig. 5.1 Classification
between males and females

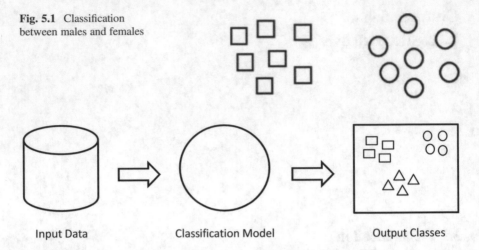

Input Data   Classification Model   Output Classes

Fig. 5.2 Classification process

unknown records based on the attribute values and classes in the training dataset.
The accuracy of the classification model depends on many factors including the size
and quality of the training dataset.

Figure 5.3 shows a generic classification model.

The accuracy of the classification model is measured in terms of the records
correctly classified as compared to a total number of records. The tool used for this
purpose is called the confusion matrix given in Table 5.3.

A confusion matrix shows the number of records of class i that are predicted as
belonging to class j. For example, f_{01} means the total number of records that belong
to class "0," but the model has assigned it the class "1."

A confusion matrix can be used for deriving a number of matrices that can be used
for classification model evaluation. Here we will present a few metrics.

Accuracy Accuracy defines the ratio of the number of classes correctly predicted as
compared to the total number of records. In our confusion matrix, f_{11} and f_{00} are the
correctly predicted classes.

$$\text{Accuracy} = \frac{\text{Number of correct predictions}}{\text{Total number of predictions}} = \frac{f_{11} + f_{00}}{f_{11} + f_{10} + f_{01} + f_{00}} \quad (5.1)$$

Error Rate Error rate defines the ratio of incorrectly predicted classes to the total
number of records. In our confusion matrix f_{10} and f_{01} as incorrectly predicted
classes.

$$\text{Error rate} = \frac{\text{Number of wrong predictions}}{\text{Total number of predictions}} = \frac{f_{10} + f_{01}}{f_{11} + f_{10} + f_{01} + f_{00}} \quad (5.2)$$

Table 5.1 Sample dataset

Name	Has garden	Occupation	Home type	Garden size	Area type	Taxation	Electric system	Class label
John	Yes	Teacher	Single story	Small	Urban	Tax payer	Solar	C1
Elizabeth	No	Manager	Double story	Small	Urban	Non-tax payer	Grid	C2
Mike	No	Manager	Double story	Big	Urban	Non-tax payer	Solar	C3
Munsab	Yes	Teacher	Single story	Big	Urban	Non-tax payer	Solar	C1
Qian	No	Salesman	Double story	Medium	Urban	Tax payer	Grid	C4
Maria	No	Manager	Double story	Small	Urban	Tax payer	Solar	C2
Jacob	Yes	Teacher	Single story	Small	Rural	Tax payer	Grid	C1
Julie	Yes	Technician	Double story	Small	Rural	Tax payer	Solar	C5
Smith	Yes	Shopkeeper	Single story	Small	Urban	Tax payer	Solar	C1
Evans	No	Manager	Single story	Big	Urban	Non-tax payer	Solar	C3
Roberts	No	Manager	Double story	Medium	Urban	Tax payer	Solar	C2
William	Yes	Technician	Double story	Medium	Urban	Tax payer	Solar	C5
Poppy	Yes	Plumber	Single story	Small	Urban	Tax payer	Grid	C1
Emma	No	Manager	Double story	Big	Urban	Non-tax payer	Solar	C3
Emily	No	Salesman	Double story	Medium	Urban	Tax payer	Grid	C4

Table 5.2 A new record

Name	Has garden	Occupation	Home type	Garden size	Area	Taxation	Electric system	Class label
Seena	No	Manager	Double story	Small	Urban	Tax payer	Grid	?

A model that provides the maximum accuracy and the minimum error rate is always desirable.

Sometimes, classes have a very unequal frequency. For example, in the case of fraud detection, 98% of transactions are valid and only 2% are fraud. Similarly, in the case of medical data, 99% of the patients don't have breast cancer and only 1% have.

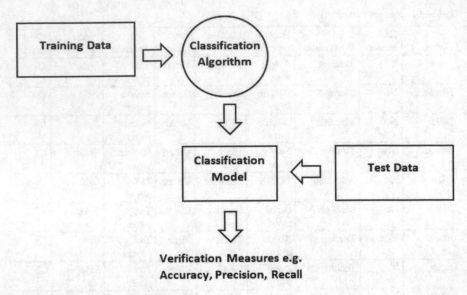

Fig. 5.3 A generic classification models

Table 5.3 Confusion matrix

		Predicted class	
		Class = 1	*Class* = 0
Actual class	*Class* = 1	*f*11	*f*10
	Class = 0	*f*01	*f*00

In most cases, the class of interest is commonly called the positive class, and the rest negative classes.

Let us consider a two-class problem:

- Number of negative examples = 9990
- Number of positive examples = 10

In such a case, accuracy can be misleading. Suppose that the model does not predict any positive example but predicts all examples belonging to the negative class. It will still have an accuracy of 99.9%.

In such a case, we need alternative measures. Two that we can use are precision and recall.

– Precision: % of tuples that the classifier labeled as positive are positive.

$$\text{Precision} = \frac{\text{True Positive}}{\text{True Positive} + \text{False Positive}} \tag{5.3}$$

– Recall: % proportion of negative tuples that are correctly identified.

$$\text{Recall} = \frac{\text{True Positive}}{\text{True Positive} + \text{False Negative}} \tag{5.4}$$

F-measure combines precision and recall into one measure. Mathematically:

$$F - Measure = (2^* Precision^* Recall)/(Precision + Recall) \qquad (5.5)$$

It is a harmonic mean of two numbers that tend to be closer to the smaller of the two. Thus, for the F-measure to be large, both precision and recall must be large.

5.2 Decision Tree

The decision tree is one of the most commonly used classifiers. We have already discussed the decision tree in previous chapters. Here we will provide details of the concept along with relevant algorithms.

Now before going into details of how decision trees perform classification, let's consider that we want to classify a person "Shelby" based on the values of the attributes presented in Table 5.1. What we can do is we can start with the first attribute and by considering that either her home has a "garden" or not; we may then consider the value of the second attribute, i.e., what is her occupation; after this, we may continue with attributes one by one until we reach the decision class of the person. This is how a decision tree exactly works.

Figure 5.4 shows a sample decision tree of the dataset given in Table 5.1.

Fig. 5.4 A sample decision tree

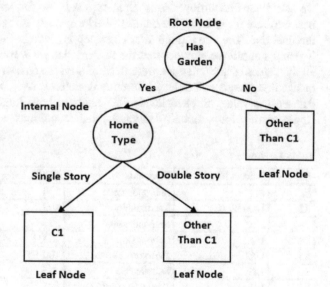

Building a decision tree is a computationally expensive job because there may be n number of potential solutions. So finding an optimal one is computationally very expensive and especially when the number of attributes increases beyond a smaller number.

So, for this purpose, we normally use the greedy search approach using local optimum. There exist many such solutions, and one of the algorithms is Hunt's algorithm. Hunt's algorithm uses the recursive approach and builds the final tree by incrementally adding subtrees.

Suppose D_t is a training set and $y = \{y_1, y_2,..., y_c\}$ are class labels, then according to Hunt's algorithm:

Step1: If there is only one class in the dataset and all records belong to this class, then there is only one leaf node labeled as y_t.

Step2: In the case of multiple classes, we select an attribute to partition the dataset into smaller subsets. We create a child node for each value of the attribute. The algorithm proceeds recursively until we get the complete tree.

For example, suppose we want to create a classification tree for determining that either a person will be permitted for registration or not. For this purpose, we consider the following dataset given in Table 5.4. We take the data of the previous students and try to determine it on this basis. The dataset is shown in Table 5.4. The attributes considered are {Student level, Student type, Marks, Registration}.

We start with an initial tree that contains only a single node for class Registration = Not-permitted. Now note that the dataset contains more than one class label. Now we select our first attribute, i.e., "Student level." So far, we assume that this is the best attribute for splitting the dataset. We keep on splitting the dataset until we get the final tree. The tree formed after each step is given in Fig. 5.5.

Hunt's algorithm assumes that the training dataset is complete which means that all the values of attributes are present. Similarly, the dataset is also consistent which means that unique values lead to unique decision classes. Note that in our training dataset, the value "Student level=PG" leads to both Registration=Permitted and Registration=Not-permitted. However, such conditions normally do not exist in

Table 5.4 Student dataset

Sid	Student level	Student type	Marks	Registration
S1	PG	Self-sponsored	130	Not-permitted
S2	UG	Scholarship	105	Not-permitted
S3	UG	Self-sponsored	75	Not-permitted
S4	PG	Scholarship	125	Not-permitted
S5	UG	Partial sponsored	100	Permitted
S6	UG	Scholarship	65	Not-permitted
S7	PG	Partial sponsored	225	Not-permitted
S8	UG	Self-sponsored	90	Permitted
S9	UG	Scholarship	79	Not-permitted
S10	UG	Self-sponsored	95	Permitted

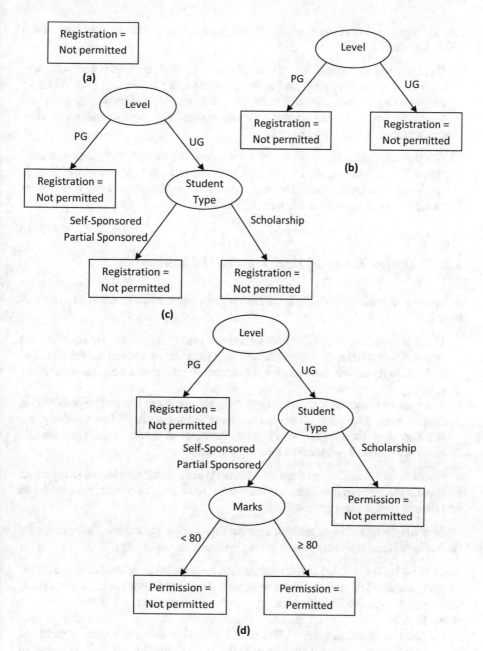

Fig. 5.5 Hunt's algorithm example

datasets due to many reasons. Here, we will present some other assumptions of the Hunts algorithm.

1. There may be a condition that the child nodes created are empty. Normally this happens when the training dataset does not have the combination of values of attributes that result in this child node. In such cases, this node is declared as a leaf node, and a class is assigned to it based on the majority class of training records that are associated with its parent.

2. If attribute values in D_t are the same with respect to attribute values but only differ in class labels, then we cannot split the records further and the node is declared as a child node. The class assigned to this node is the one having maximum occurrences in the training dataset.

5.2.1 Design Issues of Decision Tree Induction

A learning algorithm for inducing decision trees must address the following two issues.

- The algorithm incrementally implements the classification tree by selecting the next attribute to split the dataset. However, as there are a number of attributes and each attribute can be used to split the dataset, there should be some criteria for selecting the attributes.
- There should be some criteria to stop the algorithm. There can be two possibilities here. We can stop when all the records have the same class label. Similarly, we can also stop when all the records have the same attribute value. There can be some other criteria such as the threshold value, etc.

It should be noted that there can be different types of attributes and splitting based on these attributes can be different in each case. Now we will consider the attributes and their corresponding splitting.

Binary Attributes Binary attributes are perhaps the simplest ones. The splitting on the basis of these attributes results in two splits as shown in Fig. 5.6a.

Nominal Attributes Nominal attributes may have multiple values, so we can have as many numbers of splits as many possible values. If an attribute has three values, there may be three splits. Consider the attribute "Student type"; it can have three possible values, i.e., "Self-sponsored', 'Partial sponsored," and "Scholarship based," so the possible splits are given in Fig. 5.6b. It should be noted that some algorithms, e.g., CART may also produce binary splits by considering all $2^{k-1}-1$ ways of creating a binary partition of k attribute values. Figure 5.6c illustrates the binary split of the "Student type" attribute that originally had three values.

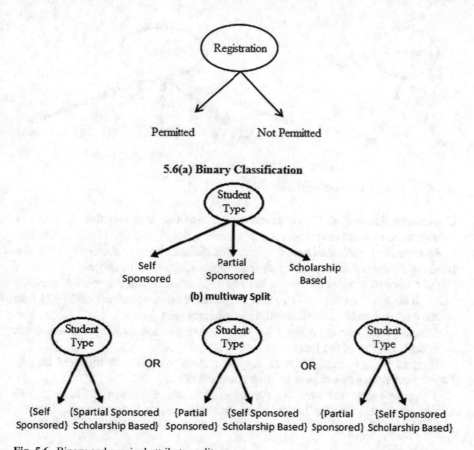

5.6(a) Binary Classification

(b) multiway Split

Fig. 5.6 Binary and nominal attributes split

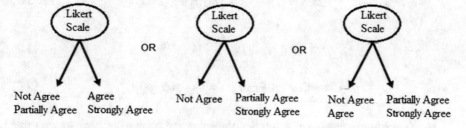

Fig. 5.7 Splitting on the basis of ordinal attributes

Ordinal Attributes Ordinal attributes have an order between the attribute values, e.g., a Likert scale may have values Not Agree, Partially Agree, Agree, and Strongly Agree. So, we can split an ordinal attribute in binary and more splits. However, it should be noted that the split should not affect the order of the attribute values. Figure 5.7 shows possible splits of the ordinal attribute Shirt Size.

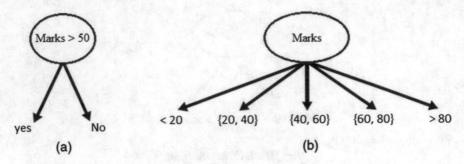

Fig. 5.8 Splitting continuous attributes

Continuous Attributes Continuous attributes are those attributes that can take any value between a specified range, e.g., temperature, height, weight, etc. In the context of decision trees, such attributes may either result in a binary or multi-way split. For binary split, the test condition for such attributes may be of the form:

$(C < v)$ or $(C \geq v)$. Here we will have two splits, i.e., either the value of attribute "C" will be less than value "V" or will be greater-than-or-equal-to "V." Figure 5.8a shows a typical test condition resulting in a binary split.

Similarly, we can have test conditions where we have multiple splits. Such conditions may be of the form:

$V_i \leq C < v_{i+1}$. Here we may have a number of splits in different ranges. Figure 5.8b shows an example of such a test condition.

There are many measures that can be used to select the best attribute that we should select next for distribution. Examples may include Entropy, Gain, Classification error, etc.

$$\text{Entropy } (t) = - \sum_{t=0}^{c-1} p(i|t) \log_2 p(i|t) \tag{5.6}$$

$$\text{Gini } (t) = 1 - \sum_{t=0}^{c-1} [p(i|t)]^2 \tag{5.7}$$

$$\text{Classification Error } (t) = 1 - \max_t [p(i|t)] \tag{5.8}$$

Here are some examples of a number of classes and corresponding measures that can be used to select the attribute:

Node N1	Count	
		Gini=$1-(0/4)^2-(4/4)^2=0$
Class=0	0	Entropy=$-(0/4)\log_2(0/4)-(4/4)\log_2(4/4)=0$
Class=1	4	Error=$1-\max[0/4,4/4]=0$

Node N1	Count	$Gini=1-(1/4)^2-(3/4)^2=0.375$
Class=0	1	$Entropy=-(1/4)\log_2(1/4)-(3/4)\log_2(3/4)=0.811$
Class=1	3	$Error=1-max[1/4,3/4]=0.25$

Node N1	Count	$Gini=1-(2/4)^2-(2/4)^2=0.5$
Class=0	2	$Entropy=-(2/4)\log_2(2/4)-(2/4)\log_2(2/4)=1$
Class=1	2	$Error=1-max[2/4,2/4]=0.5$

5.2.2   Model Overfitting

A decision tree can have three types of errors called training errors, i.e., the mis-classifications that a tree performs on the training dataset; test errors, i.e., the misclassification performed on the training dataset; and generalization errors, i.e., the errors that are performed on the unseen dataset. The best classification model is the one that avoids all of these types of errors. However, it should be noted that for small trees, we may high test and training errors. We call it model underfitting because the tree is not trained well to generalize all the possible examples mainly because of insufficient test and training data. Model overfitting on the other hand results when the tree becomes too large and complex. In this scenario, although the training error decreases but the test error may increase. One reason behind this may be that the training dataset contains noise and the classes are accidentally assigned to data points which may result in misclassification of the data in the training dataset. Model overfitting can result due to many factors here we will discuss a few.

It can be due to noise in the training dataset. For example, consider Tables 5.5 and 5.6 representing the training and test datasets. Note that the training dataset has two misclassifications representing the noise. It means that the tree resulting from the training dataset will not have any error, but when the test dataset will run the

Table 5.5 Training dataset

Name	Has garden	Home type	Taxation	Electric system	Registration
Qian	Yes	Single story	Tax payer	Grid	Permitted
Smith	Yes	Single story	Tax payer	Solar	Permitted
Jacob	Yes	Single story	Non-tax payer	Grid	Not-permitted[a]
Munsab	Yes	Single story	Non-tax payer	Solar	Not-permitted[a]
Emily	No	Double story	Tax payer	Grid	Not-permitted
Maria	No	Double story	Tax payer	Solar	Not-permitted
Elizabeth	No	Double story	Non-tax payer	Grid	Not-permitted
Mike	No	Double story	Non-tax payer	Solar	Not-permitted
Emma	Yes	Double story	Non-tax payer	Solar	Not-permitted
John	No	Single story	Non-tax payer	Solar	Not-permitted

[a]Represents the two misclassifications

Table 5.6 Test dataset

Name	Has garden	Home type	Taxation	Electric system	Registration
John	Yes	Single story	Non-tax payer	Solar	Permitted
Elizabeth	Yes	Double story	Non-tax payer	Solar	Not-permitted
Mike	Yes	Single story	Tax payer	Solar	Permitted
Evans	No	Single story	Non-tax payer	Solar	Not-permitted
Roberts	No	Double story	Tax payer	Solar	Not-permitted
William	No	Double story	Non-tax payer	Solar	Not-permitted
Emma	No	Double story	Non-tax payer	Solar	Not-permitted
Julie	Yes	Single story	Non-tax payer	Solar	Permitted
Smith	Yes	Double story	Tax payer	Grid	Permitted
Jamaima	No	Double story	Tax payer	Grid	Not-permitted

incorrect class assignments, it will result in the misclassification of training data. Figure 5.9 shows these classification trees.

Similarly, the models that are developed from smaller training data may also result in model overfitting mainly because of the insufficient number of representative samples in training data. Such models may have zero training error, but due to the immaturity of the classification model, the test error rate may be high, e.g., consider the dataset given in Table 5.7. Few training samples are resulting the classification error for the test dataset given in Table 5.6. Figure 5.10 shows the resulting classification tree.

5.2.3   Entropy

Now we will provide details of the measure called Information Gain to explain how this measure can be used to develop a decision tree. But before we dwell on this measure, let's discuss what is entropy.

Entropy defines the purity/impurity of a dataset. Suppose a dataset D_t has a binary classification (e.g., positive and negative classes), the entropy of D_t can be defined as:

$$\text{Entropy}(D_t) = -p + \log2\, p + - p\log2\, p - \tag{5.9}$$

Where:

$p+$ is the proportion of positive examples in D_t
$p-$ is the proportion of negative examples in D_t

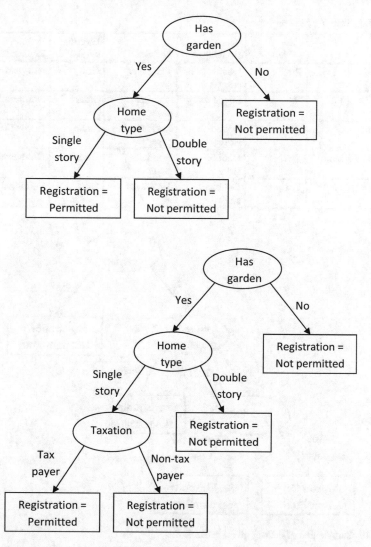

Fig. 5.9 Tree of dataset given in Tables 5.4 and 5.5

Suppose D_t is a collection of 14 data points having binary classification, including 9 positive and 5 negative examples, then the entropy of D_t will be:

$$\text{Entropy } (D_t) = -(9/14)\log 2 \ (9/14) - (5/14)\log 2 \ (5/14) = 0.940$$

It should be noted that entropy is zero (0) when all data points belong to a single class and entropy is maximum, i.e., 1 when the dataset contains an equal number of both of the examples in binary classification.

Table 5.7 Training dataset

Name	Has garden	Home type	Taxation	Electric system	Registration
Salamander (Emily)	No	Double story	Tax payer	Grid	Not-permitted
guppy (John)	No	Single story	Non-tax payer	Solar	Not-permitted
Eagle (Emma)	Yes	Double story	Non-tax payer	Solar	Not-permitted
Poorwill (Waeem)	Yes	Double story	Non-tax payer	Grid	Not-permitted
Platypus (Razaq)	Yes	Double story	Tax payer	Grid	Permitted

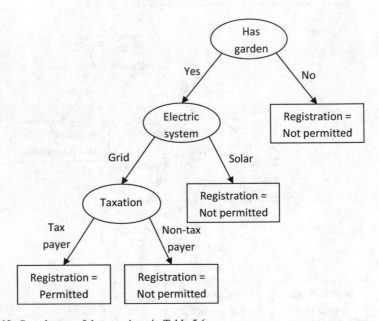

Fig. 5.10 Sample tree of dataset given in Table 5.6

5.2.4 Information Gain

Given that entropy defines the impurity in datasets, we can define the measure of Information Gain (IG) as follows:

Information Gain defines the expected reduction in entropy caused by partitioning the examples according to this attribute. Now we will create a classification tree with IG as attribute selection criteria. Consider the following dataset given in Table 5.8.

Table 5.8 Sample dataset for Information Gain-based decision tree

D	X	Y	Z	C
D1	x1	y1	z2	1
D3	x2	y2	z1	1
D3	x3	y2	z1	0
D4	x2	y2	z1	0
D5	x3	y1	z2	0

First, we calculate the entropy of the entire dataset as follows:

$$E(D) = -p + \log 2\, p + -p\log 2\, p -$$

$$E(D) = -(2/5)\,\log_2(2/5) - (2/5)\,\log_2(3/5) = 0.97$$

Now we will calculate IG of each attribute using this entropy; the attribute having maximum information gain will be placed at the root of the tree.

The formula for Information Gain is as follows:

$$\text{IG}(D, X) = E(D) - \frac{|D_{x1}|}{|D|} E(S_{x1}) - \frac{|D_{x2}|}{|D|} E(S_{x2}) - \frac{|D_{x3}|}{|D|} E(S_{x3}) \qquad (5.10)$$

Here $|D_{x1}|$ is the number of times the attribute "X" takes the value "$x1$." So, in D:

$$|D| = 5$$
$$|D_{x1}| = 1$$

In our example:

$E(D_{x1}) = 0$
$E(D_{x2}) = 1$
$E(D_{x3}) = 0$

Putting all these values in IG(D,A):

IG(D,X) = 0.57
IG(D,Y) $-$ 0.02
IG(D,Z) = 0.02

This means that attribute "X" has the maximum information gain, so it will be selected as a root node. Figure 5.11 shows the resulting classification tree.

It should be noted that the tree correctly classifies the samples $D1$, $D3$, and $D4$. For the remaining samples, we will perform another iteration by considering the other two attributes and only the remaining two samples.

IG(D',Y) = 0
IG(D',Z) = 1

Fig. 5.11 Entropy-based decision tree after the first iteration

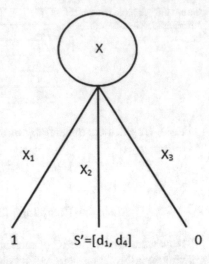

Fig. 5.12 Entropy-based decision tree after the second iteration

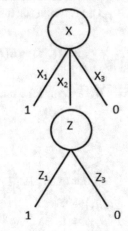

So, here attribute Y has the highest information gain, so we will consider it as the next candidate to split for the remaining two records. The decision tree is shown in Fig. 5.12.

Now all the records have been successfully classified; we will stop our iterations and the decision tree in Fig. 5.15 will be our final output.

Now we will discuss some other classification techniques.

5.3 Regression Analysis

Before we discuss regression, it's worth mentioning that analysis involves identifying the relationship between different variables. There are two types of variables in this regard. One type of variable is called dependent variable, and other ones are called independent variables. Consider the dataset given in Table 5.9.

Table 5.9 A sample dataset

X	Y
1	2
2	5
3	10
4	17
5	26

Here we have shown values of two variables, i.e., "X" and "Y." The value of "X" is used to determine the value of "Y." In other words, "Y" is dependent on "X," so, "Y" will be the dependent variable and X will be the independent variable. The relationship between them can be mentioned by the model:

$$Y = (X)^2 + 1$$

Regression analysis is one of the most common techniques used for performing the analysis between two or more variables, i.e., how the value of the independent variable affects the value of the dependent variable. The main purpose of regression analysis is to predict the value of the dependent variable for some unknown data. The value is predicted on the basis of the model developed using the available training data. So, overall regression analysis provides us following information:

1. What is the relation between independent variables and dependent variables? Note that independent variables are also called predictors and dependent variables are called outcomes.
2. How good are the predictors to predict the outcomes?
3. Which predictors are more important than the others in predicting the outcome?

There are different types of regression analysis:

- Simple linear regression
- Multiple linear regression
- Logistic regression
- Ordinal regression
- Multinomial regression

For the sake of this discussion, we will concise only to simple linear regression.

Simple Linear Regression Simple linear regression is the simplest form of regression analysis involving one dependent variable and one independent variable. The aim of the regression process is to predict the relation between the two.

The model of the simple linear regression comprises a straight line, passing through the pints on a two-dimensional space. Now the analysis process tries to find the line that is close to the majority of the points to increase accuracy. The model is expressed in the form:

$$Y = a + bX \tag{5.11}$$

Table 5.10 Heights and weights dataset

Height	Weight
5.5	44.5
6	48
4.5	37.5
5.4	?

Fig. 5.13 Weights vs Height

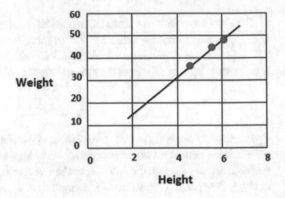

Here:

Y – Dependent variable
X – Independent (explanatory) variable
a – Intercept
b – Slope

If you remember, this is the equation of a simple linear line, i.e., $Y=mx+c$. Now we will explain it with the help of an example. Consider the simple dataset comprising of two variables, i.e., "Height" and "Weight" as shown in Table 5.10.

Now we have to predict the weight of the person with a height of 5.4 in. First, we draw the first three points on a two-dimensional space as shown in Fig. 5.13.

Now by using the model discussed above, we try to predict the weight of the person with a "Height" of 5.4 in.

If we consider $a = 6$ and $b = 7$, the predicted weight of the person will be 43.8 kg as shown in the Fig. 5.14.

5.4 Support Vector Machines

Support vector machines (SVM) are another classification technique commonly used for classifying the data in N-dimensional space. The aim of the classifier is to identify the maximal margin hyperplane that classifies the data. Before discussing support vector machines, let's discuss some prerequisite concepts.

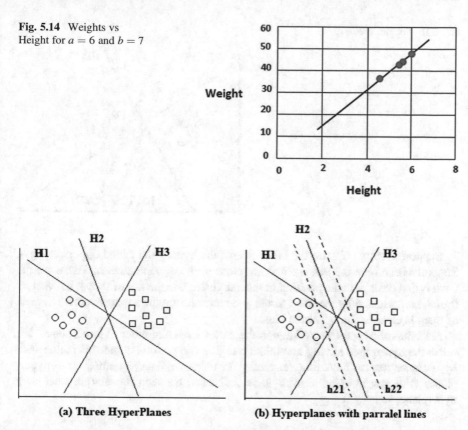

Fig. 5.14 Weights vs Height for $a = 6$ and $b = 7$

(a) Three HyperPlanes   **(b) Hyperplanes with parralel lines**

Fig. 5.15 Hyperplanes

Hyperplane: A hyperplane is a decision boundary that separates the two or more decision classes. Data points that fall on a certain side of a hyperplane belong to a different decision class. If the points are linearly separable, then the hyperplane is just a line; however, in the case of three or more features, the hyperplane may become a two-dimensional space. A linearly separable dataset is one where data points can be separated by a simple linear (straight) line. Consider the figure below showing a dataset comprising two decision classes "Rectangle" and "Circle." Figure 5.15a shows three hyperplanes H1, H2, and H3.

Note that each hyperplane can classify the dataset without any error. However, the selection of a particular hyperplane depends upon its margin distance. Each hyperplane H_i is associated with two more hyperplanes h_{i1} and h_{i2} which are obtained by drawing two parallel hyperplanes close to each decision class. In Fig. 5.15b, there are two more dotted lines h_{21}, and h_{22} are parallel hyperplanes for H_2. Parallel hyperplanes are obtained such that each parallel hyperplane is closed to a decision class. The Fig. 5.16 shows the hyperplane drawn close to the filled circle and rectangle.

Fig. 5.16 Support vectors

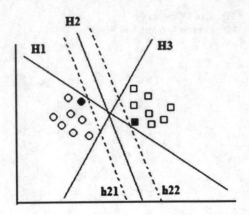

Support vectors point on the other side of the hyperplane called opponent class. Support vectors are the vectors that are close to the opponent class. These are the vectors that determine the position of the margin line. In the figure above, the vectors (points) represented by the filled circle and rectangle are the support vectors as each of them is close to the opponent class.

Selection of hyperplane: Although there may exist a number of hyperplanes, we select the hyperplane having maximum margin. The reason behind this is that such hyperplanes tend to have more accuracy for unknown data as compared to hyperplanes with low margins. A margin is a distance between the margin lines of a hyperplane, so,

$$D = d_1 + d_2$$

Where d_i is the distance of the margin line from the hyperplane.

For linearly separable problems the model for decision boundary is the same as we discussed in the case of linear regression, i.e.:

$$W \times Ps + b = h \text{ for } h < 0$$

$$\text{And } W \times Ps + b = h' \text{ for } h > 0$$

So, if we label all the Squares with S and Circles with C, then the prediction model will be:

$$f(x) = \begin{cases} C, & x < 0 \\ S, & x > 0 \end{cases}$$

Nonlinear support vector machines: So far, we discussed support vector machines for linearly separable problems where a simple straight line can be used as a hyperplane. However, this may not be the case all the time. For example, consider the dataset shown in Fig. 5.17.

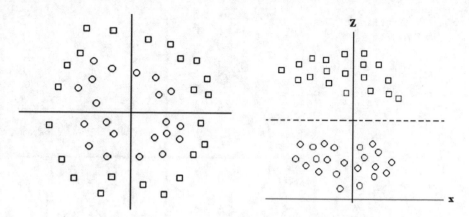

Fig. 5.17 Data points distributed nonlinearly

As we can see, the above data points cannot be classified using a linear SVM, so, what we can do here is we can convert this data into one with higher feature space. We can introduce a new dimension classed Z-Feature as follows:

$$Z = x^2 + y^2$$

Now the dataset can be classified using a linear SVM. Now suppose that the hyperplane is "K" along Z-axis, then:

$$k = x^2 + y^2$$

So, we can represent the transformed hyperplane in the original data space as shown in Fig. 5.18.

5.5   Naïve Bayes

A Naïve Bayes classifier is a probabilistic classifier that predicts the class of data based on the previous data by using probability measures. It assumes conditional independence between every pair of features given the value of the class variable. The technique can be used both for binary and multiclass classification.

The model of the classifier is given below:

$$P(c|x) = \frac{P(c|x)P(c)}{P(x)} \tag{5.12}$$

Here:

$P(c|x)$ is the probability of class c given feature x and $P(c)$ is the probability of class c.

Fig. 5.18 Hyperplane in original dataset

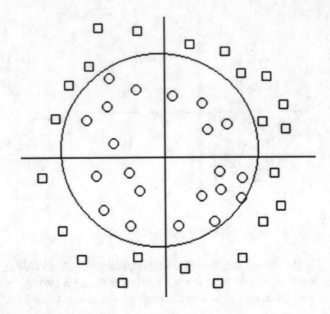

Listing 5.1 shows the pseudo code of the algorithm:

Listing 5.1 Naïve Bayes Pseudo Code

```
Input: Dataset with C features and D decision class
1) Create a frequency table
2) Create a Likelihood table
3) Calculate the probability of each class using the Naïve Bayes model
```

Now we will explain it with an example. Consider the dataset below which shows the result of each student in a particular subject. Now we need to verify the following statement:

The student will get high grades if he takes "Mathematics."

$$P(\text{High} \mid \text{Math}) = (P(\text{Math} \mid \text{High})^* \, P(\text{High}))/P(\text{Math})$$

Here we have P (Math | High) = 3/9 = 0.33, P(Math) = 5/14 = 0.36, P(High) = 9/14 = 0.64

Now, P (High | Math) = 0.33 * 0.64 / 0.36 = 0.60

So it is likely that he will have high grades in mathematics. Table 5.11a–c show the dataset, frequency, and likelihood tables for this example.

Now we will discuss some advantages and disadvantages of the Naïve Bayes approach.

Table 5.11 Dataset, frequency, and likelihood tables

(a) Dataset

Student	Subject	Grades
X_1	Math	Low
X_2	English	High
X_3	Physics	High
X_4	Math	High
X_5	Math	High
X_6	English	High
X_7	Physics	Low
X_8	Physics	Low
X_9	Math	High
X_{10}	Physics	High
X_{11}	Math	Low
X_{12}	English	High
X_{13}	English	High
X_{14}	Physics	Low

(b) Frequency table

	Low	High
English		4
Physics	3	2
Math	2	3
Grand total	5	9

(c) Likelihood table

	Low	High		
English		4	4/14	0.29
Physics	3	2	5/14	0.36
Math	2	3	5/14	0.36
Grand total	5	9		
	5/14	9/14		
	0.36	0.64		

Advantages:

- Simple and easy to implement.
- It is a probabilistic-based approach that can handle both continuous and discrete features.
- Can be used for multiclass prediction.
- It assumes that features are independent, so works well for all such datasets.
- Requires less training data.

Disadvantages:

- Assumes that features are independent which may not be possible in a majority of datasets.

- It may have low accuracy.
- If a categorical feature is not assigned a class, the model will assign it zero probability and consequently will be unable to make a prediction.

5.6 Artificial Neural Networks

Scientists were always interested in making machines that can learn just like humans learn. This resulted in the development of artificial neural networks. The first neural network was "Perceptron." In this topic, we will discuss multilayer perceptron with backpropagation. A simple neural network model is shown in Fig. 5.19.

As shown, a simple neural network model comprises three layers, i.e., input layer, hidden layer, and output layer.

Input layers comprise the n features given as input to the network, as shown in the diagram above; the input comprises the value $\{x_0, x_1, x_2, \ldots, x_m\}$. Now each input is multiplied by the corresponding weight. The weight determines how important the input is in the classification. All the inputs are provided to the summation function which is shown below:

$$Y = \sum_{i=0}^{m} w_i x_i + \text{biase} \tag{5.13}$$

The value of "z" is then provided to the activation function which will enable to neuron to be activated based on the summation value. One of the most common activation functions is sigmoid:

$$z = \text{Sigmoid}(Y) = \frac{1}{1 + e^{-y}} \tag{5.14}$$

The sigmoid activation function returns the value between zero and one as shown in Fig. 5.20.

Fig. 5.19 A simple neural network model

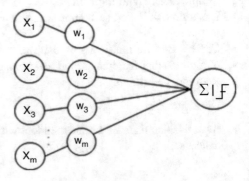

Fig. 5.20 Sigmoid function

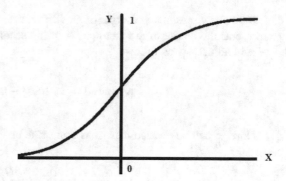

Fig. 5.21 A sample neural network

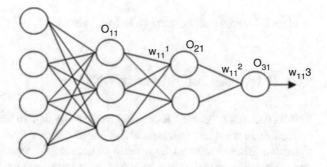

A value above 0.5 will activate the neuron and a value below 0.5 will not activate it. Sigmoid is the simplest one; however, there are many other functions as well like rectified linear unit (Relue); however, we will not discuss them.

Loss function: The loss function determines how much the predicted value is different from the actual value. This function is used as a criterion to train a neural network. We calculate the loss function again and again by using backpropagation and adjust the weights of the neural network accordingly. The process continues until there is no decrease in the loss function. There are multiple loss functions. Here we will discuss only the Mean Squared Error (MSE).

MSE finds the average squared difference between the predicted value and the true value. Mathematically:

$$MSE = \sum_{i=1}^{m} (y_i - \widehat{y}_i) \tag{5.15}$$

Here y_i is the actual output and $\widehat{y}_i$ is the predicted output. We calculate the loss function and adjust weights accordingly in order to minimize the loss. For this purpose, we use backpropagation.

In this process, we calculate the derivate of the loss function with respect to the weights of the last layer.

Consider the following neural network model as shown in Fig. 5.21.

Here O_{11} represents the output of the first neuron of the first hidden layer, and O_{21} represents the output of the first neuron of the second layer. With backpropagation, we will adjust w_{11}^3 as follows:

$$w_{11}^3(\text{new}) = w_{11}^3(\text{old}) - \eta \frac{\partial L}{\partial w_{11}^3} \tag{5.16}$$

Now $\frac{\partial L}{w_{11}^3}$ can be defined using the chain rule as follows:

$$\frac{\partial L}{\partial w_{11}^3} = \frac{\partial L}{\partial O_{31}} * \frac{\partial O_{31}}{\partial w_{11}^3}$$

Here ∂L represents the change in loss value.

5.7   K Nearest Neighbors (KNN)

KNN is called lazy learner. It does not require any data to learn. In other words, it has no requirement for any training data. It learns directly from it. This makes the KNN algorithm faster than other algorithms which require training. Since KNN requires no training data, new data can be added in real time and will not impact the accuracy of the algorithm.

KNN has only two parameters

- The value of K
- The distance function

Suppose we have two features, i.e., age and height, and based on these features we have classified data as shown in Fig. 5.22.

Now within that, we have a new data point represented by a star. The question is "Should it be classified as the square or as the circle based on the two features?"

Fig. 5.22 Classification based on age and height

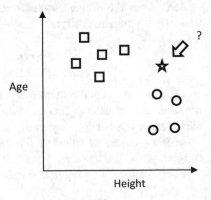

Fig. 5.23 Calculating distances

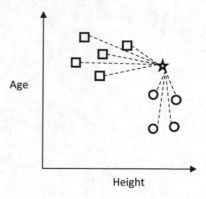

Fig. 5.24 Three closest points

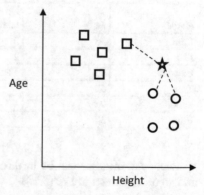

This can be solved by first calculating all the distances between the star and the rest of the data points as shown in Fig. 5.23.

We then find the k closest points or neighbors. Suppose in the above example, we set the value of k to be 3. We then find the three closest points as shown in Fig. 5.24.

In this case, the two closest neighbors are the circle and one closest point is the square. Since the majority class here is the circle, therefore we will assign the new data point represented by the star as the circle.

Of course, we need to measure the distance between the points, and for that, we can use various techniques such as Euclidean or Manhattan. The most common technique is Euclidean shown in Eq. (5.17):

$$d = \sqrt{(x - x_1)^2 + (y - y_1)^2} \tag{5.17}$$

In our example, the distance (d) will be calculated as shown in Fig. 5.25.

Fig. 5.25 Euclidean distance

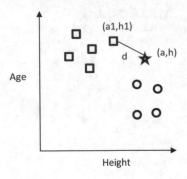

Table 5.12 KNN example

Name	Age	No of loans taken	Total amount borrowed	Bank response
John	35	3	£35,000	No
Bill	22	2	£50,000	Yes
Edward	63	1	£200,000	No
Tom	59	1	£170,000	No
Sarah	25	4	£40,000	Yes
Elliot	37	2	£50,000	?

Euclidean distance can be in three dimension, four dimension, five dimension, and so on as shown in Eq. (5.18).

$$d = \sqrt{(x - x_1)^2 + (y - y_1)^2 + (z - z_1)^2} \tag{5.18}$$

Let us consider a very simple working example of KNN. Consider Table 5.12.

In this example, we have various customers, which have taken loans from a bank and when they have applied for a new loan based on their data of Age, Number of loans previously taken, and Total amount already borrowed; the bank will make a decision as to give them a new loan or not.

Now, in the example, we have to find if Elliot whose age is 37, has already taken two loans, and has so far borrowed £50,000, will the bank give him a new loan or not? We will now apply KNN to see if the bank will give Elliot a new loan or not. Let us suppose that the value of k is set to 3 and we are using the Euclidean distance. By calculating the Euclidean distance, we will get Table 5.13.

The three nearest neighbors for Elliot are John, Bill, and Sarah. For John, the bank response is "No," whereas for both Bill and Sarah, it is "Yes." Thus for Elliot, the bank's response will also be "Yes."

There are disadvantages of KNN too. Firstly it is not suitable for large datasets as the cost of calculating the distance between the new point, and each existing point can be very large which degrades the performance of the algorithm. Also, the user

Table 5.13 KNN example with Euclidean distance from Elliot

Name	Age	No of loans taken	Total amount borrowed	Bank response	Euclidean distance from Elliot
John	35	3	£35,000	No	15.2
Bill	22	2	£50,000	Yes	15
Edward	63	1	£200,000	No	152.2
Tom	59	1	£170,000	No	122
Sarah	25	4	£40,000	Yes	15.7
Elliot	37	2	£50,000	?	

has to set the value of k. Generally, the value of k is set to be an odd number. KNN can be sensitive to noisy data, missing values, and outliers. In this case, manual preprocessing may be required.

5.8 Ensembles

Ensembles are aggregate predictions made by multiple classifiers. Just like individual classifiers, they are used to predict class labels of previously unseen records. Multiple classifiers can be aggregated by using multiple techniques including:

- Majority voting
- Weighted voting
- Bagging

Each individual or single classifier has some limitations as previously discussed. This introduces the concept of the "No Free Lunch" Theorem, i.e., no single algorithm wins all the time!

5.8.1 Assembling an Ensemble of Classifiers

There are two important factors for assembling an ensemble of classifiers:

- Selection of classifiers
- Combination of classifiers

For a good ensemble model, it should consist of two main points: diversity and accuracy, i.e., different learning operators can address a problem (accuracy) and different learning operators make different mistakes (diversity). That means predictions on a new example may differ, if one learner is wrong, others may be right.

Thus we can say is:

- Accurate classifier: Accurate classifier has an error rate better than the random when guessing new examples.
- Diverse classifiers: Two classifiers are diverse if they use different training data or different learning algorithms and produce predictions independently.

The main purpose of an ensemble model is to maximize accuracy and diversity. There are different ways to combine the output of single classifiers for maximum accuracy such as majority voting, weighted voting, bagging, stacking, etc.

Here we shall have a look at some of these techniques.

5.8.2 Majority Voting Ensemble

The majority voting ensemble will output the class which has the highest number of votes.

Suppose we have five classifiers, i.e., Naïve Bayes, decision tree, artificial neural network, support vector machine, and K nearest neighbor that have predicated the following classes for a test instance which is either 1 or 0.

Naïve Bayes (NB) = 0
Decision tree (DT) = 0
Artificial neural network (ANN) = 1
K nearest neighbor (KNN) = 0
Support vector machine (SVM) = 1

Class 0: NB+ DT + KNN → 3 votes
Class 1: ANN + SVM → 2 votes

Thus the final output will be class 0.
Majority voting is the simplest way of combining predictions.

5.8.3 Boosting/Weighted Voting Ensemble

This method uses classifiers combination using a weighted scheme. The main idea is that a high accuracy classifier will attain high weight and vice versa.

The weighted voting ensemble will output the class which has the highest weight, in this case, accuracy, associated with it.

Suppose we have five classifiers, i.e., Naïve Bayes, decision tree, artificial neural network, support vector machine, and K nearest neighbor. Classifier training is performed on training data, to calculate the accuracy of each classifier. Suppose the accuracies are as follows:

Naïve Bayes = 70%
Decision tree = 75%
Artificial neural network = 80%
K nearest neighbor = 82%
Support vector machine = 87%

The resultant weights will be as follows: Naïve Bayes = .7, decision tree = .75, artificial neural network = .8, K nearest neighbor = .82, and support vector machine = .87.

Now suppose the following classes are predicted for a test instance by each classifier:

Naïve Bayes (NB) = 1
Decision tree (DT) = 0
Artificial neural network (ANN) = 1
K nearest neighbor (KNN) = 0
Support vector machine (SVM) = 1

> Class 0: DT + KNN → 0.75+0.82 → 1.57
> Class 1: NB + ANN + SVM → 0.7+0.8+0.87 → 2.37

Thus the final output will be class 1.

We may use multiple parameters for voting such as precision and recall or F-measure which is a harmonic mean of two numbers and tends to be closer to the smaller of the two. Thus, for the F-measure score to be large, both precision and recall must be high.

Suppose we have five classifiers, i.e., Naïve Bayes, decision tree, artificial neural network, support vector machine, and K nearest neighbor. Classifier training is performed on training data, to calculate the F-measure of each classifier. Suppose the F-measure is as follows:

Naïve Bayes = 60%
Decision tree = 70%
Artificial neural network = 80%
K nearest neighbor = 85%
Support vector machine = 90%

The resultant weights are as follows: Naïve Bayes = .6, decision tree = .7, artificial neural network = .8, K nearest neighbor = .85, and support vector machine = .9.

Suppose the following classes are predicted for a test instance:

Naïve Bayes (NB) = 0
Decision tree (DT) = 0
Artificial neural network (ANN) = 1
K nearest neighbor (KNN) = 0
Support vector machine (SVM) = 1

Class 0: NB + DT + KNN → 0.6+0.7+0.85→ 2.15
Class 1: ANN + SVM → 0.8+0.9 → 1.7

Thus the final output will be class 0.

5.8.4 Bagging

The idea of bagging is derived from the fact that models may differ when learned on different data samples. In such a case, we create samples by picking examples and train a model on each sample. Once that has been done, we combine models to create a single model as shown in Fig. 5.26.

Usually, the same base learner is used.

Random forest is a variation of bagging with decision trees. The idea is to train a number of individual decision trees each on a random subset of examples and then combine the output.

5.9   Methods for Model Evaluation

There are three main methods for estimating the performance measures. These are

- Holdout method
- Random subsampling
- Cross-validation

Fig. 5.26 Bagging ensemble model

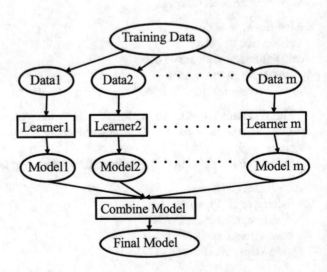

5.9.1 Holdout Method

The holdout method reserves a certain amount of the labeled data for testing and uses the remainder for training. Usually, one-third is used for testing and the rest for training. However, this is not fixed and can be varied.

However, for unbalanced datasets, this may not be the most appropriate method as random samples might not be representative of the whole dataset. Records that are important for learning may always be present in test cases. This problem can be solved by using stratified sampling, in which we sample each class independently so that records of the minority class are present in each sample.

5.9.2 Random Subsampling

Holdout estimate can be made more reliable by repeating the process with different subsamples. The number of subsamples can vary. Although this improves, still records that are important for learning may always be present in test cases.

5.9.3 Cross-validation

The aim of cross-validation is to avoid overlapping test sets. It is a two-step process:
- Data is split into k subsets of equal size.
- Each subset in turn is used for testing and the remainder for training.

It is called *k-fold* cross-validation where the most common value of k is 10 or tenfold cross-validation.

5.10 Summary

In this chapter, we discussed the concept of classification in detail. The reason behind this was the classification tree is one of the most commonly used classifiers. All the concepts necessary for developing a classification tree were also presented using examples. It should be noted that most of the concepts related to the classification tree also relate to other classification techniques.

Further Reading

Following are some valuable resources for further reading:

- Data Classification: Algorithms and Applications (Chapman & Hall/CRC Data Mining and Knowledge Discovery Series) Part of: Chapman & Hall/CRC Data Mining and Knowledge Discovery by Charu C. Aggarwal, CRC Press, 2020. Classification is one of the important machine learning techniques used in various machine learning applications. The book discusses the concept of classification, different classification algorithms, and their applications. Firstly, the book describes the classification techniques including decision trees, probabilistic methods, instance-based methods, rule-based methods, neural networks, and support vector machine (SVM). After this, different enhancements in classification methods are discussed, and finally, the book covers domain-specific methods.
- Machine Learning with scikit-learn Quick Start Guide: Classification, regression, and clustering techniques in Python 1st Edition, Kindle Edition by Kevin Jolly, Packt Publishing, 2018. This book covers the scikit machine learning library including its installation and how it can be used for developing machine learning models. Book discusses both the supervised and unsupervised machine learning models.
- Classification Problem in Data Mining Using Decision Trees by Ashima Gambhir, LAP LAMBERT Academic Publishing, 2019. This book in-depth details about decision trees and how they can be used for performing classification tasks for data analysis applications.
- Machine Learning with Random Forests And Decision Trees: A Visual Guide For Beginners Kindle Edition by Scott Hartshorn, 2016. This book provides an overview of random forests and decision trees. It provides details about how the decision trees can be effectively used in random forest algorithms in order to perform predictions.
- Tree-Based Methods for Statistical Learning in R (Chapman & Hall/CRC Data Science Series) by Brandon M. Greenwell, Chapman and Hall/CRC, 2022. This book presents mathematical concepts along with coding examples. It explains R statistical language and its use for the implementation of random forest and gradient tree boosting functions. The book also provides practical problems for the readers to solve. Book presents the state-of-the-art machine learning methodologies and statistical developments. Few of the topics covered in this book include conditional inference trees, random forests, boosting, and bagging.
- Tree-Based Machine Learning Methods in SAS by Dr. Sharad Saxena, SAS Institute, 2022. This book includes how to build decision trees using SAS viya. The book discusses predictive tree structures, pruning methodologies, gradient boosted trees, and forests. Each chapter provides detailed concepts along with state-of-the-art trends. After completing this book, the reader will be able to build regression trees, classification trees, Poisson, tweedy gradient boosted regression models, run isolation forest, dimension reduction, and decision trees for exploratory data analysis.

Exercises

Exercise 5.1: What is a confusion matrix, and how can it measure the accuracy and error rate of other machine learning models?

Exercise 5.2: Consider the following sample dataset given in the table below, and find the accuracy and error rate using the confusion matrix.

ID	Sick		Output
	actual	Predicted	
1	1	0	FP
2	0	1	FN
3	1	1	TP
4	0	0	TN
5	0	1	FN
6	1	1	TP
7	1	0	FP
8	0	1	FN
9	1	0	FP
10	0	0	TN

Exercise 5.3: In the decision tree, leaf nodes and parent nodes represent which type of information from the given dataset.

Exercise 5.4: In a decision tree, how can you identify which column has the highest information gain?

Exercise 5.5: Consider the following sample dataset of students, use entropy and information gain, and construct the decision tree. The registration column is used as the goal.

ID	Degree	Finance	Marks	Registration
1	Postgraduate	Self-finance	140	Not-permitted
2	Undergraduate	Scholarship	115	Not-permitted
3	Undergraduate	Self-sponsored	55	Not-permitted
4	Postgraduate	Scholarship	135	Not-permitted
5	Undergraduate	Partial sponsored	110	Permitted
6	Undergraduate	Scholarship	75	Not-permitted
7	Postgraduate	Partial sponsored	205	Not-permitted
8	Undergraduate	Self-sponsored	100	Permitted
9	Undergraduate	Scholarship	89	Not-permitted
10	Undergraduate	Self-sponsored	105	Permitted

Exercise 5.6: Explain the terms given below, and mention the differences in all of the given terms with the help of a table.

- Binary attributes
- Nominal attributes
- Ordinal attributes
- Continuous attributes

Exercise 5.7: A decision tree can have three types of errors that are given below. Explain each type of error.

- Training errors
- Test errors
- Generalization errors

Exercise 5.8: A few types of regression analysis are given below. Provide a brief introduction of each of these types.

- Simple linear regression
- Multiple linear regression
- Logistic regression
- Ordinal regression
- Multinomial regression

Exercise 5.9: Consider the given dataset of cricket players. The dataset provides runs made by each player and the wickets taken by him.

Player	Runs	Wickets	Type
P1	7404	368	Batsman
P2	2327	418	Bowler
P3	5213	249	Batsman
P4	2185	387	Bowler
P5	6718	57	Batsman
P6	2477	440	Bowler
P7	4887	413	Batsman
P8	6364	274	Batsman
P9	2143	253	Bowler
P10	5999	95	Batsman
P11	9887	203	Batsman
P12	6520	57	Batsman
P13	1127	103	Bowler

Using SVM, classify the player P14 as having runs 2512 and 352 wickets.

Exercise 5.10: What are the advantages and disadvantages of the Naïve Bayes classifier?

Exercise 5.11: Consider the dataset of students given below. Create the frequency table and the likelihood table in the context of the Naïve Bayes classifier.

Student	Subject	Grades
S1	English	High
S2	English	Low
S3	Physics	High
S4	Math	High
S5	Physics	Low
S6	English	Low
S7	Math	High
S8	English	Low
S9	English	High
S10	Physics	High
S11	English	Low
S12	English	High
S13	Math	High
S14	Math	High

Exercise 5.12: Two classifiers (A and B) are designed to predict patients' disease. The tests result in the following two confusion matrices, one for each classifier:

Classifier A

		Known class	
		Disease	Normal
Predicted class	Disease	80	80
	Normal	44	332

Classifier B

		Known class	
		Disease	Normal
Predicted class	Disease	222	89
	Normal	61	386

Calculate the accuracy, precision, and recall for each classifier. Which is better?

Exercise 5.13: Consider the following confusion matrix:

	Predicted negative	Predicted positive
Negative cases	9221	1222
Positive cases	361	41

What are the values of:

(a) Precision
(b) Recall
(c) Accuracy
(d) F-Measure

Exercise 5.14: Consider the following three attribute (A, B, C) dataset for a binary classification problem. Show your classifier tree, with class predictions in its leaf nodes.

Using the tree, classify the following query: $A = 2, B = 0, C = 0$.

A	B	C	Class
0	0	0	–
1	0	0	–
1	0	1	+
2	1	1	+
2	2	2	+
2	0	2	–

Exercise 5.15: Consider the following three attributes (A, B, C) dataset for a binary classification problem. Calculate information gains obtained when splitting on A and B, respectively. Which attribute would the decision tree induction algorithm choose and why?

A	B	C	Class
0	0	0	–
1	0	0	–
1	0	1	+
2	1	1	+
2	2	2	+
2	2	1	–

Exercise 5.16: Consider the following perceptron where I = [2.0, 0.5, 1.0, 1.5] and W = [0.3, 0.8, 1, 0.2]. The threshold for activation is $t = 5.0$.

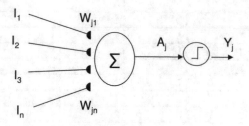

Will the neuron fire or not for the above values?

Exercise 5.17: Consider the following decision table.

	C1	C2	C3	C4	D
X1	0	0	b	0	Yes
X2	0	1	a	1	Yes
X3	0	0	a	1	Yes
X4	0	1	a	0	No
X5	0	1	b	1	Yes
X6	1	1	b	1	Yes
X7	1	0	a	0	Yes
X8	1	1	b	1	Yes
X9	1	1	a	0	No
X10	1	1	b	0	Yes

The testing and training should be 70/30, i.e., first seven are for training and the last three are for testing. Using the testing data, make an ensemble model which consists of three classifiers, i.e., KNN (where $K = 5$), KNN (where $k = 3$), and Naïve Bayes.

(a) Carrying out boosting on the ensemble model for the three training records. For each testing record, if the classifier has produced the correct decision, give it a boost of +1, and if the wrong decision then give it −1.
(b) Classify a C1 = 1, C2 = 1, C3 = a and C4 = 1 using the boosted ensemble created in part.

Exercise 5.18: Consider the following dataset.

A	B	C	Class
0	0	0	−
1	0	0	−
1	0	1	+
1	1	1	+
1	1	2	+
2	0	2	−
2	1	2	−
2	1	1	−
2	0	2	+

You are to train Naïve Bayes for the dataset. However, the aim here is to see which model of evaluation is most accurate. You are to compare the following models in terms of accuracy for Naïve Bayes.

(a) Hold out method (use the first seven for training and the rest for testing)
(b) Random subsampling (use 50% for training and 50% for testing)

Chapter 6
Clustering

6.1 Cluster Analysis

A cluster is a group of data that share some common properties, e.g., in Fig. 6.1 two clusters are given that share the same properties, e.g., all the objects in the left cluster are closed-shaped, whereas all the objects in the rightmost cluster are open-shaped. Cluster analysis is the process of clustering the objects in groups on the basis of their properties and their relation with each other and studying their behavior for extracting the information for analysis purposes. Ideally, all the objects within a cluster have similar properties which are different from the properties of objects in other groups.

Figure 6.1 shows some sample clusters of a dataset.

However, it should be noted that the idea of dividing objects into groups (clusters) may vary from application to application. A dataset that is divided into two clusters may also be divided into more clusters at some refined level, e.g., consider Fig. 6.2 where the same dataset is divided into four and six clusters.

Clustering can also be called unsupervised classification in that it also classifies objects (into clusters) but here the classification is done on the basis of the properties of the objects and we do not have any predefined model developed from some training data. It should be noted that sometimes the terms like partitioning or segregation are also used as a synonym for clustering but technically they do not depict the true meaning of the actual clustering process.

6.2 Types of Clusters

Clusters can be of different types based on their properties and how they are developed. Here we will provide some detail about each of these types.

U. Qamar, M. S. Raza, *Data Science Concepts and Techniques with Applications*, https://doi.org/10.1007/978-3-031-17442-1_6

Fig. 6.1 Two sample
clusters

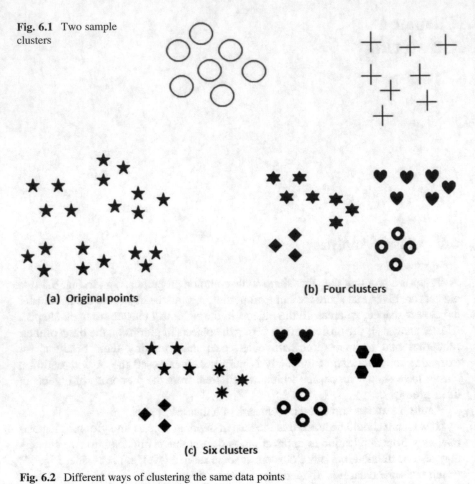

(a) Original points

(b) Four clusters

(c) Six clusters

Fig. 6.2 Different ways of clustering the same data points

6.2.1 Hierarchical Clusters

A hierarchical cluster in simple terms is one having sub-clusters. It may be possible that we have some clusters inside another cluster, e.g., cluster of the items sold within the cluster of the winter items. So at the outermost level, we have a single cluster that is further divided into sub-clusters. The outermost level can be referred to as the root cluster, whereas the innermost clusters (that are not further clustered) can be referred to as the leaf clusters as shown in Fig. 6.3:

Fig. 6.3 Two sample
clusters

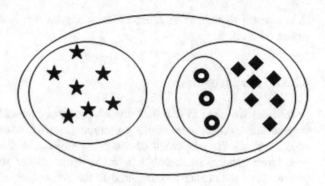

6.2.2 Partitioned Clusters

Unlike hierarchical clusters, a partitioned cluster is simply the division or grouping
of objects into separate (nonoverlapping clusters). So, the objects in one cluster are
not part of any other cluster. The two clusters in Fig. 6.2 form two nonoverlapping
partitioned clusters.

6.2.3 Other Cluster Types

We may have some other cluster types like fuzzy clusters. In fuzzy theory, an object
belongs to a set by some degree defined by its membership function. So the objects
in fuzzy clusters belong to all clusters to a degree defined by their membership
function. The value is taken between zero and one.

Similarly, we may have complete and partial clustering. In complete clustering,
every object is assigned to some cluster, whereas in partial clustering the objects may
not. The reason behind this is that the objects may not have well-defined properties
to assign them to a particular cluster. Note that we will discuss only the partitioned
and hierarchical clusters.

There are some other notations used for different types of clusters. Here we will
discuss a few of them:

6.2.3.1 Well Separated

As in a cluster, the objects share common properties, so the well-separated clusters
are the ones in which objects in one group (cluster) are more close to each other as
compared to the objects in some other cluster. For example, if we talk about the
distance of objects from each other as a proximity measure to determine how close
the objects are, then in well-separated clusters, objects within a cluster are more close

as compared to objects in other clusters. Figure 6.1 is an example of two well-separated clusters.

6.2.3.2   Prototype-Based

Sometimes we need to consider one object from each cluster to represent it. This representative object is called the prototype, centroid, or medoid of the cluster. This object has all the properties of the other objects in the same cluster. So, in a prototyped cluster, all the objects in the same cluster are more similar to center (centroid) of the cluster as compared to the center (centroid) of the other clusters. Figure 6.4 shows four center-based clusters.

6.2.3.3   Contiguous Clusters

In contiguous clusters, objects are connected to each other because they are at a specified distance and the distance of an object with another object in the same cluster is closer to its distance with some other object in another cluster. This may lead to some irregular and intertwined clusters. However, it should be noted that if there is noise in data, then two different clusters may appear as one cluster as shown in the figure below due to a bridge that may appear due to the presence of noise. Figure 6.5 shows examples of contiguous clusters.

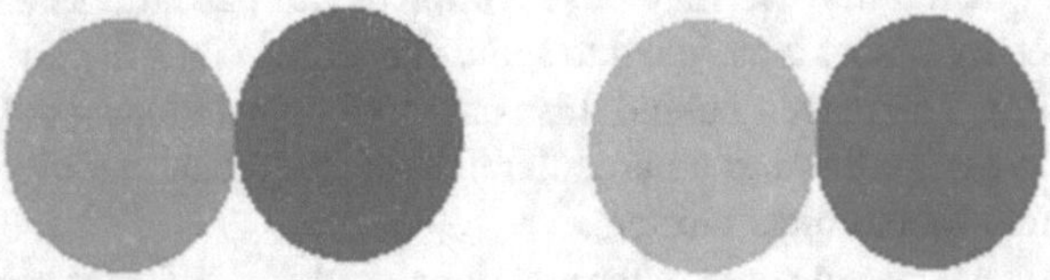

Fig. 6.4   Two center-based clusters

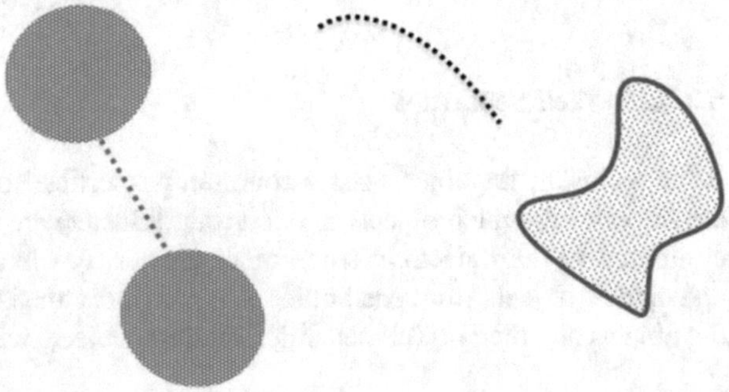

Fig. 6.5   Contiguity-based clusters

6.2.3.4 Density Based

In density-based clustering, the clusters are formed because the objects in a region are dense (much closer to each other) than the other objects in the region. The surrounding area of the density-based clusters is less dense having a lesser number of objects as shown in Fig. 6.6. This type of clustering is also used when clusters are irregular or intertwined.

6.2.3.5 Shared Property

These are also called conceptual clusters. Clusters of this type are formed by objects having some common properties. Figure 6.7 shows the example of shared property or conceptual clusters.

6.3 K-Means

Now we will discuss one of the most common clustering algorithms called K-Means. K-Means is based on prototype-based clustering, i.e., they create prototypes and then arrange objects in clusters according to those prototypes. It has two versions, i.e.,

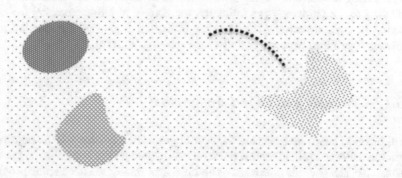

Fig. 6.6 Density-based clusters

Fig. 6.7 Shared property clusters

K-Means and K-Medoid. K-Means define prototypes in terms of the centroids which are normally the mean points of the objects in clusters. K-Medoid, on the other hand, defines the prototypes in terms of medoids which are the most representative points for the objects in a cluster.

Algorithm 6.1 shows the pseudo code of the K-Means clustering algorithm:

Algorithm 6.1 K-Means Clustering Algorithm

---

1. Select value of K (number of initial centroids).
2. Randomly select centroids.
3. Assign objects to the closest centroids.
4. Update centroids.
5. Repeat steps 3 to 4 until there are no more updates in centroids.

Here K is the number of centroids and defines the number of clusters that will be resulted. The value of K is specified by the user.

The algorithm initially selects k centroids and then arranges the objects in such a way that each object is assigned to the centroid closer to it. Once all the objects are assigned to their closer centroids, the algorithm starts the second iteration and the centroids are updated. The process continues until centroids do not change anymore or no point changes its cluster. Figure 6.8 shows the working of the algorithm.

The algorithm completes the operation in four steps. In the first step, the points are assigned to the corresponding centroids. Note that here the value of K is 3, so three centroids are identified. Centroids are represented by "+" symbol. Once the objects are identified as centroids, in the second and third steps, the centroids are updated according to the points assigned. Finally, the algorithm terminates after the fourth iteration because no more changes occur. This was a general introduction to the K-Means algorithm. Now we will discuss each step in detail.

6.3.1 Centroids and Object Assignment

As discussed above, centroids are the central points of a cluster. All the objects in a cluster are closer to their central points as compared to the centroids of the other clusters. To identify this closeness of an object, normally we need some proximity measure. One of the commonly used measures is the Euclidean distance. According to Euclidean distance, the distance between two points in the plane with coordinates (x, y) and (a, b) is given by:

$$\text{dist}((x,\ y)(a,\ b)) = \sqrt{(x-a)2 + (y-b)^2} \qquad (6.1)$$

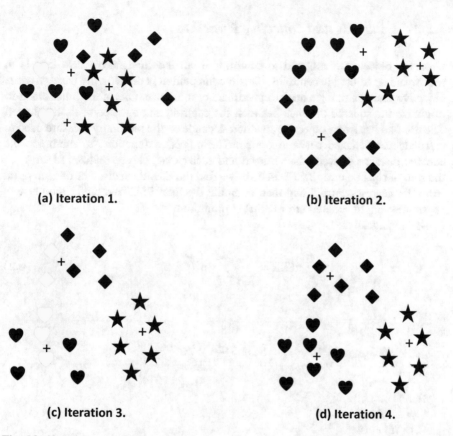

(a) Iteration 1.   (b) Iteration 2.

(c) Iteration 3.   (d) Iteration 4.

Fig. 6.8 Shared property clusters

As an example, the (Euclidean) distance between points $(2, -1)$ and $(-2, 2)$ is found to be:

$$\text{dist}((2, \ -1)(-2, \ 2)) = \sqrt{(2-(-2))^2 + ((-1)-2)^2}$$

$$\text{dist}((x, \ y)(a, \ b)) = \sqrt{(2+2)^2 + (-1-2)^2}$$

$$\text{dist}((x, \ y)(a, \ b)) = \sqrt{(4)^2 + (-3)^2}$$

$$\text{dist}((x, \ y)(a, \ b)) = \sqrt{16 + 9}$$

$$\text{dist}((x, \ y)(a, \ b)) = \sqrt{25}$$

$$\text{dist}((x, \ y)(a, \ b)) = 5$$

6.3.2   Centroids and Objective Function

Once the objects are assigned to centroids in all the clusters, the next step is to recompute or adjust the centroids. This recomputation is done on the basis of some objective function that we need to maximize or minimize. One of the functions is to minimize the squared distance between the centroid and each point in the cluster. The use of a particular objective function depends on the proximity measure used to calculate the distance between points. The objective function "to minimize the squared distance between the centroid and each point" can be realized in terms of the sum of squared error (SSE). In SSE, we find the Euclidean distance of each point from the nearest centroid and then compute the total SSE. We prefer the clusters where the sum of squared error (SSE) is minimum.

Mathematically:

$$SSE = \sum_{j=1}^{n} \sum_{x \in C_i} dist(C_i, x)^2 \tag{6.2}$$

where:

$$SSE = \frac{1}{m_i} \sum_{x \in C_i} X$$

Here:

$x =$ an object.

$C_i = i$th cluster.

$c_i =$ centroid of cluster C_i.

$c =$ centroid of all points.

$m_i =$ number of objects in the ith cluster.

$m =$ number of objects in the dataset.

$n =$ number of clusters.

The centroid c_i of the ith cluster can be calculated as follows:

The proximity measure that we have shown above was related to data in Euclidean space. However, we may have document-oriented data as shown in Table 6.1:

For such data, our objective function may be to maximize the similarity between the documents. So we cluster the documents having maximum similar terms. One of the objective functions in all such cases may be:

$$SSE = \sum_{j=1}^{n} \sum_{x \in C_i} consine(C_i, x) \tag{6.3}$$

There may be a number of proximity functions that can be used depending on the nature of data and requirements. Table 6.2 shows some sample proximity measures:

Table 6.1 Document-term matrix

	To	The	a	an	I	am	He	It	She	They
Document 1	4	4	6	2	2	2	3	6	2	1
Document 2	3	2	3	1	5	1	1	5	4	3
Document 3	6	1	5	1	2	2	4	2	3	2

Table 6.2 Some sample proximity measure

Proximity function	The measure used by the function
Bregman divergence	Mean
Cosine	Mean
Manhattan	Median
Squared Euclidean	Mean

6.4 Reducing the SSE with Post-processing

Whenever we use SSE as a proximity measure, the SSE may increase due to outliers. One of the strategies may be to increase the value of K, i.e., produce more number of clusters; thus the distance of the points from centroids will be minimum resulting in minimum SSE. Here we will discuss two strategies that can reduce SSE by increasing the number of clusters.

6.4.1 Split a Cluster

We can use different strategies to split a cluster for we can split a cluster having the largest SSE value; alternatively, the standard deviation measure can also be used, e.g., the cluster having the largest standard deviation with respect to a standard attribute can be split.

6.4.2 Introduce a New Cluster Centroid

We can introduce a new cluster centroid. A point far from the cluster center can be used for this purpose. However, for selecting such a point, we will have to calculate the SSE of each point.

6.4.3 Disperse a Cluster

We can remove the centroid and reassign the points to other clusters. To disperse a cluster, we select the one that increases the total SSE the least.

6.4.4   Merge Two Clusters

To merge two clusters, we choose clusters that have the closest centroids. We can also choose the clusters which result in a minimum SSE value.

6.5   Bisecting K-Means

Bisecting K-Means is a hybrid approach that uses both K-Means and hierarchical clustering. It starts with one main cluster comprising of all points and keeps on splitting clusters into two at each step. The process continues until we get an already specified number of clusters. Its Algorithm 6.2 gives the pseudo code of the algorithm.

Algorithm 6.2   Bisecting K-Means Clustering Algorithm

---
1. Specify the resulting number of cluster.
2. Make all the points as one cluster.
3. Split the main cluster with k=2 using the K-Means algorithm.
4. Calculate the SSE of resulting clusters.
5. Split the cluster with a higher SSE value into two further clusters.
6. Repeat steps 4 to 5 until we get the desired number of clusters.

We will now explain the algorithm with a simple and generic example. Consider the dataset P= $\{P_1, P_2, P_3, P_4, P_5, P_6, P_7\}$ which we will use to perform clustering using bisecting K-Means clustering algorithm.

Suppose we start with three resulting clusters, i.e., the algorithm will terminate once all the data points are grouped into three clusters. We will use SSE as split criteria, i.e., the cluster having a high SSE value will be split into the further cluster. So our main cluster will be C = $\{P_1, P_2, P_3, P_4, P_5, P_6, P_7\}$ as shown in Fig. 6.9:

Now by applying K = 2, we will split the cluster C into two sub-clusters, i.e., C_1 and C_2, using the K-Means algorithm; suppose the resulting clusters are as shown in Fig. 6.10:

Fig. 6.9   Main cluster

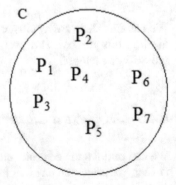

Fig. 6.10 The main cluster splits into two sub-clusters

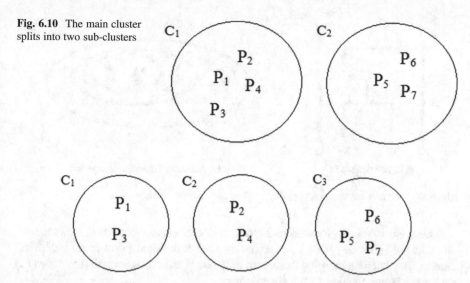

Fig. 6.11 C_1 split into two further sub-clusters

Now we will calculate the SSE of both clusters. The cluster that has the higher value will be split further into sub-clusters. We suppose cluster C_1 has a higher SSE value. So it will be split further into two sub-clusters as shown in Fig. 6.11:

Now algorithm will stop as we have obtained our desired number of resultant clusters.

6.6   Agglomerative Hierarchical Clustering

Hierarchical clustering also has widespread use. After partitioned clustering, these techniques are the second most important clustering techniques. There are two common methods for hierarchical clustering:

Agglomerative These approaches consider the points as individual clusters and keep on merging the individual clusters on the basis of some proximity measure, e.g., the clusters that are closer to each other.

Divisive These approaches consider all the data points as a single all-inclusive cluster and keep on splitting it. The process continues until we get singleton clusters that we cannot split further.

Normally the hierarchical clusters are shown using a diagram called dendrogram which shows the cluster-sub-cluster relationship along with the order in which clusters were merged. However, we can also use a nested view of the hierarchical clusters. Figure 6.12 shows both views.

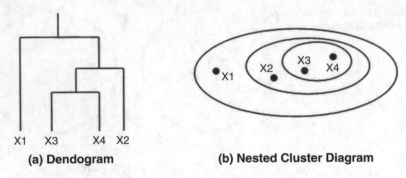

(a) Dendogram   **(b) Nested Cluster Diagram**

Fig. 6.12 Dendrogram and nested cluster view of hierarchical clusters

There are many agglomerative hierarchical clustering techniques; however, all work in the same way, i.e., by considering each individual point as an individual cluster, they keep on merging them until only one cluster is remained. Algorithm 6.3 shows the pseudo code of such approaches:

Algorithm 6.3 Basic Agglomerative Hierarchical Clustering Algorithm

1. Calculate proximity measure.
2. Merge the two closest clusters.
3. Update proximity matrix.
4. Repeat steps 2 to 3 until there remains only one cluster.

All the agglomerative hierarchical clustering algorithms use some proximity measure to determine the "closeness" of the clusters. The use of a particular type of proximity measure depends on the type of clustering. For example, for graph-based clustering, MIN, MAX, and group average are common proximity measures.

MIN proximity measure defines the distance between two closest points in different clusters as shown in Fig. 6.13a. MAX proximity measure defines the distance between two farthest points in two different clusters as shown in Fig. 6.13b. Group average defines the average of the distance of all points in one cluster with all other points in the other cluster as shown in Fig. 6.13c.

6.7   DBSCAN Clustering Algorithm

DBSCAN is a density-based clustering approach. It should be noted that the density-based approach is not as common as the distance-based measure. Furthermore, there are many definitions of density, but we will consider only center-based density measure. In this measure, the density for a particular point is measured by counting the number of points (including the point itself) within a specified radius called Eps (epsilon). Figure 6.14 shows the number of points within the radius of Eps of point P is 6 including point A itself.

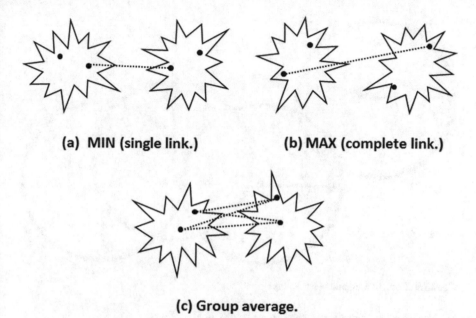

(a) MIN (single link.)   (b) MAX (complete link.)

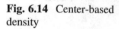

(c) Group average.

Fig. 6.13 MIN, MAX, and group average proximity measures

Fig. 6.14 Center-based
density

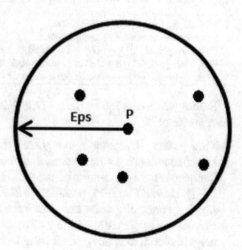

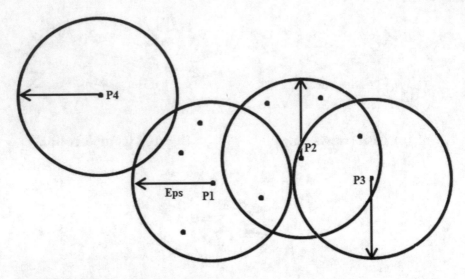

Fig. 6.15 Core, border, and noise points

Algorithm 6.4 shows the pseudo code of the algorithm.

Algorithm 6.4 Pseudo Code of Density-Based Clustering Algorithm

---
1. Find all core, border, or noise points.
2. Remove points marked as noise points.
3. Find core points that are in Eps of each other and mark and edge between them.
4. Find each group of connected core points and mark them as a separate cluster.
5. Allocate border points to one of the clusters of their associated core points.

Before we discuss the working of DBSCAN, let's discuss some terminologies of the algorithm:

- *Core points:* If a point P has greater than or equal to a minimum number of specified points in its Eps radius, then the point P will be considered a core point. Suppose if we consider a minimum number of points as five (including the point itself), then in Fig. 6.15, points P1 and P2 are core points.
- *Border points:* If a point does not have a minimum number of specified points (five points in our case) within its Eps radius but has at least one core point in its neighborhood, then this point is called a border point. In Fig. 6.15, point P3 is the border point.
- *Noise points:* If a point is neither a border point nor a core point, then it is considered a noise point. In Fig. 6.15, point P4 is a noise point.

The algorithm works as follows:

Any two core points which are within the Eps of each other are selected and merged into a cluster. Similarly, the border points are merged into the cluster of their closest core point. Noise points, however, are simply discarded.

6.8 Cluster Evaluation

As in classification models, once the model is developed, the evaluation is an essential part to justify the model; however, in clustering, the evaluation process is not well defined. However, the available measures used to evaluate the performance of clustering algorithms can broadly be categorized into three categories as follows:

- *Unsupervised:* Unsupervised measures measure the quality of clustering structures without using any class labels. One such common measure is SSE.

- *Supervised:* Supervised measures measure the quality of clustering structure with respect to some external structure. One of the examples of such measures is entropy.
- Relative: This compares different clustering structures. The comparison can be based on any of the measures, e.g., entropy or SSE.

6.9 General Characteristics of Clustering Algorithms

So far we have discussed different clustering algorithms. Now we will discuss some general characteristics of clustering algorithms.

Order Dependence For some algorithms, the order and quality of the resulted clustering structure may vary depending on the order of processing of data points. Although sometimes desirable, generally we should avoid such algorithms.

Nondeterminism Many clustering algorithms like K-Means require initial random initialization. Thus the results they may produce may vary on each run. For all such algorithms, to get an optimal clustering structure, we may need to have multiple runs.

Scalability It is common to have datasets with millions of data points and attributes. Therefore, their time complexity should be linear or near to linear in order to avoid performance degradation.

Parameter Setting Clustering algorithms should have minimum parameter settings and especially those parameters that are provided by the user and significantly affect the performance of the algorithm. It should be noted that the minimum number of parameters that need to be provided is always better.

Treating Clustering as an Optimization Problem Clustering is an optimization problem, i.e., we try to find the clustering structures that minimize or maximize an objective function. Normally such algorithms use some heuristic-based approach to optimize the search space in order to avoid exhaustive search which is infeasible.

Now we will discuss some important characteristics of the clusters.

Data Distribution Some techniques may assume distributions in data where a cluster may belong to a particular distribution. Such techniques, however, require a strong conceptual basis of statistics and probability.

Shape Clusters can be of any shape: some may be regularly shaped, e.g., in a triangle or rectangle, while some others may be irregularly shaped. However, in a dataset, a cluster may appear in any arbitrary shape.

Different Sizes Clusters can be of different sizes. Same algorithms may result in different clusters of different sizes in different runs based on many factors. One such factor is random initialization in the case of the K-Means algorithm.

Different Densities Clusters may have different densities, and normally clusters having varying densities may cause problems for methods such as DBSCAN and K-Means.

Poorly Separated Clusters When clusters are close to each other, some techniques may combine them to form a single cluster which results in the disappearance of true clusters.

Cluster Relationship Clusters may have particular relationships (e.g., their relative position); however, the majority of techniques normally ignore this factor.

Subspace Clusters Such techniques assume that we can cluster the data using a subset of attributes from the entire attribute set. The problem with such techniques is that using a different subset of attributes may result in different clusters in the same dataset. Furthermore, these techniques may not be feasible for the dataset with the large number of dimensions.

Summary

Clustering is an important concept used for analysis purposes in data science applications in many domains including psychology and other social sciences, biology, statistics, pattern recognition, information retrieval, machine learning, and data mining. In this chapter, we provided an in-depth discussion on clustering and related concepts. We discussed different cluster types and relevant clustering techniques. Efforts were made to discuss the concepts through simple examples and especially in pictorial form.

Further Reading

The following are some valuable resources for further reading:

- *Clustering Methods for Big Data Analytics Techniques, Toolboxes and Applications* (Editors: Olfa Nasraoui, Chiheb-Eddine Ben N'Cir), Springer & Co, 2019. The book provides complete details about the state-of-the-art methods that are used for performing clustering in big data. The book also provides applications of these methods in different artificial intelligence-based systems. The first chapter shows how we can apply deep learning in the case of clustering. The book provides the details about blockchain data clustering and different cybersecurity applications that can be used for detection of the threats. The book then discusses different distributed clustering methods for big data. The book also provides a discussion on the use of clustering approaches for large-scale streams of data. The heuristics-based algorithms like Particle Swarm Optimization are also discussed. The book also provides details of the tensor-based clustering approach for web graphs, sensor streams, and social networks.

- *Multicriteria and Clustering Classification Techniques in Agrifood and Environment* (Authors: Zacharoula Andreopoulou, Christiana Koliouska, Constantin Zopounidis), Springer & Co, 2017. The book provides an introduction to operational research methods and their application in the agri-food and environmental sectors. It explains the need for multicriteria decision analysis and teaches users how to use recent advances in multicriteria and clustering classification techniques in practice.

- *Data Analysis in Bi-partial Perspective: Clustering and Beyond* (Authors: Jan W. Owsiński), Springer & Co, 2020. The book discusses the bi-partial approach for performing data analysis. The approach is a general approach that can be used to develop different techniques for solving the problems related to data analysis including the models and the algorithms. The book provides a strong base for all the people who wish to extend their careers as data scientists because the book helps them address different issues that are often overlooked by other books.

- *A Primer on Cluster Analysis: 4 Basic Methods That (Usually) Work* by James C. Bezdek, Design Publishing, 2017. This book describes four conventional clustering methods in static and small datasets. The recent developments in cluster analysis have their wide use, especially in social networks, big data, wireless sensor networks, streaming video, etc. The book is a worthy resource written by the founding father of fuzzy clustering principles and is considered a valuable contribution in the domain of clustering analysis. The book provides in-depth details of the clustering analysis along with real-life problems and their applications. Overall the book is an excellent resource for anyone who wants to get specialized information about cluster analysis.

- *Reverse Clustering: Formulation, Interpretation and Case Studies (Studies in Computational Intelligence, 957)* by Jan W. Owsiński et al., Springer & Co, 2021. The book provides new solutions for data analysis problems faced by organizations these days. The solutions include different components of machine learning including cluster analysis, classification, outlier detection, feature selection, and even factor analysis as well as the geometry of the dataset. The book is useful for all those who are looking for new, nonconventional approaches to their data analysis problems.

- *Evolutionary Data Clustering: Algorithms and Applications (Algorithms for Intelligent Systems)* by Ibrahim Aljarah, Hossam Faris, and Seyedali Mirjalili, Springer & Co, 2021. This book provides a deep analysis of the state-of-the-art evolutionary clustering techniques. It discusses some of the most commonly used clustering methods. The book discusses the single-objective and multi-objective clustering solutions. Furthermore, the book also discussed the fitness function and other measures used for the evaluation of clustering algorithms. This book provides literature reviews about single-objective and multi-objective evolutionary clustering algorithms. In addition, this book gives a comprehensive assessment of the health capabilities and assessment measures that are utilized in the maximum of evolutionary clustering algorithms. Furthermore, the book provides many recent nature-inspired algorithms for data clustering. These algorithms include the genetic algorithm, particle swarm algorithm, ant colony algorithm, etc. The book also discusses the applications of data clustering in many other domains like image processing, medical applications, etc.

Exercises

Exercise 6.1: What is the difference between clustering and cluster analysis?

Exercise 6.2: Give a graphical example of two, three, and four clusters. Use the shapes of different colors for the identification of different clusters.

Exercise 6.3: What are the differences between the following types of clustering?

- Hierarchical clustering
- Partitioned clustering
- Well-separated clustering
- Prototype-based clustering
- Contiguous clusters clustering
- Density-based clustering
- Shared property clustering
- K-Means clustering
- Fuzzy C-Means clustering

Exercise 6.4: Consider the following pseudo code of the K-Means clustering algorithm and implement it on any dataset using any programming language.

1. Select the value of K (number of initial centroids).
2. Randomly select centroids.
3. Assign objects to the closest centroids.
4. Update centroids.
5. Repeat steps 3 to 4 until there are no more updates in centroids.

Exercise 6.5: While using the K-Means clustering algorithm, the value of the K is decided in the first step. What does the value of this K represent? What is the mechanism to decide this value? Is there any algorithm that can help us with the efficient selection of this value? Give detail of that algorithm.

Exercise 6.6: Consider the data given below and cluster it using the K-Means clustering method. Assume the value of k=2.

Individual	X	Y
1	7	9
2	8	1
3	4	7
4	5	1
5	6	2
6	6	2

(continued)

Individual	X	Y
7	3	5
8	8	9
9	7	2
10	1	5
11	5	5
12	3	2
13	9	1
14	5	7
15	1	7
16	2	7

Exercise 6.7: Consider the data given below and cluster it using the K-Means clustering method. Assume the value of k=3.

Individual	X	Y
1	3	8
2	5	8
3	3	7
4	8	1
5	8	2
6	1	9
7	6	3
8	8	5
9	9	3
10	7	3
11	2	8
12	2	6
13	1	6
14	9	8
15	1	4
16	3	1

Exercise 6.8: In Exercises 6.6 and 6.7, same data is clustered with different values of k. What do you observe for clustering the same data with different values of k? Which of them gives better results? Briefly describe.

Exercise 6.9: Consider the following pseudo code of bisecting K-Means clustering algorithm. Implement it on any dataset using any programming language.

1. Specify the resulting number of cluster.
2. Make all the points as one cluster.
3. Split the main cluster with k=2 using the K-Means algorithm.
4. Calculate the SSE of resulting clusters.
5. Split the cluster with a higher SSE value into two further clusters.
6. Repeat steps 4 to 5 until we get the desired number of clusters.

Exercise 6.10: Consider the data given below and cluster it using bisecting K-Means clustering method.

Individual	X	Y
1	1	2
2	7	1
3	8	5
4	9	5
5	1	2
6	5	2
7	4	6
8	9	2
9	2	4
10	9	3
11	8	8
12	3	4
13	9	3
14	6	8
15	7	4
16	5	2

Exercise 6.11: Consider the data given below and cluster it using the hierarchical clustering agglomerative method. After constructing the clusters, show them using a dendrogram and nested cluster diagram.

	X1	X2
A	3	7
B	6	5
C	3	8
D	4	1
E	4	5
F	4	1
G	7	6
H	6	9
I	2	8
J	2	1
K	1	3
L	7	7
M	2	4
N	1	7
O	1	8
P	9	3

Exercise 6.12: Consider the following given pseudo code of hierarchal clustering algorithm and implement it on any dataset using any programming language.

1. Calculate proximity measure.
2. Merge the two closest clusters.
3. Update proximity matrix.
4. Repeat steps 2 to 3 until there remains only one cluster.

Exercise 6.13: What is the MIN, MAX, and group average in clustering, and how do we measure them?

Exercise 6.14: Consider the following given pseudo code of density-based clustering algorithm and implement it on any dataset using any programming language.

1. Find all core, border, or noise points.
2. Remove points marked as noise points.
3. Find core points that are in Eps of each other and mark and edge between them.
4. Find each group of connected core points and mark them as a separate cluster.
5. Allocate border points to one of the clusters of their associated core points.

Exercise 6.15: Consider the data given below and cluster it using the DBSCAN clustering algorithm and show a diagram of generated clusters.

	X	Y
P1	5	3
P2	3	6
P3	9	4
P4	4	8
P5	1	4
P6	3	9
P7	4	4
P8	1	7
P9	2	7
P10	6	5
P11	4	8
P12	4	1
P13	8	2
P14	4	6
P15	4	7
P16	7	4

Exercise 6.16: Consider the following general and important characteristics of clustering algorithms and briefly explain each characteristic.

- Order dependence
- Nondeterminism
- Scalability
- Parameter setting
- Treating clustering as an optimization problem
- Data distribution

- Shape
- Different sizes
- Different densities
- Poorly separated clusters
- Cluster relationship
- Subspace clusters

Exercise 6.17: Cluster the following eight points with (x,y) representing the location of each point. Your task is to cluster this data into three clusters.

- A1 (2, 8)   • B2 (7, 7)
- A2 (3, 7)   • B3 (4, 6)
- A3 (4, 2)   • C1 (3, 1)
- B1 (5, 7)   • C2 (2, 4)

Initially, you can use A1, B1, and C1 as the center of each cluster, respectively. Use the K-Means algorithm to show the final three clusters. There is no limit on the number of iterations.

Exercise 6.18: Consider the following dataset.

A	B	C	Y
0	0	0	1
1	1	0	1
1	1	0	0
0	1	0	0
0	1	1	1

You need to cluster it. Firstly, remove the class label, i.e., Y. After that, perform clustering using K-Means (where k=3). Now check the accuracy of the algorithm by comparing it to the original data. What is the accuracy of K-Means?

Exercise 6.19: Differentiate between classification and clustering. From the following two datasets, which one will be used for which task?
Dataset-A:

	Width	Length
S1	5	9
S2	6	1
S3	9	7
S4	8	5
S5	1	3
S6	4	6

Dataset-B

	Width	Length	Score
S1	5	9	Good
S2	6	1	Good
S3	9	7	Very Good
S4	8	5	Very Good
S5	1	3	Satisfactory
S6	4	6	Satisfactory

Chapter 7
Text Mining

7.1 Text Mining

One of the main characteristics of textual data is that it exists in unstructured form, i.e., you only have streams of text having no metadata, no proper formats, and nothing about the semantics. This makes the processing of the textual data a bit complicated as compared to other types of data.

There are many other factors that add to the complexity of the processing of the data; a few include vagueness, ambiguity, uncertainty, etc. For example, consider the following sentence:

Old cats and dogs were taken to the hospital.

The sentence is ambiguous as it is not clear that the adjective "Old" either applies to cats or both the cats and the dogs. Similarly, there may be other types of ambiguities that we will discuss later on.

All this requires to have a set of specialized methods that could process the textual data based on its inherent characteristics for the discovery of knowledge from the text. Text mining is the process of processing text data to extract valuable information for decision-making.

Some main uses of text mining are:

- Retrieval of information from the textual data, for example, reply to a user query about a certain medicine.
- Extraction of the information from the textual data, for example, extraction of the price of an asset or the documents discussing a certain event, etc. Segregation of the documents on a certain topic, e.g., to identify all the documents that discuss the 9/11 incident.
- Title suggestions for the text, for example, suggesting the subject of an email.

All this requires the extraction of the textual data, processing it, and then applying certain techniques such as classification, clustering, etc. to extract the information.

U. Qamar, M. S. Raza, *Data Science Concepts and Techniques with Applications*, https://doi.org/10.1007/978-3-031-17442-1_7

This means that the domain of text mining not only includes text processing but incorporates different methods from other domains as well.

7.2   Text Mining Applications

There are different text mining applications; we will discuss a few here.

7.2.1   Exploratory Text Analysis

Before processing any data, it is important to get an initial understanding of it. This is what the exploratory text analysis is all about. You get the initial understanding of your text. This is essential so that you could select the appropriate models for further processing of your data. The exploratory text analysis can be taken as a first step toward the further analysis of textual data.

7.2.2   Information Extraction

Information extraction is one of the main goals of text mining. When the textual data is small in amount, it is possible to manually process this data; however, with the ever-growing size of the textual data, we need special text mining methods to process and extract information. Text mining helps us extract the relevant information from the textual data.

Consider, for example, the reviews of a product on a single web page. You can easily find the overall perspective of people about that product. But what about the scenario when there are hundreds and thousands of reviews? This is where text mining helps you extract the relevant information. You can process large volumes of textual data to extract meaningful information from the data.

7.2.3   Automatic Text Classification

Text classification is the process of assigning a class to a document on the basis of its contents. Just like other machine learning techniques, we provide the training data to our model which then performs classification for the unseen documents. Figure 7.1 shows the pictorial representation of a typical text classification model.

We have already discussed that text mining deals with the process of extracting information from textual data. In many decision support systems, the major source of data is in the textual form. So, for all such systems to work, they need to perform the

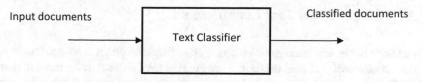

Fig. 7.1 A generic text classification model

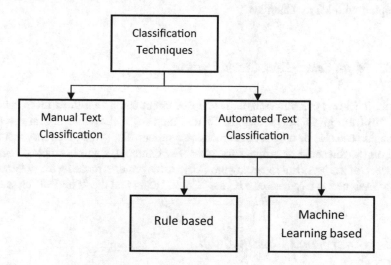

Fig. 7.2 Anatomy of text classification techniques

automatic classification of the text. This is where the process of text classification helps them a lot. Keeping all this in mind, many techniques have been proposed for automatic text classification. We will discuss a few of them later on. Figure 7.2 shows the anatomy of text classification techniques.

7.3 Text Classification Types

Now we will discuss different text classification techniques. Mainly there are three types of text classification techniques as given below:

- Single-label text classification
- Multi-label text classification
- Binary text classification

Now we discuss each of these techniques one by one.

7.3.1   Single-Label Text Classification

When each document is assigned only one class from the given class set, then this is called single-label text classification. However, in total there may be more than one class. For example, on the basis of the quality of a text in an essay, the document can be assigned only one class from the given classes, e.g., "Accepted," "Rejected," and "Accepted with Minor Changes."

7.3.2   Multi-label Text Classification

In multi-label text classification, each text document can be assigned more than one class from the given set of classes. In this method, a threshold and any scoring mechanism can also be used for document assignment. The score given to a document based on the threshold determines the class. For example, a single article discussing the speech of the president regarding the Covid arrangements made by the government may be assigned to two classes, i.e., the "Covid" class and the "Political" class.

7.3.3   Binary Text Classification

Binary text classification is a special type of single-label classification where a single text document can be assigned to a class from a set of only two classes. For example, on the basis of the contents, a message can be classified either as spam or non-spam. Similarly, an article can either be "Published" or "Not published" at the same time.

7.4   Text Classification Approaches

Text categorization involves two approaches which are:

1. Rule based
2. Machine learning

7.4.1   Rule Based

Rule-based text classification approaches classify the text on the basis of some already specified rules. As discussed in earlier chapters, rules are mentioned in the form of if...then structures. For example, a simple rule is given below:

If (spelling-mistakes > 10) $\rightarrow$ (publish$=$false)

Table 7.1 List of a few keywords for "Politics" and "Sports" class

Class=Politics	Class=Sports
1. President	1. Cricket
2. Prime minister	2. Football
3. Democracy	3. Stadium
4. Parliament	4. Match
5. Elections	5. Score

The rule specifies that if the document contains more than 10 spelling mistakes, the document will not be published. Rule-based text classification is one of the simplest methods for classifying text. Here we will give a simple example.

Consider a set of documents where we need to classify the category of the document, e.g., we may have two categories "Sports" and "Politics." To perform such classification, one of the techniques may be to find out the threshold of the words related to each category. We may define the keywords, e.g., "Democracy," "Elections," "President," "Prime minister," etc. for the category "Politics." Similarly, we may define the keywords for the "Sports" category as well. Table 7.1 shows the list of a few words belonging to each category.

Now for each document, we will count the words related to each category. The category which has a large number of related words will be considered as the assigned class of the document. As you can see, classification depends on the already defined keywords, so a small list of keywords may affect the accuracy.

7.4.2   Machine Learning

Machine learning-based text classification approaches are more accurate as compared to simple rule-based approaches. In machine learning approaches, a text classifier is first provided with the training data, and then on the basis of learning from the training data, it performs the classification of the unseen text. In the case of machine learning-based text classification approaches, the training data may comprise the text and the assigned classes, i.e., labels or tags. So far, in literature, a number of machine learning-based text classification approaches have been proposed. Here we will discuss a simple naïve-based approach with the help of an example.

Consider the dataset given in Table 7.2:

Table 7.2 Training and test documents

	Document	Words			Class
Training data	d1	Potato	Tomato	Cucumber	Vegetable
	d2	Cucumber	Beans	Banana	Vegetable
	d3	Banana	Mango	Tomato	Fruit
	d4	Banana	Tomato	Potato	Fruit
	d5	Apple	Banana	Mango	Fruit
Test data	d6	Banana	Tomato	Beans	?

Now we will apply a simple naïve-based approach (with "add 1 smoothing") for the classification of the test data. We will perform the following steps:

1. Find the probability of each class.
2. Find the probability of each word in test data given a class.
3. Assign the class label on the basis of the entire document.

First, we will find the probability of each class as given below:

P(Vegetable) = 2/5
P(Fruit) = 3/5

Now we will find the probability of each word in the dataset given a class:

P(Banana|Vegetable) = (1+1)/(6+7)
P(Tomato|Vegetable) = (1+1)/(6+7)
P(Beans|Vegetable) = (1+1)/(6+7)
P(Banana| Fruit) = (3+1)/(9+7)
P(Tomato| Fruit) = (2+1)/(9+7)
P(Beans| Fruit) = (0+1)/(9+7)

Now we will find the probability of the test data for both classes.

P(Fruit|d6) = 3/5 * 4/16 * 3/16 * 1/16 = 0.001758
P(Vegetable|d6) = 2/5 * 2/13 * 2/13 * 2/13 = 0.001457

Since P(Fruit|d6) > P(Vegetable|d6), so predicted class of d6 is "Fruit."

7.5   Text Classification Applications

Text classification is one of the main tasks in text mining. From a simple software application to a high-level decision support system, text classification is used at various stages. Here we will discuss a few applications of text calcination.

7.5.1   Document Organization

There are many scenarios where we have to deal with a huge number of documents for their automatic classification in order to properly arrange the documents. For example categories like "Weather" and "Sport" are automatically assigned to documents in a company like a newspaper. It is resource-intensive and time-consuming when the same tasks need to be performed manually.

7.5.2   Text Filtering

Text filtering is also one of the main applications of automatic text classification. For example, email services normally use automatic text classification to filter spam and non-spam emails. Consider the following list of words that can be assigned a "Spam" category:

- Lottery
- Call back
- Prize
- Visa

Now each email text can be scanned against these words, and if the threshold of occurring of these words is more than a specified value, the message is automatically classified as spam.

7.5.3   Computational Linguistics

For the processing of natural language grammar and linguistics, computational linguistics is used. Text categorization can also be used in different applications of computational linguistics which include content-sensitive spelling correction, prepositional phrase attachment, word choice selection, part of speech (POS) tagging, and word sense disambiguation.

7.5.4   Hierarchical Web Page Categorization

The World Wide Web comprises hundreds of thousands of web pages these days. People search and visit these websites on daily basis. Now for any query, the search engines need to know about the category of the websites they will have to search. For this purpose, the web pages are hierarchically classified in order to facilitate the search.

For example, Fig. 7.3 shows a sample hierarchy of web pages.

7.6   Representation of Textual Data

Computers cannot understand the data in the natural textual form. So, we need to convert the textual data into some numerical form so that computers could understand it.

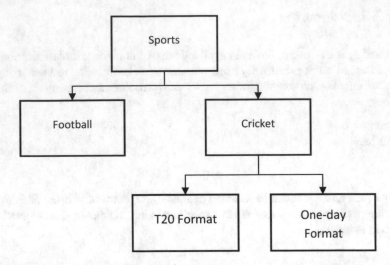

Fig. 7.3 A sample hierarchical web page classification

7.6.1 Bag of Words

Bag of words is one of the models to represent a text in numerical format for computers to understand. The bag of words model works in different steps. We will explain these steps one by one.

The first step is to define the vocabulary of the document. For example, our document comprises three sentences given below:

This is a beautiful cat.
This is a beautiful dog.
Both cat and dog are beautiful.

Our vocabulary will be as follows:

- Beautiful
- Cat
- Dog

We will call each word a feature. Note that we have removed some words from the vocabulary that do not provide much information (we will discuss this in the upcoming topic). Table 7.3 shows the frequency of each term in the document.

Now we will convert each sentence into the form of a binary vector. In binary representation of each sentence, "1" means a vocabulary term is present, whereas "0" means that the vocabulary term is not present. Table 7.4 shows the binary representation of each sentence in our document.

So, the binary representation of our first sentence will be "110" and the binary representation of our second sentence will be "101."

Table 7.3 Vocabulary terms and their frequencies

Term	Frequency
Beautiful	3
Cat	2
Dog	2

Table 7.4 Vocabulary terms and their frequencies

Term	Beautiful	Cat	Dog
Sentence 1	1	1	0
Sentence 2	1	0	1
Sentence 3	1	3	1

Table 7.5 Vocabulary terms and their frequencies

Term	Sentence 1	Sentence 2	Sentence 3
Beautiful	$\frac{1}{2}$	$\frac{1}{2}$	$\frac{1}{3}$
Cat	$\frac{1}{2}$	0	$\frac{1}{3}$
Dog	0	$\frac{1}{2}$	$\frac{1}{3}$

7.6.2 Term Frequency

From the bag of words, we can derive different measures that can be used to perform further analysis. Term frequency (TF) of a specific term "t" is the ratio of the number of occurrences of t and the total number of terms in the sentence.

$$tf(t, d) = \frac{\text{Total number of occurrences of t in d}}{\text{Total number of terms in d}} \qquad (7.1)$$

So, the term frequency of the term "cat" will be:

$$tf(t, d) = \frac{1}{2} = 0.5$$

Table 7.5 shows the term frequency of each term in the above sentences.

7.6.3 Inverse Document Frequency

Inverse document frequency (IDF) specifies how common a word is in a document. It can be obtained by the following formula:

$$tf(t, d) = \left(\frac{\text{Number of Sentences}}{\text{Number of Sentences Containing the word}} \right) \qquad (7.2)$$

Table 7.6 shows the IDF of the given terms.

Table 7.6 IDF of the terms

Term	IDF
Beautiful	$\log\left(\frac{3}{3}\right)$
Cat	$\log\left(\frac{3}{3}\right)$
Dog	$\log\left(\frac{3}{2}\right)$

Table 7.7 Vocabulary terms and TF-IDF

Term	Sentence 1	Sentence 2	Sentence 3
Beautiful	$\frac{1}{2} * \log\left(\frac{3}{3}\right)$	$\frac{1}{2} * \log\left(\frac{3}{3}\right)$	$\frac{1}{3} * \log\left(\frac{3}{3}\right)$
Cat	$\frac{1}{2} * \log\left(\frac{3}{2}\right)$	0	$\frac{1}{3} * \log\left(\frac{3}{2}\right)$
Dog	0	$\frac{1}{2} * \log\left(\frac{3}{2}\right)$	$\frac{1}{3} * \log\left(\frac{3}{2}\right)$

7.6.4   Term Frequency-Inverse Document Frequency

Term frequency-inverse document frequency (TF-IDF) gives the importance of a term in a document. A term is considered to be more important if its TF-IDF value is higher and vice versa. Mathematically, TF-IDF is obtained by multiplying the value of TF of a term with its IDF value. The TF-IDF of the term "Beautiful" will be as shown in Table 7.7.

7.6.5   N-Gram

In the above bag of words problem, each term was taken as a single term and is called a gram. However, we can combine two or more terms called N-grams. If two terms are combined, it is a 2-gram (also called bigram). Similarly, if three terms are combined, it is called 3-gram. So, N-gram is a sequence of N-terms or tokens.

7.7   Natural Language Processing

We have already discussed that most of the text in the real world exists in the form of text. Natural language processing (NLP) is the branch of computer science that deals with the processing of text in natural language so that computers could understand and process it. These days many tools are available to process natural language text. We will discuss natural language processing in Python using spaCy. spaCy is a Python-based library for processing natural language text.

The core of natural language processing is the text processing pipeline. We can also call it text preprocessing. It is the set of tasks that are performed on each textual data before using it for further processing and analysis. Figure 7.4 shows the tasks performed as part of the text preprocessing.

Now we will discuss each of these tasks one by one.

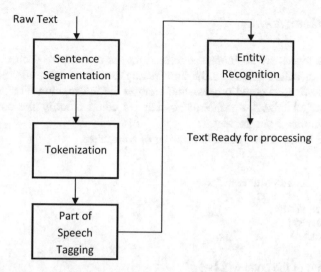

Fig. 7.4 Text processing pipeline

7.7.1 Sentence Segmentation

The textual data exists in the form of a stream of sentences. So, before processing the text, one of the essential tasks that need to be performed is to break the entire text into individual sentences so that we could handle each sentence separately. The process is called sentence segmentation. The following is the code that performs the sentence segmentation in Python using the spaCy library.

```
import spacy
nlp = spacy.load("en_core_web_sm")
MyText = "I am a good boy. My name is John."
Txt = nlp(MyText)
for sen in Txt.sents:
  print(sen)
```

The first line imports the spaCy library. Then we created an "nlp" object from the dataset "en_core_web_sm." Now our nlp object is ready to perform the preprocessing tasks. We then defined a two-sentence text and stored it in the variable "MyText." The "nlp" object is then initialized with this text. Finally, the sentences are segmented in the loop. Each sentence is then printed on the console. The following will be the output:

I am a good boy.
My name is John.

7.7.2   *Tokenization*

Once the sentences are segmented, the next task is to tokenize each sentence into separate works called tokens. In the information extraction process, we need to get the tokens so that we could process and analyze the information. For example, we may need to tokenize the sentence so that we could identify the names of the different countries from the text.

The code below shows the tokenization of a sentence.

```
import spacy
nlp = spacy.load("en_core_web_sm")
MyText = "I am a good boy."
Txt = nlp(MyText)
for token in Txt:
  print(token)
```

The output of the code will be:

I
am
a
good
boy
.

Note that each word in the sentence has been tokenized including the full stop symbol.

7.7.3   *Part of Speech Tagging*

Part of speech tagging is one of the important tasks of natural language processing. In POS tagging, each token is categorized in terms of its part of speech. For example, the word "boy" will be categorized as a noun. The code below shows each token and its POS tag.

```
import spacy
nlp = spacy.load("en_core_web_sm")
MyText = "I am a good boy."
Txt = nlp(MyText)
for token in Txt:
  print(token,otken.pos_)
```

The following will be the output of the code:

```
I PRON
am AUX
a DET
good ADJ
boy NOUN
. PUNCT
```

The token "I" is tagged as "PRON," i.e., pronoun. Similarly the word "am" is categorized as an auxiliary verb.

7.7.4 Named Entity Recognition

A named entity is a real-life object that can be identified and tagged. In NLP, named entity recognition (NER) is the process of identifying the named entities from the text. The identified entities can then be used for further processing. The following code identifies the named entities of each token in the sentence.

```
import spacy
nlp = spacy.load('en_core_web_sm')
sentence = "Asia is the biggest subcontinent in the world"
Txt = nlp(sentence)
for ent in Txt.ents:
  print(ent.text, ent.label_)
```

The output of the code will be:

Asia LOC

As you can see, the token "Asia" has been identified as "LOC" which refers to a location object.

7.8 Case Study of Textual Analysis Using NLP

Now we will discuss a case study to show how the NLP can be used to perform textual analysis. We will take some text and try to extract the components of the object-oriented class diagram from the sentences. A class diagram represents an object-oriented structure of the system. It shows different objects present in the system and their relationships with each other, e.g., association, multiplicity, etc. Figure 7.5 shows the class diagram of the sentence:

A student can register for multiple seminars.

Fig. 7.5 A sample class diagram

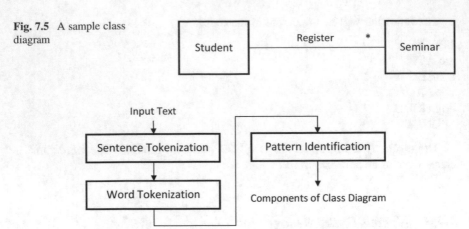

Fig. 7.6 A sample class diagram

In the above diagram, "Student" and "Seminar" are classes. "Register" is the association between these two classes, whereas the asterisk symbol represents the multiplicity. This symbol states that a student can register for more than one seminar.

To extract the components of the class diagram, we will use the Abbott textual analysis technique. Abbott textual analysis technique is one of the most common techniques used to extract class diagram components. This technique uses the parts of speech as heuristics to identify different components. For example, a common noun in a sentence specifies a "Class." A proper noun, on the other hand, may specify an object. Similarly, a verb may specify a method or an association.

We will see how different tasks of the NLP can help us identify the components from the textual data using the Abbott technique. For this purpose, we will use the methodology shown in Fig. 7.6.

First, we will tokenize the paragraph into sentences. After this, each sentence will be tokenized into separate words. We will then identify the POS tag of each token, and based on its POS tag, we will consider it a particular component of the class diagram. However, note that before doing all this, we will have to define the patterns to extract the components. A pattern is a rule that specifies that if a sentence fulfills a certain pattern, which components are present in it? For example, consider the following rule:

NN:C, VB:ASOC, NN:C

Here:

NN:C specifies that the first noun will be identified as a "Class."
VB:ASOC specifies that the verb after the first noun will be specified as an association.
NN:C specifies that the noun after the verb will be specified as a "Class."

Similarly, we will have to identify different patterns by keeping in mind the different structures of the sentences.

Table 7.8 Sample patterns

PID	Primary pattern	Secondary pattern
1	NN:C,VB:Mth	–
2	NN:C,VB:ASOC,JJ:M-R,NNS:C	–
3	JJ:M-L,NNS:C,VB:ASOC,JJ:M-R,NNS:C	–
4	NN:C,VBZ:Mth	–
5	NNS:C,VB:ASOC,NNS:C	NN:C,[Multiple:M-L],VB:ASOC, [Multiple:M-R],NN:C
6	NN:C,VB:ASOC,NN:C	–
7	NN:C,NNP:DM,NN:DM,NN:DM	–

It should be noted that sometimes a pattern may not be explicitly present in a sentence. For example, consider the following sentence:

Students can register for seminars.

Here the multiplicity is many-to-many, which means that a student can register for more than one seminar and a seminar can be registered by more than one student. We will have to define an alternate (secondary) pattern (to annotate the multiplicity information) so that if a sentence follows this structure, the alternate pattern (containing the multiplicity information) is applied. The pattern that is fulfilled by the above sentence is as follows:

NNS:C, VB:ASOC, NNS:C

"NNS" specifies the plural noun. A plural noun indicates a "Many" multiplicity. So, we can define the alternate pattern as follows:

NN:C,[Multiple:M-L],VB:ASOC,[Multiple:M-R],NN:C

Here:

"[Multiple:M-L]" specifies the left multiplicity and "[Multiple:M-R]" specifies the right multiplicity.

Table 7.8 shows some primary and secondary patterns.

These patterns are stored in a database table. We will read the patterns from the database and apply them to each sentence one by one.

Now we will provide the Python source code of the above case study. We will try to explain some important tasks performed by the code.

The source code is divided into two files named "pparser.py" and "main.py." The "pparser.py" file implements the parser that will be used to parse the sentence and apply the patterns. "main.py" provides the entry point of the execution along with some functionality for creating the interface.

Following is the source code from "main.py":

```
import pparser
import spacy
import pyodbc
import Tkinter
from Tkinter import scrolledtext
```

The above statements import the required Python libraries. "pyodbc" library is used to interact with the database as we have stored our patterns in a database table. We will use "Tkinter" library for the creation of the GUI.

```
nlp = spacy.load("en_core_web_sm")
```

The above statement creates the "nlp" object that will be used to perform NLP-related tasks.

```
conn = pyodbc.connect(
  r'Driver={Microsoft Access Driver (*.mdb, *.accdb)};DBQ=C:\Users
\hp\PycharmProjects\basitcode\Patterns.accdb;')
cursor = conn.cursor()
```

As discussed earlier, we have stored our patterns in the database. The above statement connects with the database and returns a cursor. We will use this cursor to extract the patterns from the database table.

```
def createposlog(sentence):
  tmp = ""
  for t in sentence:
    tmp = tmp + t.text + ":" + t.tag_ + ", "
  tmp = tmp + "\n"
  a = sentence.text + "\n" + tmp
  poslog.insert(Tkinter.INSERT, a)
  poslog.insert(Tkinter.INSERT, '\n')
```

In the above statements, we have defined a user-defined function that takes one sentence as input and applies POS tagging to it. The POS tags along with the sentence are then displayed in the textbox named "poslog."

```
def createclassdiagram(sentence):
  cursor.execute('select * from Pat order by ID')
  plength = 0
  foundedpatterns = 0

  rows = cursor.fetchall()

  for row in rows:
    txt = row[1]
    templ = len(txt.split(","))
```

```
if templ > plength:
  isfound, p = pparser.parseprimary(sentence, row[1])
  if isfound:
    plength = templ
    foundedpatterns = foundedpatterns + 1
    p1 = p
    if(row[2]) != "-":
      p1 = pparser.parsesecondary(p1,row[2],nlp)

# print("Matched Pattern: ", p)
if foundedpatterns > 0:
  cd.insert(Tkinter.INSERT, p1)
  cd.insert(Tkinter.INSERT, "\n")
  # print(p1)
```

In the above statements, we have defined another function named "createclassdiagram." The function takes one sentence as input. First of all, the function executes an SQL query and fetches all the patterns stored in the database table named "Pat." It then iterates through each pattern and parses the sentence against it to find out the pattern the sentence matches with. In case a pattern has a secondary pattern, the function parses the sentence against the secondary pattern.

Note that in case a sentence matches more than one pattern, the pattern having a large length is considered. This is based on the concept that the pattern having a large length provides more information as compared to the one having a small length.

```
def Convert():
  poslog.delete('1.0', Tkinter.END)
  cd.delete('1.0', Tkinter.END)
  txt = nlp(req.get("1.0", Tkinter.END))
  for sentence in txt.sents:
    createposlog(sentence)
    createclassdiagram(sentence)

master = Tkinter.Tk()
master.title("N Language")
master.geometry("2000x1500")
```

The above function is related to the GUI. When the user clicks the "Convert" button, this method is called. This method first erases the previous contents from the textboxes named "poslog" and the "cd." It then iterates through each sentence. For tokenization, it calls the "createposlog" function and for extraction of the class components, it calls the "createclassdiagram" function.

```
btnconvert = Tkinter.Button(master, text="    Click    ",
command=Convert)
btnconvert.place(x=730, y=15)

req = scrolledtext.ScrolledText(master, width=90, height=45,
wrap="word")
req.place(x=25, y=50)
```

```
poslog = scrolledtext.ScrolledText(master, width=90, height=25,
wrap="word")
poslog.place(x=770, y=50)

cd = scrolledtext.ScrolledText(master, width=90, height=19,
wrap="word")
cd.place(x=770, y=470)

master.mainloop()
```

The above code simply creates a GUI using the "Tkinter" Python library and starts the execution of the main loop. The main loop activates the main window and starts receiving messages related to GUI events.

Now we will explain the source code from the "pparser.py" file.

```
import spacy
```

In the above statement, again we have imported the spacy library.

```
def parseprimary(req_, pat_):
  r = []
  pcounter = 0
  pat = pat_.split(",")
  for tok in req_:
    if pcounter < len(pat):
      pa, pc = getpatant(pat[pcounter])
      if tok.tag_ == pa:
        pcounter = pcounter + 1
        r.append(tok.lemma_ + "->" + pc)

  if pcounter == len(pat):
    return True, r # pattern found
  else:
    return False, None    # Pattern not found
```

The above statements define a function named "parseprimary." This function takes the sentence and the current pattern as input. It then parses the sentence against that pattern. For this purpose, the function uses POS tagging. Once a sentence matches a pattern, the function returns the value "True" along with components of the class diagram annotated in the sentence.

```
def getpatant(pp):
  p = pp.split(":")
  return p[0], p[1]
```

The function shown above splits a part of the pattern on the basis of the ":" symbol. It then returns both parts.

```
def parsesecondary(pcd,sp,nlp):
  r=[]
  sptemp = sp.split(",")
  pcdcounter=0

  for i in range(0,len(sptemp)):
    temptxt1 = sptemp[i]
    if temptxt1[0] != "[":
      sptemp1 = sptemp[i].split(":")    #NN:C
      x=pcd[pcdcounter].split("->")           #Student->C
      if nlp(x[0])[0].tag_ == sptemp1[0]:
        r.append(nlp(x[0])[0].lemma_ + "->" + x[1])
        pcdcounter = pcdcounter+1
    else:
      totext = sptemp[i]
      totext = totext[1:len(totext)-1]
      sptemp1 = totext.split(":")
      r.append(sptemp1[0] + "->" + sptemp1[1])

  return r
```

In case a pattern has a secondary pattern, this function is called to traverse the secondary pattern. The secondary pattern may have some hard-coded annotations as shown in the case of pattern 5. The hard-coded annotations also become part of the identified class components.

The overall sequence works as follows:

1. User enters a sentence and clicks the "Convert" button.
2. The "Convert" method is called. This method shows POS tags of the sentence by calling the "createposlog" function.
3. The "Convert" method then calls the "createclassdiagram" function which identifies the components of the class diagram from the sentence.
4. To identify the components of the class diagram, the "createclassdiagram" function calls "parseprimary" and "parsesecondary" functions.

Figure 7.7 shows a sample sentence and the generated class components.

Summary

Text mining involves the digital processing of text for the production of novel information and incorporates technologies from natural language processing, machine learning, computational linguistics, and knowledge analysis. In this chapter, we discussed different concepts related to text mining and especially from the classification point of view. We also discussed some basic concepts related to natural language processing with the help of examples. The overall intention was to give a brief introduction to text mining.

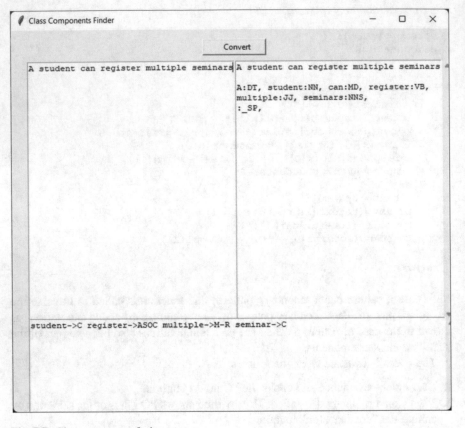

Fig. 7.7 Class component finder

Further Reading

Following are some valuable resources for further reading:

- *Intelligent Natural Language Processing: Trends and Applications* (Editors: Khaled Shaalan, Aboul Ella Hassanien, Mohammed F. Tolba), Springer & Co, 2018. The book is a one-point source for anyone working in the domain of natural language processing right from students to advance level practitioners. The book discusses different state-of-the-art approaches in the domain of natural language processing along with their applications and future directions in the same domain.
- *Natural Language Processing of Semitic Languages* (Editor: Imed Zitouni), Springer & Co, 2014. The book discusses an interesting research problem related to natural language processing, i.e., the semantic languages. The book provides details about both statistical-based approach for natural language processing and the rule-based approaches. Both of these approaches have been proved useful in certain cases.

- *Deep Learning in Natural Language Processing* (Editors: Li Deng, Yang Liu), Springer & Co, 2018. The book discusses the latest machine learning approaches and their application in tasks related to natural language processing. The tasks may include speech recognition and understanding, lexical analysis and parsing, etc. The book discusses various other tasks like machine translation, auto-answering of the questions, and generation of the natural language from the given images.
- *Advanced Applications of Natural Language Processing for Performing Information Extraction* (Authors: Mário Jorge Ferreira Rodrigues, António Joaquim da Silva Teixeira), Springer & Co, 2015. The book discusses the details of information extraction applications and how these applications can be developed to extract the information from the available sources in natural language. The sources may include web pages, documentation, books, newspapers, social media feeds, etc.
- *Text Mining with Machine Learning Principles and Techniques* by Jan Žižka, František Dařena, and Arnošt Svoboda, CRC Press, 2019. The book discusses the machine learning concepts related to knowledge discovery from textual data in the form of natural language. We can draw implicit conclusions from the textual data by processing it. These conclusions are normally hidden within the data and become apparent only through computational analysis. The book discusses various text mining algorithms using a step-by-step explanations approach. The book guides how you can use these algorithms for extracting the semantic contents from the data using the R programming language. The book has a wide range of audiences including IT specialists and any other person interested in the domain of text mining using machine learning techniques. The book provides a systematic approach. It starts from the basic concepts related to the introduction of text-based natural language processing. It then discusses various machine learning algorithms along with their strengths and weaknesses. The book uses the R programming language for implementation purposes. The main intention behind using the R language was that it incorporates many libraries for the implementation of machine learning algorithms, so the reader can concentrate on the concepts and theoretical foundation.
- *Text Mining: Concepts, Implementation, and Big Data Challenge* by Taeho Jo, Springer & Co, 2018. The book presents various text mining concepts and details about how these concepts can be used for the extraction of knowledge from textual data. The author explains how we can implement text mining systems in Java. The book starts with basic text mining concepts and keeps on moving to advance concepts. Some of the topics covered in the book include text summarization, text segmentation, basic text processing, topic mapping, and automatic text management.

Exercises

Exercise 7.1: Explain the term text mining and mention the perspectives that are considered while mining the text.

Exercise 7.2: Describe the text mining terms given below.

- Exploratory text analysis
- Information extraction
- Automatic text classification

Exercise 7.3: In the context of text mining, mention two types of text classifications.

Exercise 7.4: What are the differences in the following terms?

- Single-label text categorization
- Multi-label text categorization
- Binary text categorization

Exercise 7.5: Discuss the following terms and how they are related to the text categorization. Explain in simple words.

- Rule based
- Machine learning

Exercise 7.6: List down at least ten classes that a newspaper organization needs for the classification of their newspaper articles.

Exercise 7.7: Discuss at least three examples of supervised text classification.

Exercise 7.8: Provide at least three examples of single-label text classification.

Exercise 7.9: Is it possible to assign multiple classes to a single document? How can we assign these classes?

Exercise 7.10: What are the applications of automated text classification? Mention a few.

Exercise 7.11: What are the following applications and how are they used for text categorization?

- Document organization
- Text filtering
- Computational linguistics
- Hierarchical web page categorization

Exercise 7.12: As a computer cannot understand natural language, so to make natural language understandable for a computer, there is a need to convert the data into a numerical format so that the computer can understand it. Discuss one method that can be used for this purpose.

Exercise 7.13: What is the use of the following terms in natural language processing?

- Sentence segmentation
- Parts of speech tagging

Exercise 7.14: Text categorization uses different weighting techniques that are given below; elaborate on every technique in detail.

- Term frequency
- Inverse document frequency
- Term frequency-inverse document frequency

Exercise 7.15: Consider the following text document preprocessing techniques, and elaborate on each technique with examples.

- Tokenization
- Named entity recognition

Exercise 7.16: Consider the data given below and analyze it using exploratory text analysis and identify the average number of sentences in the following provided reviews:

Name	Review	Rating
Product-1	These flannel wipes are OK but in my opinion not worth keeping. I also ordered some Imse Vimse Cloth Wipes-Ocean Blue-12 count which are larger, had a nicer, softer texture, and just seemed higher quality. I use cloth wipes for hands and faces and have been using Thirsties 6 Pack Fab Wipes, Boy for about 8 months now and need to replace them because they are starting to get rough and have had stink issues for a while that stripping no longer handles.	3
Product-2	It came early and was not disappointed. I love planet-wise bags and now my wipe holder. It keeps my OsoCozy wipes moist and does not leak. Highly recommend it.	5
Product-3	Very soft and comfortable and warmer than it looks...fit the full-size bed perfectly...would recommend it to anyone looking for this type of quilt.	5
Product-4	This is a product well worth the purchase. I have not found anything else like this, and it is a positive, ingenious approach to losing the binky. What I love most about this product is how much ownership my daughter has in getting rid of the binky. She is so proud of herself and loves her little fairy. I love the artwork, the chart in the back, and the clever approach of this tool.	5
Product-5	All of my kids have cried nonstop when I tried to ween them off their pacifiers until I found Thumbuddy To Love\'s Binky Fairy Puppet. It is an easy way to work with your kids to allow them to understand where their pacifier is going and help them apart from it. This is a must-buy book and a great gift for expecting parents!! You will save them so many headaches. Thanks for this book! You all rock!!	5

(continued)

Name	Review	Rating
Product-6	When the Binky Fairy came to our house, we didn't have any special gift or book to help explain to her about how important it is to stop using a pacifier. This book does a great job to help prepare your child for the loss of their favorite item. The doll is adorable and we made lots of cute movies with the Binky Fairy telling our daughter about what happens when the Binky Fairy comes. I would highly recommend this product to any parent trying to break the pacifier or thumb-sucking habit.	5
Product-7	Lovely book, it's bound tightly so you may not be able to add a lot of photos/cards aside from the designated spaces in the book. Shop around before you purchase, as it is currently listed at Barnes & Noble for 29.95!	4

Exercise 7.17: Consider the data given below and analyze it using automatic text classification. You can define your own classes for this purpose.

Username	Screen name	Location	Tweet at	Original tweet
1	44953	NYC	February 03, 2020	TRENDING: New Yorkers encounter empty supermarket shelves (pictured, Wegmans in Brooklyn), sold-out online grocers (FoodKick, MaxDelivery) as #coronavirus-fearing shoppers stock up (https://t.co/Gr76pcrLWh https://t.co/ivMKMsqdT1)
2	44954	Seattle, WA	February 03, 2020	When I couldn't find hand sanitizer at Fred Meyer, I turned to #Amazon. But $114.97 for a 2 pack of Purell??!!Check out how #coronavirus concerns are driving up prices. (https://t.co/ygbipBflMY)
3	44955		February 03, 2020	Find out how you can protect yourself and loved ones from #coronavirus. ?
5	44957	Melbourne, Victoria	March 03, 2020	#toiletpaper #dunnypaper #coronavirus #coronavirusaustralia #CoronaVirusUpdate #Covid_19 #9News #Corvid19 #7NewsMelb #dunnypapergate #Costco One week everyone buying baby milk powder the next everyone buying up toilet paper. (https://t.co/ScZryVvsIh)

(continued)

Username	Screen name	Location	Tweet at	Original tweet
6	44958	Los Angeles	March 03, 2020	Do you remember the last time you paid $2.99 a gallon for regular gas in Los Angeles? Prices at the pump are going down. A look at how the #coronavirus is impacting prices. 4pm @ABC7 (https://t.co/Pyzq8YMuV5)
8	44960	Geneva, Switzerland	March 03, 2020	@DrTedros "We canÂ't stop #COVID19 without protecting #healthworkers.Prices of surgical masks have increased six-fold, N95 respirators have more than trebled; gowns cost twice as much"-@DrTedros #coronavirus

Exercise 7.18: Consider the data given below and find the term frequency of the following terms:

- Planet
- Largest
- Earth

Documents	Text
A	Jupiter is the first largest planet
B	Saturn is the second largest planet
C	Uranus is the third largest planet
D	Neptune is the fourth largest planet
E	Earth is the fifth largest planet
F	Venus is the sixth largest planet
G	Mars is the second smallest planet
H	Mercury is the first smallest planet

Chapter 8
Deep Learning

8.1 Applications of Deep Learning

We cannot perceive the importance of anything until we know its practical applications. The domain of deep learning has been involved in our daily life routine for the last few years. Here we will discuss some important applications of deep learning.

8.1.1 Self-Driving Cars

Deep learning is bringing the idea of self-driving cars practical. Data is fed to the system on the basis of which efforts are made to build the models which could safely drive the cars. This data is taken from different sensors, cameras, Global Positioning Systems (GPS), etc. which is then combined by the model to guide the vehicle on roads. A lot of work is already under process in dedicated facilities and labs, e.g., Uber AI Labs, Google self-driving cars project, etc. Once the idea of self-driving cars is securely implemented, it can be extended to further applications like self-delivery of food items and other products, etc.

8.1.2 Fake News Detection

Dealing with fake news has become a major problem in this era of the internet and social media where anyone can propagate fake news. Models can be developed that can detect fake or biased news on the basis of the data and can filter them out. Similarly, the relevant news can be filtered out on the basis of the interest of individuals using their history and liking/disliking. This enhances the reader's user experience and saves a lot of his time spent on searching the relevant news.

© The Author(s), under exclusive license to Springer Nature Switzerland AG 2023
U. Qamar, M. S. Raza, *Data Science Concepts and Techniques with Applications*,
https://doi.org/10.1007/978-3-031-17442-1_8

8.1.3   Natural Language Processing

Natural language processing has become a major area these days for the application of deep learning models. Models help remove the ambiguity inherent with natural language and thus ensure accuracy. Furthermore, translation from one language to another language, generating captions from videos, and virtual assistants are a few of the important applications of deep learning in natural language processing (NLP).

8.1.4   Virtual Assistants

Virtual assistants are the most popular applications of deep learning models. Examples include Alexa, Siri, and Google Assistant. These applications not only interact with people but also learn from this interaction and thus enrich their data for enhancing the future user experience. These applications learn the commands issued by users using natural language and then execute them.

8.1.5   Healthcare

Deep learning has transformed the total healthcare domain. From medical image processing to deep DNA analysis, deep learning models have come up as core to help the healthcare industry. From the early prediction of life-threatening diseases to predicting future risks, deep learning models have totally transformed the healthcare practices from those that were followed a few years back. Here we have tried to show some important applications of deep learning models. Few more applications include:

Visual face recognition	Image processing
Fraud detection	Pixel restoration
Automatic game playing	Photo descriptions
Marketing campaigns	Voice recognition

Now onward we will start a discussion on the core of the deep learning models, i.e., neural networks. We start our discussion with neurons.

8.2   Artificial Neurons

An artificial neuron is the base of the artificial neural network. You can call it a basic processing unit of any neural network. Figure 8.1 shows a simple artificial neuron. An artificial neuron receives the input at the input layers. There may be multiple

Fig. 8.1 Artificial neuron

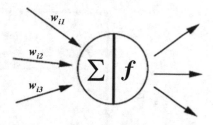

input points, one for each feature. The weights are then applied to input after which mathematical functions like ReLU or sigmoid, etc. are used to process it. Finally, the output is transferred to the next neuron.

The mathematical functions here are called the activation functions used to emulate the neurons in the brain in that they are activated (or not activated) on the basis of the strength of input signals.

From the above figure, it is clear that the inputs, i.e., $X_1, X_2, \ldots X_n$, are given, which are then multiplied with the weights, i.e., $W_{i1}, W_{i2} \ldots, W_{1n}$. The sum is then calculated and given as input to activation function f. The sum is calculated as:

$$f = \sum_{i=1}^{n} w_i x_i$$

The calculated value is then applied to the activation function.

8.3 Activation Functions

The output of the neuron is applied to the activation function. The activation functions normally convert the input value to a specific limit, e.g., from 0 to 1. One of the important activation functions is sigmoid which transforms the output from 0 to 1. Here we will discuss a few important activation functions.

8.3.1 Sigmoid

Sigmoid is the most common activation function used in a majority of artificial neural network applications.

The function is given as follows:

$$f(x) = 1/(1 + \exp(-x)) \tag{8.1}$$

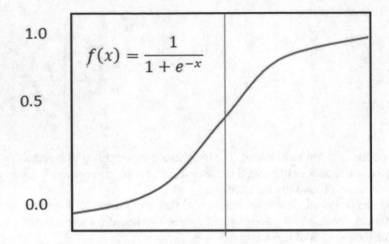

Fig. 8.2 Graph of the sigmoid function

The graph of the sigmoid function is an S-shaped curve as shown in Fig. 8.2.

The sigmoid function takes values from $-\infty$ to $+\infty$ as input and produces the output in the range from 0 to +1. It should be noted that $f(0) = 0.5$. Although commonly used, there are a few inherent problems with the function that its output becomes almost-approaching-to-zero for values less than -10 and almost-equal-to-one for values greater than +10. Note that the sigmoid function is also called the squashing function as it takes real numbers as input and generates output in a range from 0 to 1.

8.3.2   Softmax Probabilistic Function

Softmax probabilistic function is also one of the commonly used activation functions. It uses probability distribution over k classes. The formula of the softmax probabilistic function is:

$$\text{Probability} = \exp\ (1)/(\exp\ (1) + \exp\ (3) + \exp\ (2) + \ldots + \exp\ (n)) \quad (8.2)$$

Here all expression values are input elements of the input vector. These can be real number values. It should be noted that the sum of the all-output values of the softmax function is up to 1.

Fig. 8.3 Graph of ReLU function

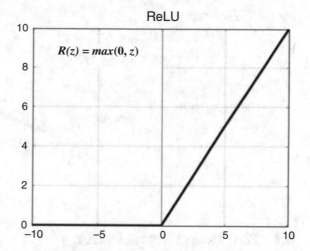

8.3.3 *Rectified Linear Unit (ReLU)*

ReLU function is given as:

$$f\,(X) = \max\,(0, X) \tag{8.3}$$

ReLU is commonly used in deep learning because it significantly improves the classification rates for speech recognition and computer vision tasks. It only activates if the output of a neuron is positive. Figure 8.3 shows the graph of the ReLU function.

8.3.4 *Leaky ReLU*

The Leaky ReLU is a modification of the ReLU function. It replaces the negative part of the ReLU domain, i.e., $[-\infty, 0]$, with some slop, alternatively helping us avoid the situation where a number of neurons become inactive (not firing at all) or do not participate in the transformation process. A small slop helps them remain active. The graph of Leaky ReLU is shown in Fig. 8.4.

Some other activation functions include Max, Argmax, hypertangent, etc.

Fig. 8.4 Graph of LReLU

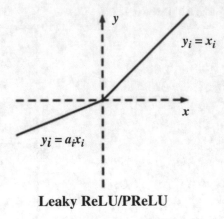

$y_i = x_i$

x

$y_i = a_i x_i$

Leaky ReLU/PReLU

8.4 How Neural Networks Learn

Neural networks use a specific algorithm to learn from the data. A generic learning algorithm of a neural network is as follows:

Step 1: Get input.
Step 2: Modify the weights.
Step 3: Calculate mean error.
Step 4: Repeat steps 2 to 3 until the mean error is minimum.
Step 5: Exit.

The first and most common learning algorithm was backpropagation. The network uses gradient descent for learning. In gradient descent, the initial weights are randomly assigned. After this, the weights are gradually updated on the basis of the error between the expected output and the actual output.

Now we will explain the above steps in detail:

8.4.1 Initialization

We need to provide the initial weights. Normally the initial weights are randomly initialized.

8.4.2 Feed Forward

Information is provided from one layer to the next. The output of the first layer becomes the input of the next layer. On each layer, the activation functions are applied.

Table 8.1 Sample dataset

Object	Feature-1	Feature-2	Feature-3	Class
X1	0.6	0.2	0.1	1
X2	0.5	0.1	0.3	1
X3	0.1	0.03	0.3	0

8.4.3   Error Calculation

The output of the network is matched with the actual output by calculating the mean error.

8.4.4   Propagation

On the basis of the error, the weights are modified. Error is propagated backward. The gradient descent of the change in error is calculated with respect to the change in weights.

8.4.5   Adjustment

Weights are adjusted using the gradients of change. The intention is to minimize the error. The weights are adjusted on the basis of the derivative of the activation function, the differences between the network output and the actual target outcome, and the neuron outputs.

Note that there are no proper guidelines for setting up the initial values of weights; however, one rule of thumb is to initialize the weights between $-2n$ and $2n$, where n is the number of input attributes.

8.5   A Simple Example

Now we will explain a simple neural network with a simple example. Consider a dataset given in Table 8.1:

Here we have three inputs for our neural network. We consider a simple activation function as follows:

$$f(x) = \begin{cases} 0, & x < 0.5 \\ 1, & x \geq 0.5 \end{cases} \tag{8.4}$$

The next step is to randomly initialize the weights. We consider the three weights as equal to "1."

Now:

$X = (0.6 \times 1) + (0.2 \times 1) + (0.1 \times 1)$
$X = 0.6 + 0.2 + 0.1$
$X = 0.9$

Applying the activation function, the output will be "1."

8.6   Deep Neural Network

A lot of work has already been done in the domain of deep learning techniques. The reason behind this is the performance of these techniques. These techniques are successfully employed in various domains (a few of which have been discussed above). Here we will provide an in-depth insight into deep neural networks.

Figure 8.5 shows the deep neural network in general. It comprises an input layer, a number of hidden layers, and finally the output layer. It is similar to a multilayer perceptron, but with these multiple layers and multiple connected neurons, the deep neural network is able to perform complex classifications and predictions.

The data is entered in the form of input features (attributes). These features pass through complex transformations in these hidden layers. The input values are then multiplied with the weights to produce the output for the next layers. The process ends at the output layers where the final classification or prediction is made. The weights are gradually updated in order to make the output close to the expected one. Once the closest match is found, the network is said to be trained and becomes ready to be used for actual data.

Fig. 8.5 Deep neural network

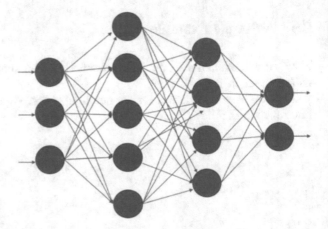

8.6.1   Selection of Number of Neurons

The selection of the number of neurons in a neural network is a very common problem when we start the implementation of a deep neural network. A simple rule of thumb may be to select the number of neurons on the basis of the patterns you want to extract. However, there may be variations in patterns due to many reasons even for the data about the same entities. The data about the quality of different products on different dates may contain different variations. Now the variations may be systematic which can be used to identify the number of classes, so we can say that the exact number of neurons required in a hidden layer may vary from problem to problem.

The more the number of neurons, the more patterns, and the finer structures can be identified; however, the problem is that increasing the number of neurons is not a simple solution as it increases the complexity of the network, consequently requiring a greater number of resources for the implementation of the network. Furthermore, by increasing the number of neurons, the model may result in the problem of overfitting, consequently resulting in a model that works well for training data but may not be accurate for test data.

8.6.2   Selection of Number of Layers

The selection of the number of layers in a deep neural network is also a problem. It may be possible that you select a model with a greater number of layers that performs well for training data but produces inaccurate results for testing datasets. So, it is a challenging problem that can be called problem specific. However, traditional techniques can also be used, e.g., using a number of experiments just to find the optimal number of layers or some heuristics-based approaches. However, adding a greater number of layers increases the complexity.

The problem of the selection of the number of layers and the number of neurons in each layer should always be guided by the principle that the performance and accuracy should enhance.

As the datasets in the real world may have hundreds and thousands of features, so building neural networks for such large datasets may require a lot of execution time. The time may be in days, especially the time taken by the network for training purposes. So, we can say that selection of the number of layers and neurons is related to your problem-solving skills and the experience of developing such networks.

Now we will explain a simple guideline for selecting the number of hidden layers and neurons in each layer. We will explain with the help of an example.

- First of all draw decision boundaries between decision classes on the basis of data.
- Represent decision boundaries in the form of lines. Now join all the lines using connections.

- The number of neurons in the first hidden layer should be equal to the number of selected lines.
- The number of selected lines represents the number of hidden neurons in the first hidden layer.
- For connecting the lines in the previous layer, add a new hidden layer. So the number of neurons in the hidden layer is equal to the number of connections needed.

Now we will explain this with the help of an example.

Consider a dataset comprising two decision classes. The classes in two-dimensional space are represented as shown in Fig. 8.6:

As shown in the above figure, we may need a nonlinear line to separate the classes as we cannot separate them using a single linear line. So, we will add a hidden layer to get the best decision boundary.

So, how many hidden layers and how much number of neurons in each layer will be there in this case?

As discussed above, firstly we will draw a decision boundary between the classes as shown in Fig. 8.7. Note that there may be more than one possible way to draw the decision boundaries; you can select any.

Following the previous procedure, the first step is to draw the decision boundary that splits the two classes. There may be more than one possible decision boundary. You can select any. Figure 8.8 shows the decision boundary that we selected.

Now we will separate the decision boundary using different lines. In our case, there will be two lines.

The main intention behind dividing the decision boundary using the lines is that a line can be linearly used to separate different classes and a single-layer perceptron is also a linear classifier that uses a line equation to separate the classes, so a greater number of neurons mean a greater number of lines for classification.

In our case there are two lines that can represent the decision boundary, so we can say that there will be two neurons in the first hidden layer.

Fig. 8.6 Graphical representation of two decision classes

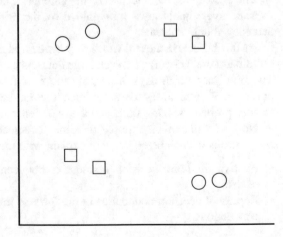

Fig. 8.7 Decision boundary
between classes

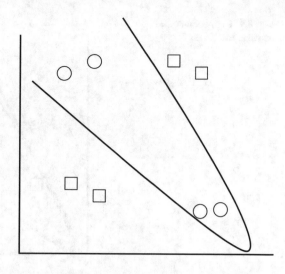

Fig. 8.8 The selected
decision boundary

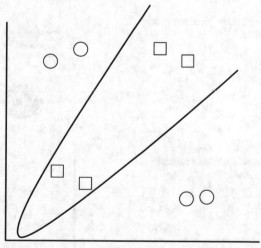

Next, we will find the connection points of the lines representing the decision boundary. In our case, there will be one connection point as shown in Fig. 8.9.

So far, we have found that there are two lines and thus two neurons in the first hidden layer. However, as we need to create one classifier, there should be some merging point to merge the output of the two neurons. For this, we will create another hidden layer with one neuron to merge the outputs of the two neurons in the previous hidden layer. The final network in our case will look like the one given in Fig. 8.10:

Now let's talk about a more complicated example as shown in Fig. 8.11. Note that here the decision boundary is more complex as compared to the previous example.

So, first, we will draw a decision boundary. Once the decision boundary is drawn, we will find out the split points. This will help us find the number of lines and hence

Fig. 8.9 Connection point between two lines

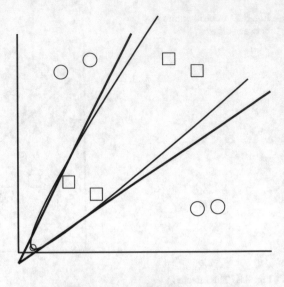

Fig. 8.10 Final network model

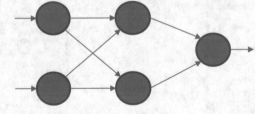

Fig. 8.11 Complex representation of decision classes

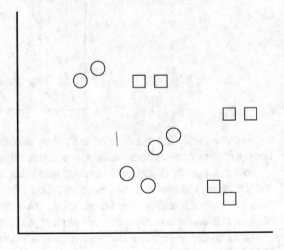

Fig. 8.12 Split points on
the decision boundary

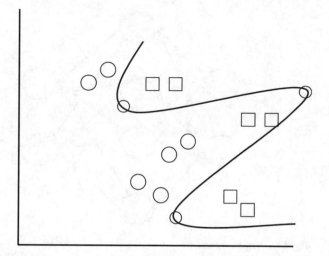

Fig. 8.13 Lines drawn on
the decision boundary

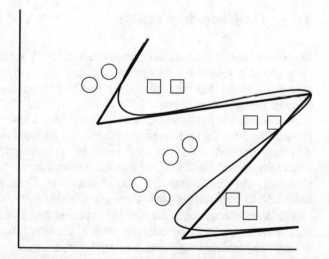

the number of perceptions. Figure 8.12 shows the decision boundary and the split
points.

Now on the basis of the split points, we will find the number of lines as shown in
Fig. 8.13:

As there are four lines, so there will be four perceptrons in the first hidden layer.
These four perceptrons will produce four classification outputs which mean we need
to have another layer to merge these outputs. Here the user can insert a new hidden
layer with two perceptrons for combining the results of the previous hidden layer.
Figure 8.14 shows what the overall architecture will look like:

There are many ways we can use to enhance the performance and reduce the
execution time. Here we will discuss a few commonly used methods.

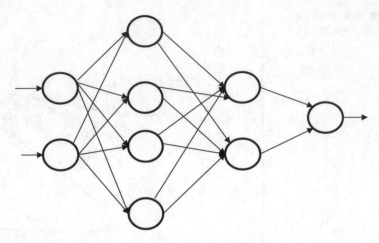

Fig. 8.14 The final model

8.6.3   *Dropping the Neurons*

Dropout means to omit certain neurons in a hidden layer. Dropping out neurons in deep neural networks has its upsides and downsides. By randomly omitting some neurons, it seems that the performance will be boosted due to the lesser number of neurons in the hidden layers.

While training the network, we randomly drop a few neurons having a certain probability. Now for each training instance, the different number of neurons will be randomly omitted. This idea may result in a model with weak classification or prediction power. However, just like the mechanism we have in the random forest algorithm, the predictions and classifications of different weak models can be combined to form strong predictions and classifications, thus resulting in a strong prediction or classification models. Different combination mechanisms can be used and one of them may be majority voting. However, the following observations should be considered regarding the dropout of neurons:

- Increasing the number of dropouts can slow down the learning process.
- Dropouts normally help for large and complex deep neural networks.
- Dropout forces a neural network to learn more robust features that are useful in conjunction with many different random subsets of the other neurons.
- Dropout reduces the training time for each epoch.

Dropout can be used on any hidden layer and even the input layer. However, it is not used on the output layer. Figure 8.15 shows two models with and without the dropout of neurons.

Fig. 8.15 A sample neural network with and without dropout

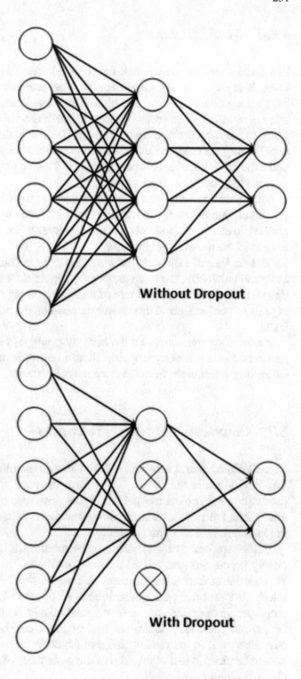

8.6.4   Mini Batching

In a simple deep neural network, the change in weights is calculated for each neuron in each layer for every single epoch. The process is known as batch learning backpropagation. This seems to be a simple mechanism for small networks having a lesser number of neurons. However, the large networks may have millions of weights for which the changes need to be calculated in each epoch. This may significantly increase the execution time and the required resources. It may even make the batch learning backpropagation infeasible for problems having a number of neurons beyond the average number.

Mini batching may be one of the solutions to reduce the number of gradients (or calculation of change in weights). In mini batching, in contrast to the stochastic gradient descent method, we calculate gradients on several instances all at once instead of each individual instance.

The number of instances, for which gradient is calculated, is called a batch. The more the batch size, the more memory is required by the model. It should be noted that so far there is no specific rule of thumb to specify the batch size that could result in optimal performance. The size may depend on the number of neurons required in a layer.

Another approach may be the early stopping of the training of the deep neural network. However, early stopping should be implemented if we can achieve considerable performance from the deep neural network.

8.7   Convolutional Neural Networks

Convolutional neural networks (CNNs) are proved more effective so far. They were first introduced in the 1980s. CNNs form the core of image processing and audio processing, e.g., object recognition, figure print recognition, voice command recognition, etc. On some occasions, the applications using the CNNs have even beaten the human image recognition ability.

CNNs are best at identifying special relationships which enables them to efficiently process images-related data. For example, the dataset of an employee may have different features, e.g., name, salary, tax deduction, etc. Now, this dataset is totally different from the spatial images of human faces. The difference is that the employee dataset features have no relationship with each other. Salary and tax deductions are independent and any of them can be processed before the other one. However, in the case of images of human faces, the data of the lips of a face should be close to that of the nose of the same face; otherwise, there will be issues in classification or prediction.

So if your dataset has spatial relationships, CNNs will be a good choice to process such data.

8.7.1 *Convolution in* n-*Dimensions*

Convolution is the main operation in any CNN. Mathematically:

$$(x * w)[n] \equiv \sum_{m=-\infty}^{\infty} x[m]w[n-m] \qquad (8.5)$$

where n is the size of the data vector and m is the size of the weight vector.

Here we will explain the idea of convolution on one-dimensional and two-dimensional data.

Consider that we have the following input vector V and their weights W.

$V = [2,3,4]$
$W = [1,3,-1]$

Now the one-dimensional convolution will be calculated as:

$H[0] = 2 \times -1 = -2$

[2,3,4]
[1,3,-1]

$H[1] = (2 \times 3) + (3 \times 1) = 6 + 3 = 9$

[2,3,4]
[1,3,-1]

$H[2] = (2 \times 1) + (3 \times 3) + (4 \times 1) = 2 + 9 + 4 = 15$

[2,3,4]
[1,3,-1]

$H[3] = (3 \times 1) + (4 \times 3) = 3 + 12 = 15$

[2,3,4]
[1,3,-1]

$H[4] = (4 \times 1) = 4 = 4$

[2,3,4]
[1,3,-1]

So the one-dimensional convolution yields:

$h = [-2,9,15,15,4]$

The highlighted text shows the weight and the vector point that will be multiplied.

Note: in Python, you can use the convolve method of numpy to get the one-dimensional convolution as follows:

```
Import numpy as np
h = np.convolve([2,3,4], [1,3,-1])
print (h)
```

The following output will be displayed:

$[-2,9,15,15,4]$

It should be noted that the complete information is obtained in the case of h [2] where the input data fully overlaps with the weights. In Python this can be obtained as follows:

```
import numpy as np
h = np.convolve([2,3,4], [1,3,-1],'valid')
print (h)
```

The following output will be displayed:

15

It should be noted that convolution does not give complete or certain information at the edges of the vectors because data and the weights do not fully overlap. In order to make the output the same as the input (e.g., to make the processing easy), the following code can be used:

```
import numpy as np
h = np.convolve([2,3,4], [1,3,-1],'same')
print (h)
```

The following output will be displayed:

[9,15,15]

So, you can use the argument "valid" when you need the information without noise and "same" when you want to make the output the same as the input.

A two-dimensional convolution is similar to a one-dimensional convolution. The two-dimensional datasets are represented as a matrix which is ideal to represent images-related data.

Mathematically:

$$(x * w)[n_1, n_2] = h[n_1, n_2] = \sum_{m_1 = -\infty}^{\infty} \sum_{m_1 = -\infty}^{\infty} x[n_1, n_2]w[n_1 - m_1, n_2 - m_2] \quad (8.6)$$

Suppose we have the following two matrices where one represents the input data and the other represents the weights (filters):

$$X = \begin{bmatrix} 1 & 3 & 2 \\ 6 & 5 & 1 \\ 1 & 2 & 3 \end{bmatrix}$$

$$w = \begin{bmatrix} 2 & 4 & 1 \\ 2 & 1 & -1 \\ 3 & 1 & -1 \end{bmatrix}$$

Now we will calculate convolutions.

H[0,0] = 3×1 = 3

$$\begin{bmatrix} 1 & 3 & 2 \\ 6 & 5 & 1 \\ 1 & 2 & 3 \end{bmatrix}$$
$$\begin{bmatrix} 1 & 3 & 2 \\ 6 & 5 & 1 \\ 1 & 2 & 3 \end{bmatrix}$$

H[0,1] = (2×1)+(3×3) = 2+9 = 11

$$\begin{bmatrix} 1 & 3 & 2 \\ 6 & 5 & 1 \\ 1 & 2 & 3 \end{bmatrix}$$
$$\begin{bmatrix} 1 & 3 & 2 \\ 6 & 5 & 1 \\ 1 & 2 & 3 \end{bmatrix}$$

Just like the above, we can calculate the convolutions till h[4,4]. You can use the numpy's convolve2d() to calculate two-dimensional convolutions in Python.

Similarly, we can calculate the convolutions for n-dimensions. Table 8.2a–c shows the original two-dimensional image and the convolved image.

Table 8.2 (a) Image data; (b) filter data; (c) convolved image

(a)				
1	1	1	0	0
0	1	1	1	0
0	0	1	1	1
0	0	1	1	0
0	1	1	0	0
(b)				
1	0	1		
0	1	0		
1	0	1		
(c)				
4	3	4		
2	4	3		
2	3	4		

The following are some characteristics of CNNs:

- CNNs can use the spatial properties of the data and thus are ideal for processing images-related data.
- Different image processing-related applications can be easily developed using CNNs.
- A lot of tools related to support are available.
- CNNs can be combined with other techniques to get the optimal performance.

8.7.2   *Learning Lower-Dimensional Representations*

Although the convolutional neural network requires a large amount of labeled data, in the real world, we may not have such plentiful labeled data. So, there should be some methods that could be used in such situations. We will see the process of embedding. The process can be summarized in Fig. 8.16:

The algorithm takes unsupervised data as input data and then produces the embeddings which are then used by the actual algorithm to produce the output.

8.8   Recurrent Neural Networks (RNN)

A recurrent neural network (RNN) is just like a conventional neural network; however the only difference is that it remembers the previous state of the network as well. For conventional networks the inputs and outputs are independent of each other; however, this is not the case with RNN. The output of the network at time t is used as input to the network at time *t*+1. So, we can say that output in RNN is

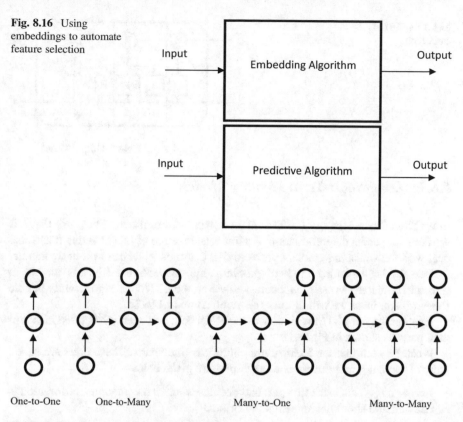

Fig. 8.16 Using embeddings to automate feature selection

Fig. 8.17 Four input/output combinations of RNN

dependent on the prior elements within the sequence. For this purpose, it has a special memory unit. On the basis of the number of inputs/outputs, there may be four possible combinations of the network:

- *One-to-One RNN:* One-to-One RNN is the simplest form of RNN which has a single input and single output.
- *One-to-Many RNN:* One-to-Many RNN, as the name indicates, has one input and many outputs. A simple example may be the automatic generation of a description of an image.
- *Many-to-One RNN:* Many-to-One RNN takes multiple inputs and produces a single output. One of the examples may be a sequence of words comprising a news article, and the output may be the prediction of the category of the article.
- *Many-to-Many RNN:* Many-to-Many RNN takes multiple inputs and produces multiple outputs. A simple example may be machine translation.

Figure 8.17 shows the pictorial representation of all of these combinations:

Now we will discuss an important RNN architecture called Long Short-Term Memory (LSTM) Model.

Fig. 8.18 LSTM
architecture

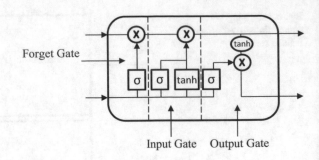

8.8.1 *Long Short-Term Memory Models*

Long Short-Term Memory (LSTM) is an advance form of neural network that lets
the network persist the information. An important feature of LSTM is that it can also
deal with the vanishing gradient problem that a normal RNN can't properly handle.

Due to the vanishing gradient problem, the conventional RNN is unable to
remember the long-term dependencies; however, the LSTM can significantly handle
this problem. Now we will discuss the architecture of LSTM.

Apparently, the LSTM is similar to the RNN cell; however, it has three independent parts as shown in Fig. 8.18.

It can be seen that the architecture comprises three parts. These parts are called
gates. The names and functions of each part are given below:

- *Forget gate:* This is the first part and decides whether the previous information is
  relevant and should be remembered or not.

- *Input gate:* This is the second part and learns the important information.

- *Output gate*: This is the third part and passes the information to the next
  timestamp.

LSTM, just like the conventional RNN, comprises hidden states as well. H_t
represents the hidden state of the current timestamp t, whereas the H_{t-1} represents
the hidden state of the previous timestamp. The state is referred to as short-term
memory.

Each cell has a cell state that is also known as log-term memory shown as follows:

C_{t-1}: cell state at timestamp $t-1$
C_t: cell state at timestamp t

Hidden states and cell states are shown in Fig. 8.19:

It is interesting to note that the cell state carries the information along with all the
timestamps.

Let us discuss the above functionalities with the help of an example.

John has won the competition. Bill will leave tomorrow. You can see that before
full stop, we are talking about John in the first sentence and when the full stop occurs

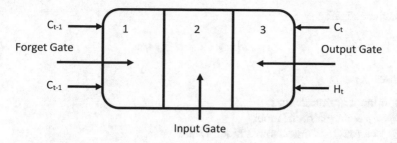

Fig. 8.19 LSTM with hidden states

the next context starts where we are talking about Bill. So after a full stop, the network should forget about John and remember the context of Bill.

8.8.1.1   Forget Gate

The role of the first cell is to decide whether to remember or forget the information. The equation for the forget gate is:

$$f_t = \sigma(x_t * H_{t-1} * w_t x_t) \tag{8.7}$$

where:

- X_t = current timestamp input
- U_f = input weight
- H_{t-1} = previous timestamp's hidden state
- W_f = weight matrix for hidden state

After this, we apply the sigmoid function. If the output of the function is 1, the information will be remembered; otherwise it will be forgotten.

8.8.1.2   Input Gate

The input gate specifies which information is important and which to discard. For example, consider the following sentence:

John is selected for the competition. After visiting his father this week, he will contest in the second week of this month.

Here the overall context is about John. The first sentence gives us the information that John is selected for the competition, whereas in the second sentence, we get the information that after visiting his father, he will contest in the second week of this month. So, our model has to decide which information is important. The information that he will contest in the second week is more important. This is the task of the input gate to prioritize the information and discard the unimportant ones.

Mathematically:

$$i_t = \sigma(x_i * U_i + H_{t-1} * w_i) \tag{8.8}$$

Here:

- $X =$ input at timestamp t
- $U_i =$ weight matrix of input
- $H_{t-1} =$ previous timestamp's hidden state
- $W_i =$ hidden state's weight matrix

After this, we will apply the sigmoid function.

8.8.1.3 Output Gate

Now consider this sentence:

John got the first position in 10th grade in physics. For winning first position _____ was given a prize.

Now when we see the words "first position," we can immediately get the idea that we are talking about John as he got the first position and not the entire 10th grade class nor the physics. So, while filling the blank, the most relevant answer will be John. This is the task of the output gate.

Mathematically:

$$O_t = \sigma(x_t * U_o + H_{t-1} * w_o) \tag{8.9}$$

So these were some details about the LSTM model. Just like LSTM, we may have other RNN models as given in Table 8.3:

Although each of the above models can be explained in further detail, here we will only explain a simple architecture of Sequence to Sequence (Seq2Seq) RNN. We will explain the encoder/decoder architecture.

Table 8.3 LSTM models

Model	Description
Vector to Vector (Vec2Vec) RNN	As the name indicates, these RNN take a single vector as input and produce a single vector as output as well
Sequence to Vector (Seq2Vec) RNN	In Sequence to Vector RNN, the input is a sequence, whereas the output is a vector, e.g., a language predictor
Vector to Sequence (Vec2Seq) RNN	The input is a vector and the output is a sequence
Sequence to Sequence (Seq2Seq) RNN	These models take input as sequence and produce output as sequence as well. It should be noted that the length of the input and output sequence may not be the same, e.g., while translating from one language to other, the lengths of both input and output may be different

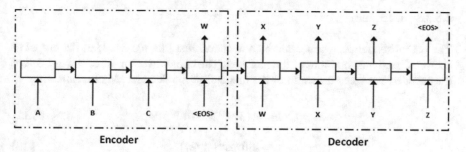

Fig. 8.20   Encoder/decoder architecture of Seq2Seq

8.8.2   Encoder and Decoder

Encoder/decoder is a classic example of Sequence to Sequence RNN. The architecture takes input a sequence of any length and produces a sequence of the same or different lengths. Figure 8.20 shows the encoder/decoder architecture of Seq2Seq model.

The architecture comprises two components called encoder/decoder. Each box comprises a cell that contains an RNN. Input is received by the encoder in the form (A, B, C, <eos>) and the decoder generates the output sequence (X, Y, X, <eos>). Here is a brief description of both encoder and decoder components.

8.8.2.1   Encoder

The encoder comprises different cells where each cell contains an RNN. A single vector is an input to a cell followed by the end-of-statement token.

The encoder is a recurrent RNN in each cell. Each cell receives a single vector followed by an end-of-statement token, <EOS>, which marks the end of the input sequence. The encoder then generates a context vector W which is then input to the decoder.

The information is passed from one cell to the next as shown by the links. Mathematically this information is calculated by the formula:

$$h_t = f\left(W^{(hh)}h_{t-1} + W^{(hh)}x_t\right) \qquad (8.10)$$

Here:

W^{hh} = weight given to the previous hidden state at timestamp $t-1$.
W^{hx} = weight given to the current input at timestamp t.

8.8.2.2   Decoder

The output sequence is generated by a decoder. Just like the encoder, the output of each cell becomes the input of the next cell. The context vector W, of the last cell of the encoder, becomes the input of the first cell of the decoder. Mathematically:

$$h_t = f\left(W^{(hh)}h_{t-1}\right) \tag{8.11}$$

$$y_t = \text{softmax}\left(W^S h_t\right) \tag{8.12}$$

The encoder/decoder architecture forms the basics of the tasks especially the language translation in natural language processing. The same can be used to convert the code generated in one programming language to some other programming language.

8.9   Limitations of RNN

Along with the characteristic of remembering the previous state, the RNN suffers two major issues as mentioned below:

- Vanishing gradient problem: This problem occurs when the algorithm learning process becomes slow due to the small gradients. The algorithm takes too much time to converge.

- Exploding gradient problem: This problem occurs when the gradient becomes large such that the model becomes unstable. Again the algorithm suffers the performance issues.

The problem can be solved someway by using the minimum number of hidden layers or some advance RNN architectures.

8.10   Difference Between CNN and RNN

Here are some differences between CNN and RNN, given in Table 8.4:

Table 8.4 Difference between CNN and RNN

CNN	RNN
Uses sparse data like images. RNN is applicable for	RNN uses time series and sequential data
Simple backpropagation is used in CNN	RNN uses backpropagation with timestamps
CNN uses a finite number of inputs and outputs	No restriction on the length of input/output
CNN is a feed-forward neural network	RNN uses a feedback loop to use the output at timestamp $t-1$ as the input at timestamp t
Normally used for image processing	Normally used for time series and sequential data

8.11 Elman Neural Networks

The Elman neural network is a type of recurrent neural network. In recurrent neural networks, the output of neurons is fed back to other neurons and the entire network works in a loop. This means that the network can keep the previous state (results) and is thus able to process the time series data, e.g., textual data.

Figure 8.21 shows a simple recurrent neural network.

Note that the context layer has a delay neuron. This delay neuron works as a memory holder which stores the activation value of the earlier time step and provides this value to the next time step.

Elman neural networks are nothing but a multilayer perceptron; however, we add a context layer(s). Each context layer contains neurons equal to the number of neurons in the hidden layer. Furthermore, the neurons in the context layer are fully connected to neurons in the hidden layers.

The values of neurons in hidden layers are stored in the neurons present in the context layers. We call them the delay neurons because the values are delayed for one step and are provided as input for the next step (to provide the previous state).

Figure 8.22 shows the generic diagram of the Elman neural network.

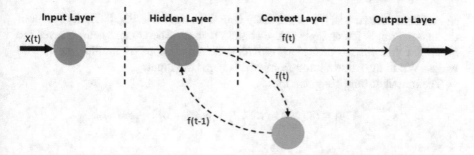

Fig. 8.21 Simple recurrent neural network with context layer

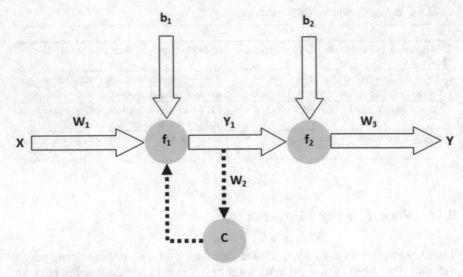

Fig. 8.22 A generic Elman neural network

Now we will explain the information flow in Elman neural network. Suppose we have a neural network comprising two layers and there is one neuron in each layer. The network has one neuron in each layer. The first neuron has two inputs, i.e., weighted input and bias. The output of the first neuron becomes the input of the second neuron along with the bias. "f1" and "f2" are the activation functions.

The output function will then be:

$$Y = f2\,(w2f1\,(w1X + b1) + b2) \tag{8.13}$$

Here:

b1 = input bias of first neuron.
b2 = input bias of second neuron.
w1 = input weight for first neuron.
w2 = input weight for second neuron.

As Elman network involves the delay neuron as shown in Fig. 8.21, the output of the first neuron is fed as input to the context neuron. The context neuron saves the input and then, in the next step, provides the same input by multiplying it with weight W2 to the same neuron from which it got the input.

The output at time t is given by:

$$Y\,[t] = f2\,(w3f1\,(w1X[t] + w2C + b1) + b2)$$

where:

$$C = Y1[t - 1]$$

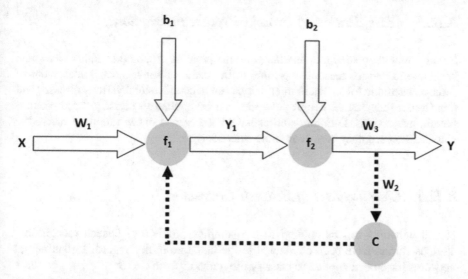

Fig. 8.23 Generic form of Jordan neural network

Similar to a conventional neural network, the weights are dynamically adjusted during each iteration on the basis of the error.

Normally the Elman neural networks are used when we need to predict the results in sequence, e.g., the next pollution level at the next hour.

8.12   Jordan Neural Networks

Jordan neural network is similar to Elman neural network. The only difference is that context neurons take input from the output layer instead of the hidden layer. Figure 8.23 shows the generic form of Jordan neural network.

So the memory, i.e., information from the previous state, is recurrently provided to context nodes.

Here we will discuss some of the applications of the Jordan neural networks. Jordan neural networks have a wide range of applications. They can model time series data and can successfully be applied for classification problems.

8.12.1   Wind Speed Forecasting

Wind speed forecasting is important, especially in coastal regions. This can help in various range of industrial and social marine activities, e.g., the prediction of wind speed can be used to determine the output of a wind turbine at a certain time. Similarly, the wind speed can be used to schedule the movements of the ship for transportation.

8.12.2 Classification of Protein-Protein Interaction

Proteins within one cell may interact with the proteins of the other cells in a human body due to various reasons, e.g., due to the result of some biochemical event or some electrostatic force. These interactions can help understand various diseases and their treatments. Jordan neural networks can be applied to classify these protein-protein interactions. Different studies have been carried out in literature where the classification accuracy was even more than 95%.

8.12.3 Classification of English Characters

Neural networks can be applied to the complex problem of speech recognition. Various studies have been conducted in the literature in this regard. Jordan neural networks have been applied to classify the numbers from 0 to 9.

8.13 Autoencoders

Autoencoder is a feed-forward neural network based on unsupervised learning. As shown in the Fig. 8.24, it is just like a perceptron and uses three layers including the input layer, a hidden layer, and one output layer.

From the above diagram, it is clear that the number of neurons in the input layer is equal to the number of neurons in the output layer. At this point, it may be different

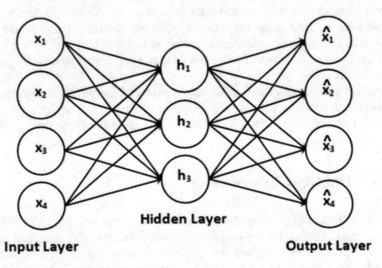

Fig. 8.24 Generic autoencoder model

from the conventional multilayer perceptron. Furthermore, instead of training where the conventional neural networks predict or classify the output on the basis of the input values, the autoencoders are trained to reconstruct the output from the given inputs.

Mainly autoencoders comprise two components called encoders and decoders consequently providing two tasks, i.e., encoding and decoding. The encoding means mapping the input to the hidden layer and decoding means mapping the output of the hidden layers to the output layer. The encoders take input in the form of feature vectors and transform them into the form of new features. The decoder then transforms these encoded features into the original feature set. For example, the original features may be X1, X2, X3, and X4. The encoder then transforms them into hidden features h1, h2, and h3. The decoder takes these features as input and converts them to x^1, x^2, x^3, and x^4 which are similar to original features, i.e., X1, X2, X3, and X4.

For developing an autoencoder, we need to have three things:

- The encoding method
- The decoding method
- A loss function

Normally autoencoders help us in dimensionality reduction. We may project the large data spaces into small ones having a minimum number of features. Here are some other properties of autoencoders:

- Data-specific: Autoencoders are data-specific which means that they can only be applied to the data that is similar to the one on which these algorithms are trained. So you cannot reduce the dimensionality of textual data using the autoencoder that has been trained on images-related data. This is totally in contrast to generic data reduction or compression algorithms.

- Lossy: Some data may be lost during the encoding and decoding process. It means that the transformed features may not be the same as the original ones. So, you can say that it is related to feature transformation rather than feature selection.

- Unsupervised: Autoencoders can work on unsupervised data which means that you don't need to provide the class labels or tagged data. They are pretty good at working with unsupervised data. Autoencoders generate their own labels during training.

8.13.1 Architecture of Autoencoders

Some basic details of autoencoders have already been discussed. Here we will see some further details.

The input is provided to encoder components of the autoencoder; the encoder produces the code. It should be noted that normally an autoencoder is a fully

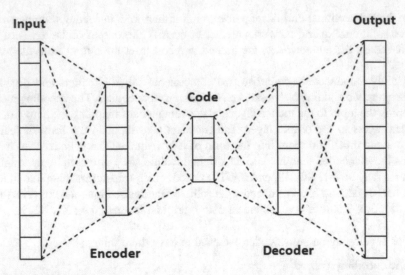

Fig. 8.25 Stacked autoencoder

connected neural network. Both the encoders and decoders implement the same neural network model. Now once the code is generated by the encoder, the decoder uses the code to generate similar features by decoding the input. The goal is to get the output similar to the input as much as possible. The decoder, which has a similar ANN structure, then produces the output only using the code. The goal is to get an output identical to the input.

For the implementation of an autoencoder, we need to consider four parameters as follows:

- *Code size:* The smaller the number of code sizes, the more the compression.

- *The number of layers:* We can have as many hidden layers, both in encoders and decoders depending on our requirements and specific problems.

- *The number of nodes per layer:* You can use any number of nodes per layer. One of the common autoencoders is called the stacked autoencoder where the number of nodes per layer keeps on decreasing as compared to the previous layer. Similarly, the number of neurons per layer keeps on increasing in the decoder.

- *Loss function:* Different loss functions can be used. A commonly used loss function is mean squared error (MSE). Similarly, another cross function may be entropy.

Figure 8.25 shows the stacked autoencoder.

Now we will implement the autoencoder using three images from MNIST dataset as shown in Fig. 8.26. We will reconstruct these images using the code size of 32. Width and height of each image are 28, so it can be represented with a vector of 784 numbers.

The following is the Python code using Keras:

```
iSize = 784
hSize = 128
cSize = 32
iImg = Input(shape=(input_size,))
fHidden = Dense(hidden_size, activation='relu')(iImg)
code = Dense(code_size, activation='relu')(fHidden)
sHidden = Dense(hidden_size, activation='relu')(code)
oImg = Dense(input_size, activation='sigmoid')(sHidden)
autoencoder = Model(iImg, oImg)
autoencoder.compile(optimizer='adam', loss='binary_crossentropy')
autoencoder.fit(xTrain, xTrain, epochs=5)
```

Figure 8.26 shows the original and the reconstructed images.

We can control our autoencoder using the abovementioned parameters. By adding more layers, adding more neurons per layer, and increasing the code size, we can make the autoencoder more powerful, and it can be used for more complex encodings. However, it will make the autoencoder itself complex. The encoder may face the issues like overfitting and copying inputs to outputs, etc.

We don't want the autoencoders to copy the input to output. For this purpose, we keep the size of the autoencoder very small which enables the autoencoder to learn the intelligent features.

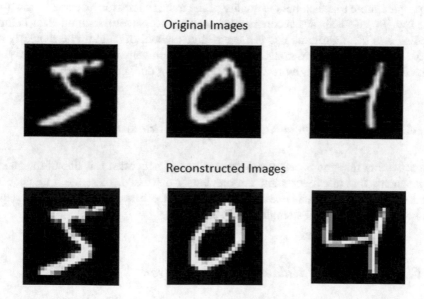

Fig. 8.26 Original and reconstructed images using autoencoder

8.13.2   Applications of Autoencoders

These days we don't have many applications for autoencoders. However, few applications may include the denoising of the data and dimensionality reduction due to which the autoencoders can be used as preprocessors for various machine learning and deep learning algorithms.

8.14   Training a Deep Learning Neural Network

In literature, there have been different practices that can be used to efficiently train a deep neural network. We will mention some guidelines that can be and should be followed. We will start with the training data.

8.14.1   Training Data

A machine learning algorithm is just like a kid. Raised in a good environment will make the kid a person with high moral values, whereas a bad environment will have the worst effect on him. The same is the case with machine learning algorithms. Lots of people ignore this fact, but the quality of the training data for machine learning is the key. So, it is always recommended to preprocess your training data before feeding it to the algorithm. You can use a large amount of data; furthermore, try to avoid corrupt data, e.g., missing, incomplete, and null values, etc. Occasionally you can create new data, e.g., by rescaling by generating derived attributes, etc.

8.14.2   Choose Appropriate Activation Functions

An activation function is one of the cores of neural networks. It is one of the major components that affect accuracy. As previously discussed, there are a number of choices and the appropriate one can be selected on the basis of requirements and the knowledge of associated strengths and weaknesses.

8.14.3   Number of Hidden Units and Layers

Selecting the number of hidden layers and the number of neurons per layer is a tricky problem that should be problem specific and should be carefully selected. In the upcoming section, we will provide a detailed discussion on it with an example. However, for the time being, it should be noted that we can't simply keep on increasing the number of hidden layers as the model may suffer the issues like overfitting. However, you should have a sufficient number of hidden layers and neurons in each hidden layer to keep the error as minimum as possible.

8.14.4 Weight Initialization

Use small random numbers for the initialization of weights. However, the question may be how small or large the weights should be and what range should be used because using small values of the weights will result in small gradients while the large values will make the sigmoid function saturate and thus result in dead neurons. However, luckily, we have a lot of research literature for selecting the appropriate values of the weights.

8.14.5 Learning Rates

The selection of the appropriate learning rate is one of the most important parameters. Here again, we have to make an appropriate choice because a slow learning rate means that the model will take a lot of time and iteration to converge whereas a high learning rate means that model will not converge at all.

In literature, we have different methods which guide us to select the appropriate learning rate value based on the value of the error function. We also have methods that let us avoid manually fixing the learning rate and with the passage of time keep on updating the learning value for the smooth convergence of the model.

8.14.6 Learning Methods

One of the most commonly used learning models is the stochastic gradient descent. However, it may not be much efficient, especially for deep neural networks suffering from two major issues called vanishing gradient descent and exploding gradient descent. However, fortunately, there are other techniques available, e.g., Adagrad, Adam, AdaDelta, etc.

8.14.7 Keep Dimensions of Weights in the Exponential Power of 2

Although the state-of-the-art deep learning models have been devised, we have powerful hardware resources; however, still, efforts should be made to help avoid the overburden a model may face. One method may be to set the values of the weights in terms of exponents of 2. This will help in sharing of weights metrics. Always keep the size of your parameters as exponential power of 2.

8.14.8   Mini-batch vs Stochastic Learning

The main goal of training a model is to learn the appropriate parameters to get the minimum error. The parameters are updated with each training sample. The major objective of training a model is to learn appropriate parameters that result in an optimal mapping from inputs to outputs. Applying different learning methods, e.g., batch, mini-batch, or stochastic learning, may affect the learning process. Each method has its own strengths and weaknesses. In the stochastic gradient approach, the weights are tuned with each training example which can introduce noise during training. It can also avoid model overfitting. Then we have a mini-batch learning process. However, the problem may be selecting the appropriate batch size to introduce an appropriate amount of noise.

8.14.9   Shuffling Training Examples

It has been observed that randomizing the order of training examples can boost the performance and can make the algorithm converge faster.

8.14.10   Number of Epochs/Training Iterations

Another parameter worth consideration is to decide the number of epochs that will be executed during training to get better results. A possible strategy may be to use fixed number of epochs or examples. After this number, compare the error and decide about further examples and epochs.

8.14.11   Use Libraries with GPU and Automatic Differentiation Support

Luckily different libraries like Keras are available for the implementation of various core functionalities of deep neural networks. These libraries provide a better way to code your mode as compared to the custom code which may not only take a lot of time but may have serious performance issues as well.

8.15   Challenges in Deep Learning

Here we will discuss some of the challenges the machine learning algorithms have to face these days.

8.15.1   Lots and Lots of Data

Deep learning algorithms need data for training. More amounts of data are needed to make sure that the algorithm is fully trained (although more training means the introduction of some errors as well). Just like human brains learn and deduce information, artificial neural networks also need data to extract the information for learning and derive the abstractions from this data. The more information you provide, the more abstractions you can derive.

A simple example may be to train a model for speech recognition. For training such a model, you may need data from multiple dialects, demographics, and time scales. Processing such a large amount of data may require more memory and powerful processing systems.

8.15.2   Overfitting in Neural Networks

A common error that almost all neural networks may face is that the difference between the predicted results and the actual result may become large. It may occur due to a large number of parameters and examples. We may train our model well on the training dataset and expect it to perform well on the actual data which may not be the actual case. As shown in Fig. 8.27, the training error (shown in blue) may

Fig. 8.27 Verification error representation

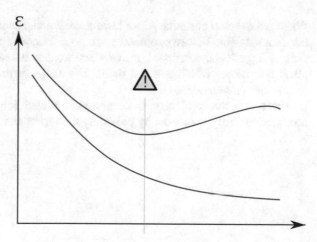

decrease after a certain time, whereas the model may suffer from overfitting (represented with the red line).

8.15.3 Hyperparameter Optimization

Another challenge is how to optimize the hyperparameters. A deep neural network model may require a lot of hyperparameters before commencement. So, a proper mechanism is needed for gradual optimization so that the model may converge with high accuracy. We should be especially careful while optimizing these parameters as a slight change in their value may result in a lot of changes in the performance and accuracy of the model.

8.15.4 Requires High-Performance Hardware

As discussed earlier, the deep learning models may require a lot of training data. These huge volumes of data require a lot of processing power for better performance. This is the very reason for using high-power GPUs having multicore architectures. However, the problem is that using such high-powered machines requires a lot of costs.

It is common for industries using deep learning applications to have their own high-powered data centers to cope with the need for implemented deep learning models. These data centers involve a lot of capital for development.

8.15.5 Neural Networks Are Essentially a Black Box

When we develop our models, we have a deep understanding of these models and the data they use; however, normally we do not know how they reached a certain state or a particular solution. For many researchers, a neural network is just like a black box model which gets the inputs and maps them to outputs. It is hard to understand these models.

This makes the verification of deep models a tough job because we may only be restricted to verify the model by providing the inputs and getting the outputs.

8.16   Some Important Deep Learning Libraries

Today, Python is one of the most popular programming languages for this task, and it has replaced many languages in the industry; one of the reasons is its vast collection of libraries.

Here we will discuss some important Python libraries.

8.16.1   NumPy

NumPy is an open-source Python programming language library. It is well known among data scientists and frequently used for data processing. Because of its ability to work with linear algebra, Fourier transformation, and processing multidimensional arrays, it is ideal for use in artificial intelligence and machine learning projects.

In contrast to list type in Python, NumPy is much faster as it allows only integer or float type arrays, and this allows speedy calculation of linear algebra operations. It is also more convenient to use and has a lower memory footprint compared to the former. It also allows performing various operations such as transposition, reshaping, operations, sorting, and many more on the data of matrices. Performance of machine learning and artificial intelligence algorithm can greatly be improved with the compiled NumPy library without much hassle.

8.16.2   SciPy

Scientific Python (SciPy) is an open-source library based on NumPy and supports advance and more complex operations such as clustering, image processing, integration, etc. It is equipped with other embedded modules for integration, optimization, linear algebra, differential equation, and many other classes of problems. Because of its applicability among multiple domains, it is considered a foundational Python library. SciPy not only includes all the NumPy functions for array manipulation, but it also provides additional tools and data structure (e.g., K-dimension trees and sparse computing) for array computing. These additional modules with their high-level syntax make SciPy a full-fledged easy-to-use scientific tool.

Though SciPy benefits from the frequent development of various modules because of its vibrant community among numerous domains, it is considered ideal for image processing, data visualization, and manipulation. Its implementation in C along with high-level syntax not only makes it faster but also very appealing to use regardless of programmer background or expertise.

8.16.3 scikit-learn

scikit-learn is an open-source and robust to use library for supervised and unsupervised machine learning algorithms. It is based on some of the commonly used NumPy, SciPy, and matplotlib Python modules. It provides efficient tools for statistical modeling and machine learning including regression, classification, clustering, model selection, and preprocessing.

scikit-learn is mainly focused on data modeling rather than loading, manipulating, and summarizing data.

8.16.4 Theano

Theano is an optimizing compiler and numerical computation Python library that is exclusively designed for machine learning. It allows us to define, optimize, and evaluate mathematical operations and multidimensional calculations for arrays required to create deep learning and machine learning models. It is frequently used by deep learning and machine learning developers because of its ability to handle large computations as required in complex deep neural network algorithms.

Theano can easily be integrated with NumPy and other Python libraries, and when deployed over a Graphical Processing Unit (GPU), it can perform operations way faster than when it is deployed over a central processing unit (CPU). It also includes the testing and debugging tools which can be used for bug and error detection.

8.16.5 TensorFlow

TensorFlow is a Python library for fast numerical computing developed and maintained by Google and released under an open-source license. Irrespective of the developer's level of expertise, Theano includes the tools and libraries which help to develop the ML and DL algorithms effectively. It is also very flexible and can easily be deployed over a range of systems including but not limited to single-core processors, multicore processor systems on chips (MPSoCs), Graphical Processing Units (GPU), Tensor Processing Unit (TPU), mobile systems, and even over the distributed systems. With this library, it is also very easy to acquire data, model, train, and serve predictions. For user convenience, it can be used with Python or JavaScript-based application programming interface (API) for application development.

For resource-constrained systems such as mobile computing and edge devices, a library that requires limited resources was also developed and referred to as TensorFlow light. For effective resource utilization, the lighter version optimizes the TensorFlow and allows users to find a trade-off between accuracy and model size. A smaller model is relatively less accurate; however reduction in memory footprint, computation requirement, and the energy consumption is significant.

8.16.6 Keras

Keras is a free and open-source Python library used for neural network development and evaluation within machine learning and deep learning models. Since it is modeled on top of Theano and TensorFlow, it is very easy and requires few lines of code to model and train a neural network. Keras contains numerous readymade building blocks for neural network models such as objective functions, layers, activation functions, and optimizers.

Keras framework flexibility and portability allow the deployment of deep models on smartphones, the web, or the Java virtual machines. For fast training, it also allows the distributed training of deep neural network models on clusters of GPUs and TPUs. Almost all neural network models including conventional and recurrent neural network models can easily be developed in Keras. When it comes to data types, Keras has the widest range as it can work with text images and images for model training and testing.

8.16.7 PyTorch

PyTorch is an open-source deep learning library developed by the Facebook AI research group, and it is based on Python and a C language framework, Torch. It is very popular among data scientists and can easily be integrated with other Python libraries such as NumPy. PyTorch is very flexible. It uses dynamic computation, because of which it is easy to develop the complex model as compared to other machine learning libraries such as TensorFlow which only allows static computing graphs. It also allows the execution and testing of code in parts so the user doesn't have to wait for the complete code implementation for testing.

Even when working with heavy and extensive graphs, PyTorch is preferred over the rest of the deep learning libraries because of its high-speed execution. In addition to GPUs and TPUs, PyTorch also operates well on simplified processors because of its flexibility. To expand on the PyTorch library, it is equipped with a set of powerful APIs and with a natural language toolkit for smoother processing. It is also very compatible with other Python IDEs and so the debugging process is very easy.

8.16.8   pandas

pandas is an open-source Python library that is used for machine learning tasks and data analysis. It is built on top of NumPy and is responsible for preprocessing datasets for machine learning. pandas is very flexible and so works well with other data science libraries. pandas supports multidimensional arrays and is used in various sectors, from science to engineering and statistics to finance.

pandas supports various input file formats such as Parquet, SQL, JSON, comma-separated values, Microsoft Excel, and database tables. It also supports various data manipulation operations such as grouping, merging, selecting, reshaping, data wrangling, and data cleaning.

8.16.9   matplotlib

matplotlib is a comprehensive Python library for data visualization and making interactive graphs and animations. It is an extension of NumPy and SciPy and for data analysis and manipulation, it uses the pandas library. As such, it is used by many developers as an open-source alternative to MATLAB. It generates high-quality graphs, charts, plots, bar charts, histograms, scatter plots, and even 3-D plots.

It is easy to use and provides full control over the plot by providing multiple scripting layer APIs, such as Pyplot and object-oriented (OO) APIs. Pyplot interface is simplistic and easy to use, whereas OO APIs are difficult to use but come with more customizable options. It also has Tkinter and PyQT toolkits, so the developer can insert the static plots in applications as well. For users with a MATLAB background, Pyplot API also makes it work like MATLAB.

8.16.10   Scrapy

Scrapy is an extensive free and open-source web scraping and crawling library. In contrast with some of the other web scraping libraries such as Beautiful Soup, it is full-fledged library and does not require additional tools or libraries. In addition to efficient system resource utilization such as CPU and memory usage, it also extracts data at an incredibly fast rate.

Scrapy uses user-defined classes referred to as spiders to scrap and crawl different web pages. These spiders define the procedure to perform crawling on a web or set of web pages, including how to perform crawl and obtain the specified data. The spider can also speed up the procedure by parallelizing the data extraction commands so that different requests don't have to wait in a queue. Scrapy also helps to process data (i.e., validating, accessing, and writing data to a certain database) by providing the item pipelines.

8.16.11 Seaborn

Seaborn is an open-source enhanced data visualization and representation library based on matplotlib. Though matplotlib provides many data visualization options, it is low level and requires extensive coding and data merging schemes when working with pandas data structures. Seaborn helps to cope with these issues by providing a high-level programming interface that requires less coding and more control over the plot options and aesthetics. This however does not mean that all the matplotlib learning is wasted as some basic working experience is still required to tweak some plot settings.

Seaborn is a very comprehensive library and is often used for generating publication quality and visually very appealing plots. A combination of matplotlib and Seaborn provides a perfect combination for quick data exploration and generating highly customizable plots.

8.16.12 PyCaret

PyCaret is an open-source modular machine learning library based on the popular Classification And REgression Training (CARET) R package. In addition to many other machine learning Python libraries such as NumPy, Keras, and PyTorch, PyCaret is known for its low-code machine learning nature. This helps the data scientist to analyze their data and execute machine learning models quickly. It is a wrapper around various machine learning libraries such as CatBoost, Optuna, scikit-learn, and many more.

PyCaret is developed in a highly modular way. Each module represents supervised learning (i.e., regression and classification) or unsupervised learning (i.e., NLP, anomaly detection, clustering, or association rule mining).

8.16.13 OpenCV

OpenCV is a huge open-source Python library developed for image processing, machine learning, and computer vision. It can be used for detecting various objects, faces, or even handwriting from different sources of inputs (i.e., images or videos). It also supports multiple languages which include C, C++, Python, and Java.

OpenCV is also available on different platforms such as Linux, Windows, iOS, Android, and OS X. OpenCV was developed with high computational efficiency in mind. To achieve this a Python wrapper referred to as OpenCV-Python was developed around its original OpenCV C++ implementation.

8.16.14 Caffe

Caffe (Convolutional Architecture for Fast Feature Embedding) is an open-source deep learning library originally written in C++ with a Python interface. It was originally developed at the University of California, Berkeley, and is widely accepted in academic research and industrial application in AI, multimedia, and computer vision.

Caffe has an expressive architecture that allows defining models and optimization with configuration files without complex coding. It also allows transporting models between systems, training models over GPU, and later testing the trained model on a variety of devices and environments. Caffe is capable of processing 60 million images in a day over a NVIDIA K40 GPU, which makes it suitable for industrial-scale deployment.

8.17 First Practical Neural Network

In this section, we will build our first deep learning model that will comprise a simple feed-forward neural network. The model will have the following properties:

- It will have a single input layer.
- It will have three hidden layers.
- We will use "ReLU" as an activation function for hidden layers.
- We will use the "sigmoid" activation function for the output layer.
- The data will comprise a simple dataset having three input features (excluding the class) and four data points.

We will use Python as our programming language. Furthermore, the NumPy and Keras will be used for defining different parts of the neural network. So, first of all, we will import both of these libraries and the necessary modules from them using the following statements:

```
1)  import numpy as np
2)  import keras
3)  from keras.layers import Dense, Input
```

In statement 1 we have imported the NumPy library. The NumPy library will be used for managing arrays. Secondly, in statement 2 we imported Keras. Keras is one of the specialized libraries commonly used for implementing deep learning models. Finally, in statement 3 we have imported the "Dense" and the "Input" module. The dense module implements the fully connected dense layer, whereas the "Input" module will be used to handle the input data.

8.17.1   *Input Data*

Every deep learning algorithm takes input some data and on the basis of this data, it trains itself. Once the model completes its training, we use the model to classify or predict the unseen data.

In our sample deep learning model, we will use a simple dataset comprising four data points and five features having some hypothetical values. The last (fifth) feature specifies the class of the data. Table 8.5 shows the data points.

As our dataset is very small, we will use a simple two-dimensional array to store the data. The values will be manually initialized. The following lines of code are doing this job:

4) Input_Features = np.array([[1,1,1], [2,2,2], [2,1,2], [1,2,2]])
5) Input_Classes = np.array([[0], [1], [0], [1]])

Statement 4 implements a two-dimensional array named "X." It is a 4×3 array that implements four data points and each data point has three features. Statement 5 implements the "Class" feature. Note that it is also a two-dimensional array comprising 4×1 dimensions, i.e., there are four rows and one column. Each row corresponds to a class for each data point. We will use these two arrays for training and testing purposes.

As it was a very small dataset, we manually initialized the arrays with the data points. However, in real life, you may come across large datasets where the loading of the data in this way may not be possible. For example, you may have your entire dataset in .CSV file. In such a case, you may load your dataset as follows:

```
data=loadtxt('dataset.csv', delimiter=',')
Input_Features = dataset[:,0:3]
Input_Classes = dataset[:,3]
```

Note that the first line loads the dataset from the "dataset.csv" file. We have specified "," as a delimiter as the data in CSV is in comma-separated format. The dataset is loaded in a variable named "data." Then we extracted the first three features and loaded them in the variable named "Input_Features." The next statement then loads the fourth column (which is actually the column at index number 3) in the variable named "Input_Classes."

Table 8.5 Input dataset

Object	Feature-1	Feature-2	Feature-3	Class
X1	1	1	1	0
X2	2	2	2	1
X3	2	1	2	0
X4	1	2	2	1

8.17.2   Construction of the Model

Next, we will have to construct the structure of our model. The model will comprise an input layer, three hidden layers, and one output layer as shown in Fig. 8.28:

The input comprises three features for each data point, where each feature is taken as input by one neuron in the first hidden layer. The first hidden layer thus has three neurons. It is important to note that the selection of the hidden layer and the neurons in each hidden layer is critical. We have discussed this in detail in previous topics. As there are three features, so we opted to have three neurons in the first hidden layer. As the model is sequential, so the output of the first hidden layer is taken as input by the second hidden layer which has two neurons. The number of neurons in the second hidden layer is selected such that the number of neurons decreases by one in each hidden layer as we proceed forward. The output of the second hidden layer is then given as input to the third hidden layer which has only one neuron. The third hidden layer then produces the output for the output layer.

To implement the above structure, you will have to write the following lines of code:

6) Model_Input = Input(shape=(3,))
7) FirstLayer = Dense(3,activation='relu')(Model_Input)
8) SencondLayer = Dense(2,activation='relu')(FirstLayer)
9) Model_Output = Dense(1, activation='sigmoid')(SencondLayer)
10) model = keras.Model(Model_Input,Model_Output)

As mentioned earlier, we will use a sequential model for our first deep learning project. For this purpose, we have used the "Sequential()" function. This function returns a sequential model object. The role of the sequential model is to arrange the layers in sequence.

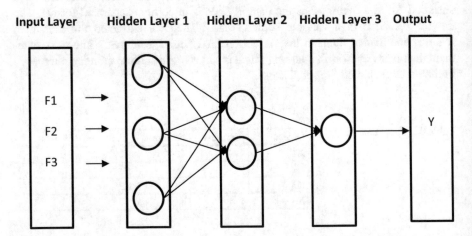

Input Layer   **Hidden Layer 1**   **Hidden Layer 2**   **Hidden Layer 3**   **Output**

Fig. 8.28   The model structure

Next, in statement 6, we have defined the input of the model. Note that for the model, we will have to provide the structure of the data. For this purpose, we have used the "Input()" function. In this function, we have specified "shape=(3,))" as input which means that there will be three inputs for the model.

Statement 7 defines the first hidden layer of the model. You should have already noted that all the layers are fully connected which means that the output of each neuron in one layer is given as input to all the neurons in the next layer. The fully connected layer is called a dense layer. For this purpose, we have used the "Dense()" function as follows:

FirstLayer = Dense(3,activation='relu')(Model_Input)

The first argument specifies the number of neurons in the layer which is equal to 3. The argument "activation='relu'" specifies the activation function used for this layer. We have used the "ReLU" activation function for this layer. There are a number of other functions that can be used. You can use any activation function based on your requirements. Note that we have specified the "(Model_Input)" in the statement as well in order to implement that the input specified by the "Model_Input" variable will be given to this dense layer. Finally, this layer is saved in the variable named "FirstLayer."

After implementing the first layer, statement 8 implements the second hidden layer. In the second hidden layer, we have placed two neurons. The activation function used in this layer is "ReLU" as well. The output of the first hidden layer is specified as the input for this hidden layer. Finally, this hidden layer is saved in the variable named "SecondLayer."

Statement 9 implements the third hidden layer. In this hidden layer, we have defined only a single neuron. The output of which becomes the output of the model. Note that in this hidden layer, we have used the "sigmoid" activation function instead of the "ReLU." The reason behind this is that we are performing the binary classification and the sigmoid function can be the best option for this purpose. The output greater than 0.5 can be considered "1" and the output less than 0.5 can be considered "0."

Statement10 constructs a sequential model and returns its object. As mentioned earlier that every deep learning algorithm requires data, so here we have provided the input and the structure of the model. So far, we have constructed the model by providing the structure, number of hidden layers, neurons per hidden layer, activation functions on each hidden layer, and the input of the model for training.

8.17.3   Configure the Model

Our next task is to actually build the model and configure it for training. For this purpose we will use the "compile()" function as shown in statement 11 below:

11) model.compile(loss=mean_squared_error, optimizer='adam', metrics=['accuracy'])

However, for the configuration of the model, we will need to provide some essential information that is necessary for the resulting model to work, for example:

- Loss function
- Optimization technique (optimizer)
- Metrics used to measure the performance of the model
- Weighted metrics (if any)
- Number of steps per execution

Besides this there is some other information as well; however, here we will provide only three things, the loss function, the optimizer, and the performance measuring metric. We have used the "mean squared error" for the loss function. There are many optimizers; however, we have used "adam" which is based on the stochastic gradient descent method. Finally, the metric used for measuring the performance of the model is "accuracy." There are many other metrics available, e.g., Binary-Crossentropy, SparseCategoricalCrossentropy, Poisson, etc. Keras provides a separate class for each of these metrics.

8.17.4 Training and Testing the Model

Our model is ready to run. We will now train the model before testing it for the test data. For training purposes, we will have to provide both the input features of each data point and its input class. We will use the "fit()" function of our model object for this purpose. Statement 12 below shows how this function is used.

12) model.fit(Input_Features, Input_Classes, epochs=100, batch_size=4)

We have provided the "Input_Features" and the "Input_Classes" variables that contain our training data as the first two arguments. Then we have provided the values of epochs and the batch_size. Epochs mean how much the forward and backward propagation will occur for the model to get accurately trained.

"batch_size" on the other hand specifies how many data points will be considered for calculating the error. As our dataset consists of only four data points, so we have specified the "batch_size" as 4. Note that in this way it becomes exactly equal to the stochastic gradient descent optimizer. For large datasets, you may specify some other number, thus making it the mini_batch algorithm.

After executing statement 12, the model gets trained and now is ready to be used for the unknown (test) data. As the purpose of this tutorial is just to give you an understanding of how things work, we will use the same dataset (the one used for training) as our test dataset and will see how the model performs. We will use the "evaluate()" function for this purpose as shown in statement 13 below:

13) loss, accuracy = model.evaluate(Input_Features, Input_Classes)

The "evaluate()" function takes the input features of the test data and the class of each data point for measuring the performance of the algorithm. Here we have specified both "Input_Fetures" and "Input_Classes" variables. The function returns the loss value and the metric specified in the training process. So, in our case, it will return the loss and the accuracy.

For our executions, the accuracy was 75%. You may have different accuracy as different weights may be assigned during each run. Furthermore, the accuracy may increase or decrease due to the number of epochs and the batch size.

Note that this was just a very simple example. The Keras provides you with the complete set of functions and models that you can use as per your requirements.

8.17.5 Alternate Method

Note that you can develop your model using different other ways, e.g., consider the following code:

1) model=Sequential()
2) FirstLayer = Dense(3,Input(shape=(3,)),activation='relu')
3) SencondLayer = Dense(2,activation='relu')
4) Model_Output = Dense(1, activation='sigmoid')
5) model.add(FirstLayer)
6) model.add(SencondLayer)
7) model.add(Model_Output)
8) model.compile(loss='mean_squared_error',   optimizer='adam',   metrics=['accuracy'])
9) model.fit(Input_Features, Input_Classes, epochs=100, batch_size=4)
10) loss, accuracy = model.evaluate(Input_Features, Input_Classes)

We have constructed a sequential model and then we have constructed three dense layers. The arguments and the provided sigmoid function are the same. Note that the input shape is provided in the first dense layer only.

Once all the layers are defined, we will have to connect these layers in the sequence we want. For this purpose, we have used the "add()" function of the model. This function adds the provided layer on top of the previous layer stack. In the provided code, the layers will be added in the same sequence in which the function is called. The function, however, may return different error values as well based on the arguments provided.

- TypeError: This error is flagged if the provided layer instance does not contain the valid layer or is empty.
- ValueError: This error is flagged if the shape of the input data is not provided.

Summary
In this chapter, we provided an in-depth discussion on deep neural networks. Starting from the basic component of the deep neural network, i.e., a perceptron, we moved up to the advance deep neural network architectures including convolutional neural networks, recurrent neural networks, Elman neural networks, etc.

We discussed the guidelines that should be considered to develop the deep neural networks along with the mathematical concepts essential to step into the domain of deep learning.

Further Reading

- *The Principles of Deep Learning Theory: An Effective Theory Approach to Understanding Neural Networks* New Edition by Daniel A. Roberts and Sho Yaida, Cambridge University Press, 2022. This book elaborates the basic theoretical concepts through an easy-to-understand approach. The book provides the maximum knowledge of deep learning to students and professionals for the use of their respective domains. The book describes the concepts and results using calculus, algebra, and informal probability theory which can be helpful for the students to enhance their analytical skills. The book is an excellent choice for researchers who want to strengthen their basic foundation in the domain of deep learning. Mathematical explanations of the related concepts help understand the results and hyperparameters. Overall, the book is an excellent resource for anyone interested in the domain of deep learning.
- *Deep Learning* (Adaptive Computation and Machine Learning series) Kindle Edition by Ian Goodfellow, Yoshua Bengio, and Aaron Courville, MIT Press, 2016. This textbook provides details of deep learning along with different deep learning techniques and the mathematical background required as a prerequisite. The book is written by three well-known authors in the field of deep learning. Deep learning is considered a type of machine learning that makes computers learn through an understanding of the world with the help of training data. As computers learn over time, there remains no need for human intervention to operate the machines. This is because the computers already have the knowledge that is required to operate computers, so it means that the computer can operate itself. Researchers, teachers, professionals, and students get many concepts that are covered in this book including machine learning, numerical computations, information theory, probability theory, linear algebra, etc.
- *Deep Learning: Fundamentals of Deep Learning for Beginners* (Artificial Intelligence Book 3) Kindle Edition by Rudolph Russell, CreateSpace Independent Publishing Platform, 2018. The book provides a great help in the field of deep learning and its related areas. It helps explain the deep learning concepts from a mathematical point of view. Starting from the basic definition, the book keeps on adding concepts. One of the best features of the book is that it provides review questions at the end of each chapter for reinforcement of the knowledge given in the chapter. As discussed earlier, the book discusses essential mathematics required for the understanding of the deep learning models, so the reader does not require any other mathematical resource for the understanding of the domain. The book provides deep learning concepts with step-by-step explanations and illustrations for a better understanding of the reader. The book is a valuable resource for students, researchers, and practitioners at the same time as it fulfills the need of each in the mentioned domain.

- *Deep Learning for Beginners: A beginner's guide to getting up and running with deep learning from scratch using Python* by Dr. Pablo Rivas, Packt Publishing, 2020. As the amount of content on the internet is increasing day by day, it is difficult to find valuable content on a specific topic. The book is intended for all those who are beginners and are interested in the domain of deep learning. The book starts from scratch and keeps on building the concepts as we proceed further in the book. The book assumes that the reader has a bit of mathematical and programming knowledge required to get started. Starting with the basic concepts of deep learning, it guides you through different Python frameworks. The book also explains preparing the data for machine learning algorithms including preprocessing and cleansing of the data. After that various deep learning algorithms are discussed in detail. The algorithms include CNNs, RNN, AEs, VAEs, GANs, etc. One of the best features of the book is that it explains the concepts with the help of practical examples. The book also provides a question-answer section at the end of each chapter to reinforce the learning of the reader.

Exercises

Exercise 8.1: Explain the working of the following deep learning applications:

- Self-driving cars
- Fake news detection
- Natural language processing
- Virtual assistants
- Healthcare
- Visual face recognition
- Image processing
- Fraud detection
- Pixel restoration
- Automatic game playing
- Photo descriptions
- Marketing campaigns
- Voice recognition

Exercise 8.2: What is the use of neurons in neural networks?

Exercise 8.3: What is the activation function and how it is used in neural networks?

Exercise 8.4: Consider the sample data given below. Create an artificial neural network and train it using this data.

Feature 1	Feature 2	Feature 3	Feature 4	Feature 5	Target
1	27	701	45	0	Class 1
0	89	753	55	1	Class 2
1	15	919	26	1	Class 2
1	76	846	34	1	Class 2
1	84	472	20	1	Class 1
1	98	268	62	0	Class 3

(continued)

Feature 1	Feature 2	Feature 3	Feature 4	Feature 5	Target
1	100	647	36	1	Class 1
0	43	960	29	0	Class 1
0	30	383	64	1	Class 1
1	40	278	80	0	Class 1
0	72	673	9	0	Class 2
0	92	600	25	0	Class 2
0	32	296	80	0	Class 3

Exercise 8.5: Why is rectified linear unit (ReLU) considered the best activation function?

Exercise 8.6: Consider the pseudo code of the generic neural network algorithm given below and implement it using any programming language and any dataset.

Step 1: Get input.
Step 2: Modify the weights.
Step 3: Calculate mean error.
Step 4: Repeat steps 2 to 3 until the mean error is minimum.
Step 5: Exit.

Exercise 8.7: Consider the following concepts related to the construction of an artificial neural network. Explain each of them in detail.

- Mini batching
- Dropping the neurons
- Selection of number of layers
- Selection of number of neurons

Exercise 8.8: Consider the data given below and calculate one-dimensional convolution.

$$V = \begin{bmatrix} 3 \\ 4 \\ 5 \end{bmatrix} \quad W = \begin{bmatrix} 2 \\ 3 \\ -1 \end{bmatrix}$$

Exercise 8.9: Consider the data given below and calculate two-dimensional convolution.

$$X = \begin{bmatrix} 2 & 2 & 3 \\ 3 & 4 & 1 \\ 1 & 3 & 2 \end{bmatrix} \quad W = \begin{bmatrix} 3 & 1 & -1 \\ 2 & 2 & 1 \\ 4 & 1 & -1 \end{bmatrix}$$

Exercise 8.10: Consider the following variations of recurrent neural networks and explain each of them with a graphical example.

- One-to-One RNN
- One-to-Many RNN
- Many-to-One RNN
- Many-to-Many RNN

Exercise 8.11: Consider the following components of Long Short-Term Memory Model architecture, and explain their use in detail.

- Forget gate
- Input gate
- Output gate

Exercise 8.12: What are the examples of Sequence to Sequence recurrent neural networks; explain each of them.

Exercise 8.13: What are the limitations of recurrent neural networks?

Exercise 8.14: What are the differences between convolutional neural networks and recurrent neural networks? Give at least five differences.

Exercise 8.15: Consider the output function given below in the context of the Elman neural network.

$$Y = f2 \left(w2f1 \left(w1X + b1 \right) + b2 \right)$$

Provide the detail of each variable in front of it in the following table:

Variable	Details
b1	
b2	
w1	
w2	

Exercise 8.16: Discuss some real-life applications of the Jordan neural networks.

Exercise 8.17: Consider the four parameters of an autoencoder and explain their working.

- Code size
- Number of layers
- Number of nodes per layer
- Loss function

Exercise 8.18: Following are the concepts related to training of a deep neural network; explain each of them.

- Training through data
- Selection of appropriate activation functions
- Selection of number of hidden neurons and layers
- Weight initialization
- Selection of learning rate value
- Mini-batch vs stochastic learning
- Shuffling of training examples
- Selection of number of epochs/training iterations
- Selection of appropriate activation function

Exercise 8.19: Following are the challenges of deep neural networks; explain the detail of every challenge.

- Requirement of large amount of training data
- Overfitting in neural networks
- Hyperparameter optimization
- High-performance hardware requirements
- Black box nature of neural networks

Exercise 8.20: Consider the sample data given below and use the following steps in the given sequence and build a practical neural network.

- Use a single input layer.
- Use a single output layer.
- Use "ReLU" as an activation function for hidden layer(s)
- Use the "sigmoid" activation function for the output layer.
- Layers should be fully connected.

Object	Feature 1	Feature 2	Feature 3	Feature 4	Class
X1	1	2	1	1	1
X2	1	1	1	2	1
X3	2	1	1	1	1
X4	2	2	1	2	1
X5	2	2	1	1	0
X6	1	2	1	2	1
X7	1	2	1	1	1
X8	2	1	1	2	1
X9	1	1	1	2	0
X10	1	2	1	2	0
X11	2	1	1	1	1
X12	1	1	1	2	0
X13	2	2	1	2	0

Chapter 9
Frequent Pattern Mining

9.1 Basic Concepts

In this section, we will discuss the market basket example and see some basic concepts that are used for association rule mining.

9.1.1 Market Basket Example

A market basket means a set of items purchased together by a customer in a transaction. As we purchase a lot of products in a single visit, all of them are considered to be a single transaction. Market basket analysis is the analysis of these items. This analysis comprises various techniques that are used to determine the associations between various items and customer behavior. The main intention of the market basket analysis is to determine which items are purchased by the customer together. The concept is that if a customer purchased a certain group of items together a certain number of times, it is likely that the customer will purchase the same again. For example, normally it is known that the customers who purchase milk are most likely to purchase bread as well. Such behaviors in purchases can help the companies to increase their sale as they can devise new marketing strategies.

Normally the retail stores are interested on which products are purchased by the customers, and they ignore the information on when they buy these products.

Consider the figure given below (Fig. 9.1):

Now if we consider each basket a transaction, then the following will be the transaction table as shown in Table 9.1:

By carefully analyzing we can see that three times the customer purchased the milk and bread together. So the container racks for these products can be placed together, or some selling discount can be collectively provided on both of these products to increase the sale.

© The Author(s), under exclusive license to Springer Nature Switzerland AG 2023
U. Qamar, M. S. Raza, *Data Science Concepts and Techniques with Applications*,
https://doi.org/10.1007/978-3-031-17442-1_9

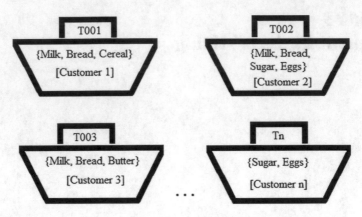

Fig. 9.1 Market baskets

Table 9.1 Transactions data

TransactionID	Products
T001	Milk, Bread, Cereal
T002	Milk, Bread, Sugar, Eggs
T003	Milk, Bread, Butter
. . .	. . .
Tn	Sugar, Eggs

Table 9.2 Sample transactional data

TID	Items
1	{Bread, Milk}
2	{Bread, Diapers, Beer, Eggs}
3	{Milk, Diapers, Beer, Cola}
4	{Bread, Milk, Diapers, Beer}
5	{Bread, Milk, Diapers, Cola}

It should be noted that market basket analysis is one of the initial techniques that tend to identify the association rules and mine frequent patterns from itemsets. Now we will discuss some of the basic concepts related to market basket analysis. For this purpose, we will consider the following dataset given in Table 9.2:

9.1.2 Association Rule

An association rule provides the information between two items. For example, the following rule can be derived from the dataset given in Table 9.2:

{Bread}→{Milk}

This rule states that if a person will purchase milk, he/she will also purchase bread. The rule is defined in the form of IF-ELSE structure. So:

If (Milk is purchased) → (Bread will be purchased)

9.1.3 Lift

Lift measure is used to find out whether an association exists or not, and if the association exists, it is a positive or negative association. If the association exists, then we calculate the support value.

$$\text{Lift} = \frac{P(A \cap B)}{P(A) * P(B)} \qquad (9.1)$$

9.1.4 Binary Representation

Instead of the conventional representation, the market basket data can be represented in the form of binary values as well. For example, Table 9.3 shows the binary representation of the dataset given in Table 9.3.

Each transaction is represented in the form of binary values, where the value of "1" means that the item is present and the value "0" means that the item is not present.

9.2 Association Rules

Association rule mining was introduced in 1996 and serves as a significant data mining model. Initially, it found its application in market basket analysis for the detection of customers' purchases of different items and their specific patterns. If/then logic provides the backbone for the rules of association which identify the relationship patterns among information repositories and apparently free-relational databases. Association rule mining (ARM) is used to identify the patterns in the non-numeric and categorical datasets. This methodology aims at noticing relationships, recurring patterns, and/or correlations from datasets present in various repositories and different kinds of databases. Association rules find their root in the "IF and THEN" relation, i.e., an antecedent followed by a consequence. Support and confidence provide the basis for the measurement of the effectiveness of the association rule. The recurrence of the "IF and THEN" relationship is directly related to the support of the items in which the relationships occur. Various studies have

Table 9.3 Binary representation of data in Table 9.2

TID	Bread	Milk	Diapers	Beer	Eggs	Cola
1	1	1	0	0	0	0
2	1	0	1	1	1	0
3	0	1	1	1	0	1
4	1	1	1	1	0	0
5	1	1	1	0	0	1

already been conducted to discuss the different aspects of ARM which are applied to observe notable relationships and patterns in different items in a dataset.

Association rules have been the focus of the database community and have received a lot of attention in the recent past. These are also among the most extensively researched domain of data mining. Agrawal introduced them for the first time in 1993 when it was tried to seek relationships in a database among a set of items. These certain relationships (functional dependencies) are not based on inherent characteristics of the data, but instead on the co-occurrence of the data items. Now we will see the use of association rules below.

Example A grocery store creates advertising supplements for its weekly specials which are published in the local newspaper. Whenever an item like pasta is decided to put on sale, staff evaluates a couple of other items which customers usually buy along with the pasta. From the sales data, they discover that customers purchase sauces with pasta 40% of the time and cheese 30% of the time. These associations help them decide to place sauces and cheese on special displays near our main sale item, pasta. These items, i.e., sauces and cheese, are not put on sale. This takes advantage of the group purchase frequency of these items by the customers and aims at higher sales volume.

Example mentions two association rules. In the first case of the purchase of pasta, sauces are purchased with it 40% of the time. In the second case, it states that cheese is also purchased in 30% of all the transactions containing pasta. Retail stores regularly use these association rules for the analysis of market basket transactions. Management uses the association rules discovered in this way to increase the effectiveness while at the same time reducing the cost of advertisement, inventory management, better marketing, and stock location. Association rules also find application in other sectors and scenarios, for example, failure prediction in telecommunication networks through identification and linkage of events occurring immediately before an actual failure.

9.2.1 Support and Confidence

Let $A = \{A_1, A_2, \ldots, A_m\}$ be a set that has m distinct characteristics, also known as literals. Suppose we have a database called B, in this database, each record (tuple) C carries a certain identifier which is unique for each record, and each tuple has an itemset in a way that $C \subseteq A$. The association rule is actually an implication of the $Y \Rightarrow Z$ form, here $Y, Z \subset A$ are sets of items known as itemsets, and $Y \cap Z = \varphi$. In this relation, Y is the antecedent, and Z is called the consequent.

Fig. 9.2 Support – circle area

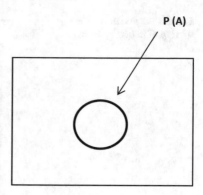

P (A)

Two main measures of ARM, support (s) and confidence (α), are defined below.

Support

Support is the number of transactions containing item A from the transaction list, i.e., dataset. It is simply considered as the probability of occurring of an item A. Support can be represented as given in Fig. 9.2.

Definition: In an association rule, the support (s) is the percent ratio of records containing Y $\cup$ Z to the total records present in the database.

Support is calculated by the given formula:

$$\text{Support (A)} = \frac{\text{Number of transactions in which A appears}}{\text{Total number of transactions}} \tag{9.2}$$

So, a statement narrating that there is 5% support for a rule means that among total records, 5% contain Y $\cup$ Z. For an association rule, support provides the measure of statistical significance. Grocery store workers would be less concerned about the relation between pasta and rice if they are purchased combined in less than 5% of total purchases.

In most of the cases for an association rule, we desire higher support values, but there can be cases otherwise. Let's study the case of failures in telecommunications switching nodes. When we use the association rule here to predict the events which occur before the failure, even though the support or recurrence of these combination events would be fairly low, still this association rule relationship will be important for us.

Confidence

Confidence is the number of transactions containing item A as well as item B. The greater overlap shows that we have greater confidence that people will also buy item B with item A. Figure 9.3 represents confidence in the overlapping area.

Definition: The ratio of records containing A $\cup$ B to the records containing A for the provided records is called confidence which is denoted by (α).

Fig. 9.3 Confidence – the
overlap of items A and B

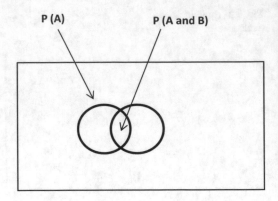

Confidence is calculated by the given formula:

$$\text{Confidence } (A \to B) = \frac{\text{Support } (A \cup B)}{\text{Support } (A)} \tag{9.3}$$

So, for an association rule, having a 90% confidence value means that of all the records in the dataset, 90% contain both A and B. The confidence value here is a measure of the degree of correlation between A and B in our dataset. Confidence measures the strength of a rule. Usually, we require a large confidence value for the association rule. If a product is purchased with pasta very rarely or an event has a rare occurrence before a failure, these relationships don't find any notable value for the management.

9.2.2   Null Transactions

A null transaction is a special type of transaction that does not contain any itemset under consideration. Typically, the number of null transactions can be more than the number of individual purchases because, for example, if we are considering the milk and coffee together, there may be a number of transactions where these items are not purchased together.

So, it is highly desirable to have a measure that has a value that is independent of the number of null transactions. A measure is null-invariant if its value is free from the influence of null transactions. Null-invariance is an important property for measuring association patterns in a large transaction database; however, if the transactional database is sparse due to the presence of "null transactions," mining performance degrades drastically.

9.2.3 Negative Association Rules

Formally, association rules are defined as follows:

Let $I = \{i1, i2, \ldots im\}$ be a set of m items. Let D be a set of transactions, where each transaction T is a set of items such that $T \subseteq I$. Each transaction has a unique identifier TID. A transaction T is said to contain X, a set of items in I, if $X \subseteq T$ where X is called an itemset.

An association rule is an implication of the form $X \rightarrow Y$ where $X, Y \subseteq I$, and $X \cap Y = \varnothing$. The rule $X \rightarrow Y$ contains the support s in the transaction set D if s% of the transactions in D contains $X \cup Y$. Similarly, for the definition of confidence, we can say that the rule $X \rightarrow Y$ holds in the D with confidence value c if c% of transactions in D that contain X also contain Y.

A negative item on the other hand is defined as $\neg i_k$. $\neg i_k$ which represents the absence of an item from the transaction. The positive item on the other hand is the item that is present in a transaction. The support of $\neg i_k$: $s(\neg i_k) = 1 - s(i_k)$.

A negative association rule is again an implication $X \Rightarrow Y$, where $X \subseteq I$, $Y \subseteq I$, and $X \cap Y = \varnothing$ and X and/or Y contain at least one negative item.

9.2.4 Multilevel Association Rules

Multidimensional databases are inherently sparse, which makes it hard to mine strong association rules at low levels of abstraction. Rules mined at high abstraction levels usually represent common sense knowledge and may prove to be helpful in good judgments. However, common sense knowledge subjectively varies between individuals and societies. On the other hand, association rules mined at multiple levels of abstraction help us discover interesting relations between data elements and find the hidden information in/between levels of abstraction.

9.2.5 Approaches to Multilevel Association Rule Mining

The support-confidence framework can be used to efficiently mine multilevel association rules using concept hierarchies. Some of the approaches based on the support-confidence framework are as follows:

9.2.5.1 Uniform Minimum Support

In a uniform minimum support approach, each level of abstraction has the same threshold support used for rules mining. Though setting the same threshold for every abstraction level greatly simplifies the search procedure, it has its shortcomings. For example, setting threshold support too high could result in some interesting associations to be remained undiscovered at lower levels of abstraction. Setting the threshold too low, on the other hand, might generate many uninteresting associations at higher levels of abstraction.

9.2.5.2   Reduced Minimum Support

Unlike uniform minimum support, this approach works by assigning each level of abstraction its minimum support threshold. As we go deeper down the abstraction levels, the threshold values become smaller and smaller.

There are several search techniques based on a reduced minimum support approach for mining multilevel associations, such as:

- *Level by level independent:* Each item is traversed in a breadth-first search manner, without utilizing any prior knowledge about frequent itemsets.
- *Level cross-filtering by a single item:* An item at a given level is traversed only if its parent node (at one level higher) is frequent.
- *Level cross-filtering by k-itemset:* An itemset at a given level is traversed only if its parent itemset (at one level higher) is frequent.

9.2.5.3   Checking for Redundant Multilevel Association Rules

In data mining, concept hierarchies are useful since they facilitate knowledge to be discovered at several levels of abstraction. For example, concept hierarchies are used in mining multilevel association rules. However, there is a downside to it, since concept hierarchies involve mining rules at several levels of abstraction which can result in redundant rules being generated. It happens because of the ancestor relationships that exist in concept hierarchies.

9.2.6   Multidimensional Association Rules

When an association rule has two or more predicates, it becomes a multidimensional association rule. Predicates in an association rule are known as dimensions.

Multidimensional association rules can either have repetitive predicates or non-repetitive attributes. Association rules with repetitive predicates are called hybrid-dimension association rules, while non-repetitive attributes are known as inter-dimensional association rules.

Qualities/attributes in the multidimensional association rule are either categorical or quantitative. Categorical attributes are also referred to as nominal attributes. They take on finite values and do not exhibit any meaningful order. Quantitative attributes are numeric and incorporate meaningful order among values. If relational data stores contain only categorical attributes, it is possible to use mining algorithms effectively for mining association rules. However, in the case of quantitative attributes, mining algorithms cannot be effectively applied because their domains are very large. There are three different approaches for mining multidimensional association rules based on how they handle quantitative attributes:

- In the first approach, predetermined static discretization of quantitative attributes is performed before mining. Discretized attributes are then treated as categorical attributes.
- In the second approach, the distribution of data determines the discretization process. Quantitative attributes are dynamically clustered into "bins." These bins are progressively discretized further during the mining process.
- In the third approach, dynamic discretization of quantitative attributes is performed, keeping the distance between data points into account. Ranges are defined based on the semantic meaning of the data.

The static discretization process is driven by predetermined concept hierarchies. Numeric values are discretized into range values. Categorical attributes, if needed, may also be discretized to more general forms. Discretization is performed before mining.

The mining process can be optimized if we store the resultant data in a relational database. By slightly modifying the mining algorithm, we would then be able to generate all frequent predicate sets. Other techniques such as hashing, partitioning, and sampling can also be used to increase the performance.

Besides relational databases, the discretized data can also be stored in data cubes. Data cubes are inherently multidimensional structures, making them a suitable choice for multidimensional association rule mining.

9.2.7   Mining Quantitative Association Rules

Quantitative association rule mining works on the principle of an association rule clustering system. Pairs of quantitative attributes are stored in a 2-D array; the array is then searched for groups of points to create the association rules.

This process constitutes two steps:

The first step is called "binning." In order to keep the size of a 2-D array manageable, a technique called binning is used. Binning reduces the cardinality of the continuous and discrete data. Binning groups the related values together in bins based on the distribution of data. The range of a quantitative attribute is partitioned into intervals known as bins.

Bins can be formed in three different ways:

- *Equiwidth binning:* Data is partitioned into equal intervals.
- *Equidepth binning:* Each interval has the same number of values.
- *Homogeneity-based binning:* Each interval has uniformly distributed values.

The second step is about "finding frequent predicate sets." Rule generation algorithm is used to search the 2-D array for frequent predicate sets. These predicate sets are then used to generate association rules.

9.2.8   Mining Multidimensional Association Rules Using Static Discretization of Quantitative Attributes

The static discretization process is driven by predetermined concept hierarchies. Numeric values are discretized into range values. Categorical attributes, if needed, may also be discretized to more general forms. Discretization is performed prior to mining.

The mining process can be optimized if we store the resultant data in a relational database. By slightly modifying the mining algorithm, we would then be able to generate all frequent predicate sets. Other techniques such as hashing, partitioning, and sampling can also be used to increase the performance.

Besides relational databases, the discretized data can also be stored in data cubes. Data cubes are inherently multidimensional structures, making them a suitable choice for multidimensional association rule mining.

9.3   Frequent Pattern Mining Methods

So far, we have discussed the concepts of the frequent patterns and how they are realized in terms of the association rules in dataset. Now we will discuss few algorithms that can be used for mining such patterns from the data stored in data repositories.

9.3.1   Apriori Algorithm

In this section, we will discuss the Apriori algorithm in detail. The algorithm was proposed by Rakesh Agarwal, and it is the first associative algorithm that was proposed, and various further tasks in association, classification, associative classification, etc. have used this algorithm in their techniques.

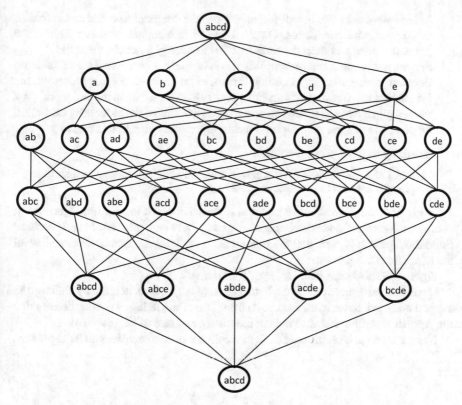

Fig. 9.4 An itemset lattice of five items

Association rule mining comprises three steps:

1. *Frequent Itemset Generation:* In this step, we find all frequent itemsets that have support ≥ predetermined minimum support count. In this step we find the interesting associations and correlations between the itemsets both in transactional and relational data. So, we can say that frequent mining shows which items normally can appear in a certain transaction. Normally the process takes several iterations to complete. We may need to traverse the dataset multiple times in order to identify the items that qualify. The process comprises two steps:

   (a) Pruning
   (b) Joining

A lattice structure can be used to represent a list of all possible itemsets. Figure 9.4 shows the lattice structure for I={a,b,c,d,e}. The figure below shows the pseudo code of the frequent itemset generation for the Apriori algorithm.

2. *Rule Generation:* We list all the rules from the frequent itemsets and calculate the support and confidence. On the basis of this support and confidence, we prune the rules and drop the ones that fail to qualify a certain threshold.
3. *Frequent Itemset Generation:* We traverse the entire database and find the frequent itemsets having a certain threshold on support. As this process scans the entire database, so this is an expensive task. As the data in the real world may exceed gigabytes and terabytes, so we need to optimize the process by excluding the itemsets that are not helpful. Apriori algorithm helps in this regard.

According to the Apriori algorithm:

Any subset of a frequent itemset must also be frequent. In other words, no superset of an infrequent itemset must be generated or tested.

In Fig. 9.4, we have shown a pictorial representation of the Apriori algorithm. It comprises k-itemset nodes and the relationship between subsets of the k-itemset. From the figure, it is clear that the bottom node contains the transaction that has all the items from the itemset.

Figure 9.5 shows the item lattice of four items.

It can be seen that it is difficult to generate a frequent itemset by finding the support and confidence of each combination. Therefore, in Fig. 9.6 it can be seen that the Apriori algorithm helps to reduce the number of sets to be generated.

If an itemset {a, b} is infrequent, then we do not need to consider all its super sets.

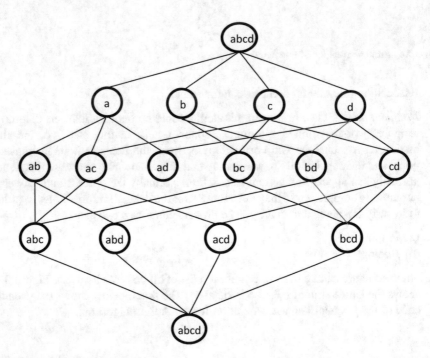

Fig. 9.5 Item lattice of four items

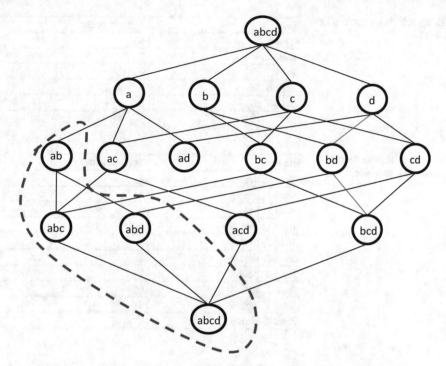

Fig. 9.6 Infrequent itemsets

Table 9.4 Transactions and the itemsets

TransactionID	Items
T001	Item-2, Item-4, Item-5
T002	Item-2, Item-4
T003	Item-2, Item-3
T004	Item-1, Item-2, Item-4
T005	Item-1, Item-3
T006	Item-2, Item-3
T007	Item-1, Item-3
T008	Item-1, Item-2, Item-3, Item-5
T009	Item-1, Item-2, Item-3

Now we will see it in the form of transactional data. Consider the items given in Table 9.4. TID represents the TransactionID and Items represent the items in a single transaction.

Step: 1

- From the transaction table, create a table containing the items along with the support count of each item in the transactional dataset as shown in the table.
- Next, we will compare the support count of each item with the minimum support count. In our case, the minimum support count is 2. So, all the items with a minimum support count of 2 will be selected. Table 9.5 shows the items and their support count.

Table 9.5 Support count of each item

Items	Support count
Item-1	6
Item-2	7
Item-3	6
Item-4	2
Item-5	2

Table 9.6 All the itemsets comprising two items

Items	Support count
Item-1, Item-2	4
Item-1, Item-3	4
Item-1, Item-4	1
Item-1, Item-5	2
Item-2, Item-3	4
Item-2, Item-4	2
Item-2, Item-5	2
Item-3, Item-4	0
Item-3, Item-5	1
Item-4, Item-5	0

Table 9.7 Itemsets having a minimum support count of 2

Items	Support count
Item-1, Item-2	4
Item-1, Item-3	4
Item-1, Item-5	2
Item-2, Item-3	4
Item-2, Item-4	2
Item-2, Item-5	2

Step: 2

- In this step, we join the items to generate other larger itemsets.
- Verify if the subset of an itemset is frequent or not. For example, the subsets of {Item-1,Item-2} are { Item-1} and { Item-2} and are frequent.
- Now search the dataset again and find the support of each itemset.

- Again check the support count. If the support count is less than the minimum support count, then ignore those items. Table 9.6 shows all the itemsets formed from the items in Table 9.5, whereas Table 9.7 shows only those itemsets that have a minimum support count of 2.

Table 9.8 Itemsets with a
minimum support count of 2

Items	Support count
Item-1, Item-2, Item-3	2
Item-1, Item-2, Item-5	2

Step: 3

- Now we have received another dataset. Again we will join the itemsets and find the support of each item. After joining items in figure six, we get the following subset:

  {Item-1, Item-2, Item-3}
  {Item-1, Item-2, Item-4}
  {Item-1, Item-2, Item-5}
  {Item-1, Item-3, Item-5}
  {Item-2, Item-3, Item-4}
  {Item-2, Item-4, Item-5}
  {Item-2, Item-3, Item-5}

- Now again we will check the subsets of these items and find if they are frequent or not. All the items that are not frequent will be removed again. Consider the itemset: {Item-1, Item-2, Item-3}; its subsets, i.e., {Item-1, Item-2}, {Item-1, Item-3}, and {Item-2, Item-3}, are frequent; however, on the other hand, the subsets of the itemset {Item-2, Item-3, Item-4} are not frequent and hence will be removed. The same will be done for all itemsets.
- Once we have removed all the non-frequent itemsets, find the support count of the remaining itemset by searching in the dataset.
- Again compare the support count of each itemset with the minimum support count, and those less than the minimum support count will be removed. Now we get the itemsets given in Table 9.8:

Step: 4

- The same process will be followed again, i.e., cross join the itemsets to form bigger itemsets and then find the support count.
- Once the itemsets are joined, find the itemset having frequent subsets. In our case, the itemsct {Itcm-1, Item-2, Item-3, Item-5} is formed after joining. If you notice, one of its subsets {Item-1, Item-3, Item-5} is not frequent. Therefore, there is no itemset left anymore.
- We will stop here because no more frequent subset is found.

Here the first step is completed. In the next step, we find all the frequent subsets and also find how strong are the association rules. The strength of each rule can be calculated by using the formula given below.

$$\text{Rule} : X \rightarrow Y$$

$$\text{Support} = \frac{\text{frq}(X, Y)}{N} \tag{9.4}$$

$$\text{Confidence} = \frac{\text{frq}(X, Y)}{\text{frq}(X)} \tag{9.5}$$

$$\text{Support} = \frac{\text{Support}}{\text{supp}(X) * \text{Supp}(Y)} \tag{9.6}$$

We take the example of a frequent itemset, e.g., {I1, I2, I3}. Now we will generate all the rules using these items. Table 9.9 shows all the rules that can be generated with these items along with the confidence level of each rule.

Here if we consider the threshold of 50%, then the first three rules will be said to have strong association as shown in Table 9.10.

Now we will see another example of frequent itemset mining. Consider the following database given in Table 9.11 containing the "TransactionID" and the corresponding Items in each transaction:

We have to find all the frequent itemset(s) with minimum support of four (i.e., $\text{Min}_{\text{Sup}} = 4$) using the Apriori method. Table 9.12 shows all the items along with the support count of each item. Table 9.13, on the other hand, shows only those items having minimum support count of 4.

Now we will form the itemsets from these items. The itemsets and their support count are given in Table 9.14.

Table 9.9 All possible rules generated from the itemset

Items	Confidence
{Item-1,I2}→{Item-3}	50%
{Item-1,I3}→{Item-2}	50%
{Item-2, I3}→{Item-1}	50%
{Item-1}→ {Item-2, Item-3}	33.33%
{Item-2}→{Item-1, Item-3}	28.57%
{Item-3}→{Item-1, Item-2}	33.33%

Table 9.10 Rules having minimum support of 50%

Items	Confidence
{Item-1, Item-2}→{I3}	50%
{Item-1, Item-3}→{I2}	50%
{Item-2, Item-3}→{I1}	50%

Table 9.11 Transactional database

TransactionID	Items
001	A, B, C, D, H, I
002	B, C, D, H, I
003	B, C, E, F
004	C, G, H, I
005	A, B, C, H, I
006	A, B, C, D, E, H, I

Table 9.12 Support count of each item

Items	Support
A	3
B	5
C	6
D	3
E	2
F	1
G	1
H	5
I	5

Table 9.13 Items with a minimum support count of 4

Items	Support
B	5
C	6
H	5
I	5

Table 9.14 Itemsets and their support count

Itemset	Support
BC	5
BH	4
BI	4
CH	5
CI	5
HI	5

Table 9.15 Itemsets and their support count

Items	Support
BCH	4
BCI	4
BHI	4
CHI	5

Table 9.16 Itemsets and their support count

Items	Support
BCHI	4

Each itemset has a support count greater than or equal to 4, so all these itemsets will be selected to form the larger itemsets. Table 9.15 shows the itemsets formed from these items and their support count.

Again each itemset has support count greater than or equal to 4, so all these itemsets will be selected to form the larger itemsets. Table 9.16 shows the itemsets formed from these items and their support count.

As no further itemsets can be formed, so we will stop here. So, {B,C,H,I} is our frequent itemset with minimum support of four (04).

9.3.2   Fuzzy Apriori Algorithm

Fuzzy Apriori algorithm is an effort to combine the fuzzy set theory with association rule mining. Fuzzy logic, in contrast to the crisp values, provides a smooth and clean conversion from one value to another. In literature the fuzzy set theory has been successfully combined with the association rule mining to generate results. The numeric data is first converted to categorical data using fuzzy logic. The transformed data can then be mined through association rule mining. In this section we will present the fuzzy Apriori algorithm. Figure 9.7 shows the flow diagram of fuzzy Apriori algorithm.

First we receive the data and convert it into a database. Then on this data, we apply the process of fuzzification. On the fuzzified data, the Apriori algorithm is applied to generate required fuzzy rules. Once the fuzzy rules are generated, rules are evaluated to ensure their correctness. Once the learning process is complete, the new rules are stored in knowledge base.

The algorithm starts with the fuzzification process. For example, consider the following dataset given in Table 9.17:

Taking the data given in Table 9.17, the fuzzification process can be applied to get the categorical data. Categorical data represents the categories instead of the numeric values. The height, e.g., can be categorized into three categories like short, medium, and tall. Similarly all the other attributes can be categorized in the same way.

Table 9.18 shows the final data after fuzzification:

Once the data is categorized, the modified Apriori algorithm can be applied to mine the data for extraction of the frequent patterns in the form of the rules.

9.3.3   FP-Tree

FP-Tree is used in FP-Growth algorithm. FP-Tree is a compact representation of transactional data. It represents each transaction in the form of a tree path. The paths may overlap with each other since different transactions may have many common items. The more the paths overlap, the more the compression in data can be achieved.

If there are small numbers of transactions, i.e., the FP-Tree is small and can fit into the memory, we can directly extract data from FP-Tree; else we will have to store it on disk and extract data in iterations.

Figure 9.8 shows the record of ten transactions:

Each node in the tree is labeled with the item number along with the number of customers who purchased that item. Note that the tree starts with the null node.

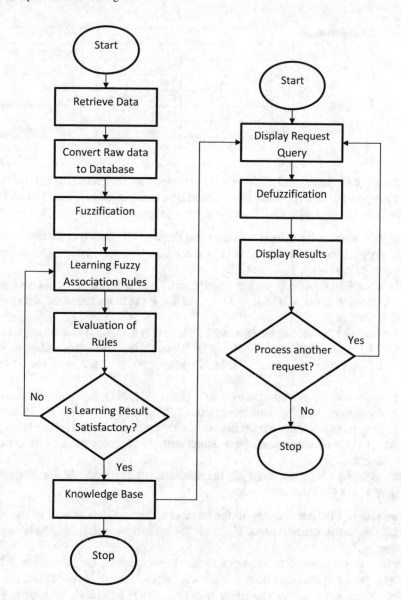

Fig. 9.7 Fuzzy Apriori algorithm

Table 9.17 Sample dataset

Person	Height	Size of house
1	6.8	240
2	4.6	350
3	7.1	100
4	5.2	50
5	4.5	25
6	5.9	200

Table 9.18 Sample dataset

Person	Height	Size of house
1	Tall	Large, medium
2	Short	Large
3	Tall	Small
4	Short, medium	Small
5	Short	Small
6	Medium, tall	Small, medium

We start with the support count of each item and then the items are sorted in order of their support count. The items from a certain threshold are discarded. To construct an FP-Tree, we start with a root node and then proceed as follows:

1. For each item, in the dataset we determine the support count and the items below the support count are discarded. For the dataset shown in Fig. 9.8, the most frequent items are J, K, L, and M.
2. The algorithm then reads the first transaction, i.e., {I,J}, and the nodes for I and J are created with the path Null → I → J. As this is the first transaction, every node has a frequency of "1."
3. Now we read the second transaction. In our case it is {J,K,L} and we map it onto the FP-Tree as Null → J → K → L. It should be noted that although both transactions share the item "J," it is still labeled as "1" because of the different prefix.
4. Then we read the third transaction, i.e., {I,K,L,M}. Now this transaction shares the same item, i.e., "I," and the prefix of "I" is also the same. So instead of mapping transaction three from the null node, we start it from the item "I" on the path of the first transaction. Now since item "I" appeared twice, its count is incremented.
5. The process continues, until all the transactions are read. At the bottom of Fig. 9.8, the final tree is shown.

The size of FP-Tree depends on the number of transactions as well as the items overlapping in these transactions. The more the items are overlapping, the lesser will the size.

In best case where all transactions share common items, the FP-Tree will have only one path. On the other hand when each transaction has unique items, the tree will have maximum paths. Generally speaking, the requirements of storage for a FP-Tree are more because it needs to store item names as labels along with count of each item. Furthermore, pointers to nodes and the overlapping paths are also stored.

It should also be noted that the size of FP-Tree also depends on the order of the support of items. The main idea behind storing the items in decreasing order of their support actually means that the maximum transactions have the same prefix for frequent items and thus the size could be reduced. If the order is reversed, the tree will be different. Figure 9.9 shows the FP-Tree with reverse order of items in terms of support count.

TID	Items
T001	{I, J}
T002	{J, K, L}
T003	{I, K, L, M}
T004	{I, L, M}
T005	{I, J, K}
T006	{I, J, K, L}
T007	{I}
T008	{I, J, K}
T009	{I, J, L}
T0010	{J, K, M}

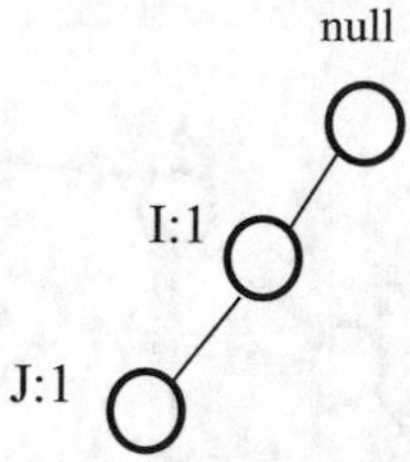

(i) After reading first transaction

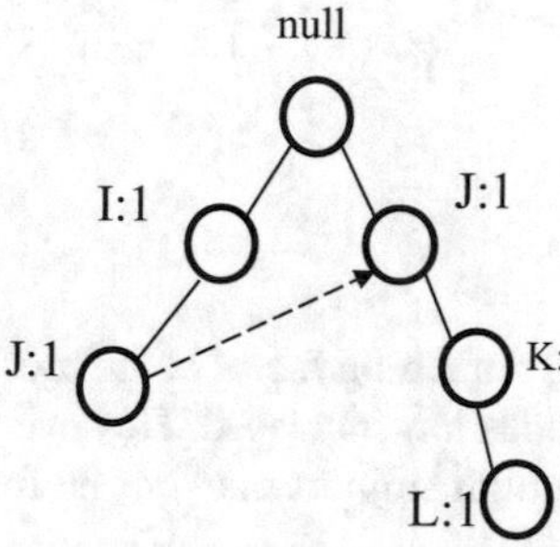

(ii) After reading Second transaction

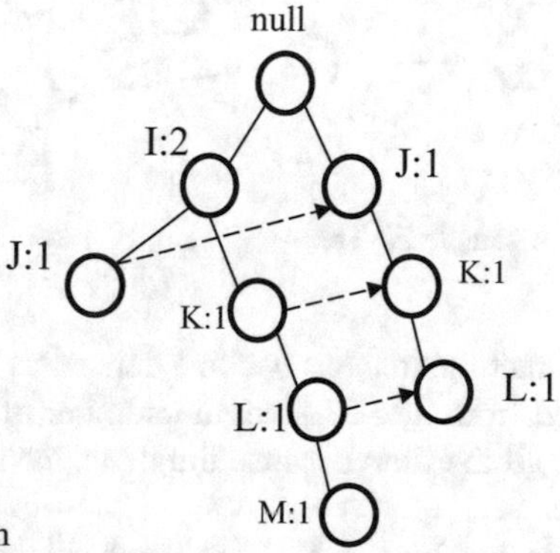

(iii) After reading third transaction

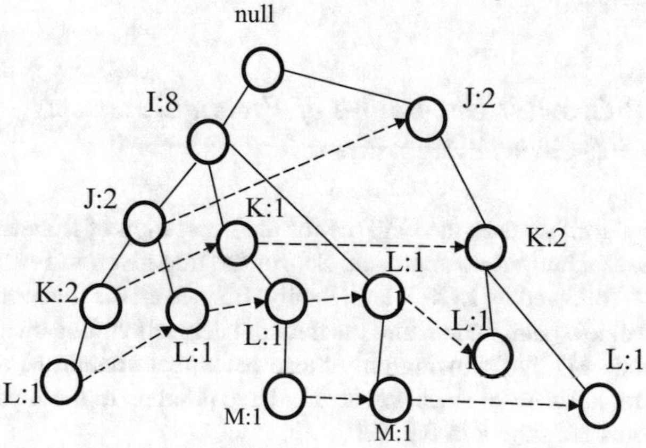

(iv) After reading tenth transaction

Fig. 9.8 A sample FP-Tree

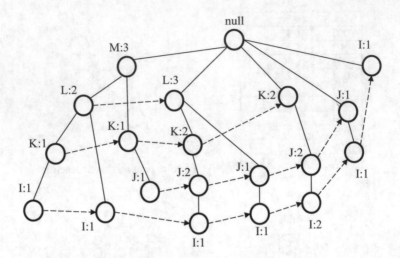

Fig. 9.9 A sample FP-Tree

Note that the tree has become denser, as the branching factor at the root node has increased from 2 to 5. Also, the number of nodes has increased. However, it is not the case all the time because the items having high support may not be frequent all the time.

The dashed lines in FP-Tree represent the pointers connecting the nodes that have same items. These pointers help access the individual items in the tree.

9.3.4   FP-Growth: Generation of Frequent Itemsets in FP-Growth Algorithm

FP-Growth algorithm uses the FP-Tree for the generation of frequent itemsets. The algorithm uses a bottom-up approach. So, for the tree given in Fig. 9.8, it will start with item M, followed by L, K, J, and, finally, I. Since all the transactions in FP-Tree are mapped onto a path, we can find the frequent itemsets ending on a particular node (e.g., the node M) by following only those paths that contain M at the end. This becomes a rapid traversal since we already have pointers to these paths.

These paths are shown in Fig. 9.10.

We can use divide-and-conquer strategy to find the frequent itemsets that end at a specific node. Suppose we want to find all the frequent items ending at M.

For this purpose first, we need to check that item {M} itself is frequent or not. If frequent, we can consider the subproblem of finding itemsets ending in LM followed by KM, JM, and IM. When we have found all the frequent items that end at M, we can merge them. Finally, we can merge all the frequent items ending at M, L, K, J, and I.

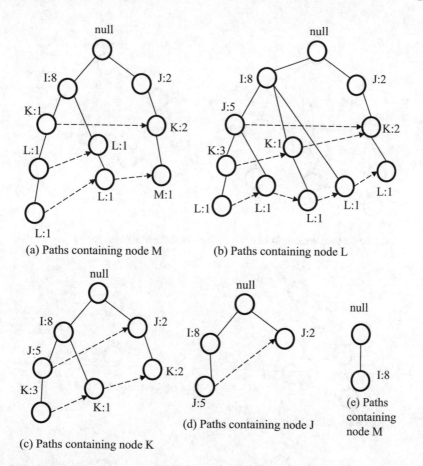

(a) Paths containing node M   (b) Paths containing node L

(c) Paths containing node K

(d) Paths containing node J

(e) Paths containing node M

Fig. 9.10 Paths extracted by using frequent itemsets ending at a particular node

For an exact example of how to solve the subproblem of finding the paths ending at M, consider the following steps:

1. First we will find all those paths that have the node M. We call these paths the prefix paths. Figure 9.11 shows these paths.
2. From these prefix paths, we try to find the support count for node M by counting the support count of all nodes with which M is associated. If the minimum support count is 2, {M} will become the frequent itemset as the support count of M is 3.
3. Since M is frequent, the algorithm will try to find the frequent itemsets that end in M. However, before this, the prefix path will be converted into the conditional FP-Tree. The structure of the conditional FP-Tree is almost similar to the FP-Tree. The only difference is that the conditional FP-Tree is used to find the paths that end at a specific suffix.

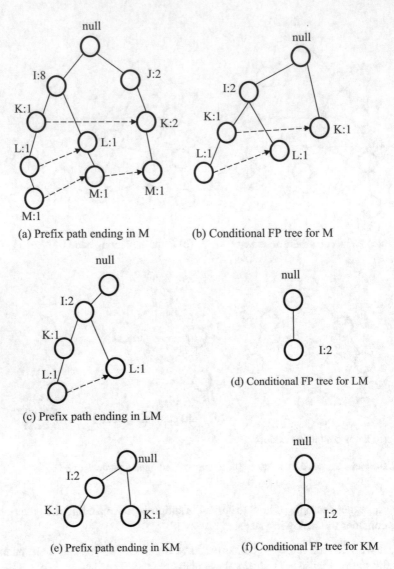

Fig. 9.11 Applying FP-Growth to find subsets ending at M

We can obtain a conditional FP-Tree by the following steps:

- First, the support counts along the prefix paths are updated. This is because some counts may contain transactions that do not include item M.
- Truncate prefix paths by removing node M. This is safe to remove because we have already updated the support count to reflect only those transactions that include node M. So, the frequent itemsets that end in LM, KM, JM, and IM do not require the information about M now.
- Once the support count along the prefix path is updated, it is a chance that some items may not remain more frequent

4. In FP-Growth we use a conditional FP-Tree for the node M in order to find the frequent itemsets that end in LM, KM, and IM. For finding those frequent itemsets that end in LM, we get the prefix paths for node L from the conditional FP-Tree developed for node M. Similarly the same process is followed for other items.

9.3.5 ECLAT

ECLAT stands for equivalence class clustering and bottom-up lattice transversal algorithm. This is another algorithm to find the frequent patterns in the transactional database. Apparently, it is one of the best methods for association rule mining. So, we can accurately use the ECLAT algorithm to find the frequent patterns. The pictorial representation of the algorithm is given in Fig. 9.12.

In order to discover the frequent patterns, ECLAT uses the depth-first search approach. Apriori algorithm on the other hand uses the breadth-first search method. In contrast to the Apriori approach, this algorithm represents the data in a vertical manner. The Apriori algorithm on the other hand represents data in a horizontal manner. This use of vertical patterns makes it faster than the Apriori. Hence we can use ECLAT for datasets having large amounts of transactions.

We use the TransactionID sets, called tid-sets, to find the support value. We try to avoid the generation of subscts that do not exist in the prefix three. First, we use all single items along with their tid-sets. The process is used recursively and in each iteration items in the tid-sets are combined. The process continues until no further item in the tid-sets can be combined.

Fig. 9.12 Flowchart of ECLAT

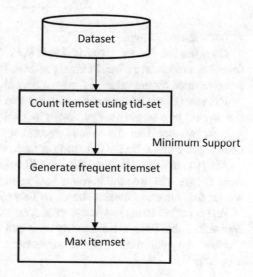

Table 9.19 Sample dataset

TID	A	B	C	D	E	F
T001	1	1	0	0	0	1
T002	0	0	1	1	1	1
T003	0	0	1	1	1	0
T004	1	0	1	0	0	0

Table 9.20 Dataset after the first recursive call

TID	Transactions
A	{T001, T004}
B	{T001}
C	{T002, T003, T004}
D	{T002, T003}
E	{T002, T003}
F	{T001, T002}

Table 9.21 Dataset after the second recursive call

TID	Transactions
{A, B}	{T001}
{A, C}	{T004}
{A, F}	{T001}
{B, F}	{T001}
{C, D}	{T002, T003}
{C, E}	{T002, T003}
{C, F}	{T002}
{D, E}	{T002, T003}
{D, E}	{T002}
{E, F}	{T002}

Let's see an example:

Consider the dataset given in Table 9.19. The data is represented in the binary format as discussed earlier. In binary format, the presence of an item in a transaction is represented by the value "1," whereas its absence is represented by the value "0."

Note that in the first recursive call, the min support is 2. Firstly we will try to find the support of a single item as shown in Table 9.20.

Now we consider the second recursive call; in the second call, we have to combine two items. So the following dataset, shown in Table 9.21, will be generated:

Now in the third recursive call, we will consider three items at a time. For all such itemsets, we will drop the itemsets having support minimum than the threshold. So we get the following dataset shown in Table 9.22:

As no further items can be combined, so we will stop here. Now from the dataset shown in Table 9.22, the following rules can be derived as shown in Table 9.23:

Now we will discuss some advantages and disadvantages of the ECLAT algorithm.

Table 9.22 Dataset after the third recursive call

TID	Transactions
{C, D, E}	{T001}

Table 9.23 Rule set

Items bought	Recommendation
C	D
C	E
D	E

Advantages

1. It uses the depth-first search approach, so it consumes less memory as compared to the Apriori algorithm.
2. As it does not generate more number of itemsets, so it is faster than the Apriori algorithm.
3. In ECLAT we don't need to traverse the dataset again and again which adds to the efficiency of the algorithm.
4. As compared to the Apriori algorithm, it can be used for small- and medium-sized datasets. Apriori on the other hand can be used for large datasets.
5. It only traverses the currently generated dataset and not the original dataset as is done by the Apriori algorithm.

Disadvantages

1. The intermediate sets on the other hand can consume memory.
2. Overall, we can say that this method can efficiently be used to implement frequent pattern mining.

9.4 Pattern Evaluation Methods

Once the patterns are identified from the data, it is important to evaluate those patterns and find out their quality in terms of the information provided by these patterns. For example, we may know that {bread} → {Eggs} rule may strongly hold in a store database; however, is this rule interesting, or does it provide any new information which could be used by the business to find some new opportunity? So, evaluating the rules just on the basis of their support level or confidence may not be sufficient. We need some other measures that could enable us to find our rules of interest.

This is not a simple task and especially in the case when the database is large and hundreds of rules are extracted. We need to filter out the rules that are of particular interest. Two types of criteria can be used. The first one comprises the subjective measures that are user and business-dependent.

The second criterion comprises the objective measures that use statistical methods to find out whether the pattern is interesting or not.

Subjective measures as the name implies are subjective in nature and depend on the input provided by the user or some custom criteria. Such information may be provided by the user through some interface. Secondly, we may also specify the templates and the patterns that fulfill those templates are extracted. Finally, the user may also specify some domain-specific interesting measure, e.g., all the patterns that contain a certain out-of-season product or a pattern that resulted in maximum profit.

9.5 Applications of Frequent Pattern Mining

Here we will review the significant applications of frequent pattern mining algorithms in various fields. Since its inception in 1993, it has attracted many data mining researchers, and various tree-based and pattern-growth-based algorithms have been designed for different areas.

However, it is one of the methods where output patterns may be large in size as compared to the input, e.g., from a database of a few thousand records, this method may identify frequent patterns which are larger in size. Therefore, it arises a need to compact the size of the result for generating a concise summary of data. In fact, these output patterns serve as an input for other subsequent tasks, i.e., classification, prediction, decision-making, etc. So, this method cannot be used as an individual method, i.e., outlier detection, classification, etc., where output itself can be examined for decision-making. Frequent pattern mining requires post-processing to formulate concise results. By keeping this aspect in view, in this chapter, we will study particularly those applications of frequent pattern mining where its importance can be best described.

Frequent pattern mining can be applied to various real-life applications which may belong to different data domains. In the context of different data problems, different versions of frequent pattern mining methods can be used, e.g., the pattern mining algorithm used for biological data is different from the mining algorithm used for the multimedia domain. Few such applications with different data domains are briefly described below:

- *Behavior Analysis:* Behavior analysis is the key application of frequent pattern mining. By analyzing customer behavior, critical business decisions can be made.
- *Intermediate Step for Other Data Mining Techniques:* Pattern mining acts as an intermediate step for other data mining techniques, i.e., clustering, classification, etc. Patterns identified can be used as a subspace cluster or to design a classifier. Moreover, patterns mined can also be used to identify outliers of the data.
- *Indexing and Retrieval:* As we have discussed earlier, the output generated by the frequent pattern mining process serves as an input to other processes; therefore, to design the techniques for indexing and data retrieval from the market basket, frequent pattern mining can be applied, but post-processing is required to make results concise for indexing.

- *Web Mining Tasks:* Pattern mining algorithms can be used to traverse web logs for identifying useful patterns. These patterns help in designing and organizing websites.
- *Bug Detection:* To identify bugs in software programs, pattern mining can be applied which discovers relevant patterns from the data.
- *Event Detection and Other Temporal Applications:* As frequent pattern mining identifies useful patterns from underlying data, it can be used in different applications for event detection. Other than that, it can be applied to perform periodic pattern mining. In these applications, pattern mining works as subroutines.
- *Spatial and Spatiotemporal Analysis:* In a data that includes spatial and nonspatial attributes of each object, frequent pattern mining can be applied to identify the relationship between the attributes. Similarly, spatiotemporal data can be studied using pattern mining algorithms.
- *Image and Multimedia Data Mining:* Image data is also spatial data that has both spatial and nonspatial attributes. Therefore, pattern mining algorithms can be applied to image data to identify the important characteristics of the image.
- *Chemical and Biological Applications:* In chemical and biological applications, frequent patterns can be applied to study the usage patterns in the graphs and structured data, e.g., RNA analysis, prediction of chemical compounds, etc.

Now we will discuss some of the abovementioned applications in detail.

9.5.1   Frequent Patterns for Consumer Analysis

Consumer analysis is the most important process that is required to predict future trends and is considered a significant application of frequent pattern mining. By considering the shopping cart and sequence of the items bought together, the pattern mining algorithms can identify the patterns of buying behavior of the customers. For example, if a pattern identifies that a customer is buying bread along with eggs, this information can be used to place these things together on the shelves. Moreover, different promotion deals can also be made for a particular group of customers based on such patterns.

Temporal mining also can be used to identify target customers by studying past buying behavior, e.g., if a customer bought a mobile phone in recent days, he may require other accessories related to the phone afterward. So, by mining customer behavior, more refined targeted customers can also be identified related to particular products.

However, frequent pattern mining generates large output, so it requires further post-processing to utilize useful patterns in decision-making. So, a more useful application of pattern mining is the construction of a classifier that makes rules based on identified patterns. By temporal mining, rules can be further refined, and sequential mining allows to make good recommendations. Some tasks which subsequently use patterns will be discussed in the aforementioned sections.

9.5.2 *Frequent Patterns for Clustering*

As mentioned earlier, frequent pattern mining can be related to other data mining techniques, e.g., by constructing the relationship between clustering technique and pattern mining, frequent items of different clusters will be identified. Subspace clustering is an extension of clustering which is used to find clusters within high-dimensional data. By definition, subspace clustering is closely related to the frequent pattern mining technique.

The CLIQUE is a subspace clustering algorithm specially designed to handle multidimensional datasets. It divides the original data into grids and uses these grids to find the relevant patterns or dense patterns. Clusters are then generated from all dense patterns using the Apriori approach. It is fastest compared to K-Means and DBSCAN algorithms, but if the size of the grid is not suitable, then correct clusters will not be found. In ENCLUS, entropy measure is used to quantify clusters in each subspace instead of using density measure. Many techniques have been designed for biclustering, projected clustering, and other types of clustering of multidimensional data using pattern mining methods.

A common problem with the clustering of biological data is that the data is presented in the form of discrete values. The sequences are too long to cluster by using a clustering algorithm only. So, the use of pattern mining can be very helpful for the clustering of such data. In CLUSEQ, clusters are designed using frequent patterns from biological data.

9.5.3 *Frequent Patterns for Classification*

Classification methods also can be closely linked with frequent pattern mining methods, especially rule-based classifiers. By analyzing the training data, rules are defined that the new instances need to fulfill in order to be part of a certain class.

As classification rules are closely related to association rule mining, therefore, pattern mining can be used to construct relevant patterns which must be discriminative. One of the most popular techniques is Classification Based on Associations (CBA) which uses pattern mining to design rule-based classifiers. Another technique is CMAR which is based on the FP-Growth method for association rule mining.

In frequent pattern mining, some methods particularly focus on discriminative pattern mining. These methods can be used for software bug detection and designing decision trees. It can be applied to different data domains, e.g., discriminative subtrees and subgraphs can be used for the classification of structural data, spatio-temporal data (in order to detect anomalies), spatial data (for medical image classification), etc.

It should be noted that we can determine frequent patterns either by using whole data or by using only class-specific data. By considering class-specific patterns, we can classify data more accurately. If a large number of rules satisfy the support and confidence constraint, then pick a small subset of rules which effectively represent training data. The method to prioritize rules may be different for different applications.

9.5.4 Frequent Patterns for Outlier Analysis

To detect outliers from the data, frequent pattern mining techniques are widely used. These methods can be applied from binary to high-dimensional data. For multidimensional data, subspace clustering is an ideal technique, but for sparse datasets, it is not possible to detect outliers using subspace methods. Some researchers have suggested to build a relationship of transactions in dense subspaces instead of sparse subspaces.

It is a typical observation that frequent patterns are less likely to occur in outliers. This is because the presence of outliers can decrease the frequency of a pattern that was otherwise frequent in case of the absence of that outlier.

9.5.5 Web Mining Applications

Web mining applications deal with web logs and the contents of websites. These applications study linkage patterns and contents and determine useful patterns. Different methods for mining web log data have been proposed to identify important patterns. Web log mining can be categorized into two types: web log mining and linkage structure mining.

Log files consist of information, e.g., username, IP address, timestamp, access request, number of bytes transferred, result status, URL that is referred, user agent, etc. Web servers maintain these web log files. We can analyze these logs to find frequent access patterns which can then be used to design a website. Furthermore, outlier analysis can be done, based on web mining, by identifying infrequent patterns. For distance learning, web log mining can be used to evaluate learners' behavior. Initially, web log mining used sequential and frequent pattern analysis to find useful web log patterns. In this method, the concept of forward reference and backward reference was used. Forward reference means when the user clicks on a link to go to a new page, whereas the backward reference means that the user returns back to the previous page. By using both forward referencing and backward referencing, the frequent patterns related to the behavior of the user can be identified.

There are a number of methods to find the traversal paths. These methods work with the assumption that irrelevant patterns may be linked to some useful patterns. Using the web graph structure, relevant patterns can be mined by ignoring irrelevant arbitrary vertices. For this purpose, a modified form of an Apriori algorithm is used. It should be noted that data preparation is an important task in web log mining, due to the presence of noise in the data. Therefore, it is complex to find relevant paths from the log data.

An important application based on association rule mining is termed as personalization in which user attitude mining is used to group the relevant items of interest and make future recommendations. Rule-based methods are considered effective methods for making recommendations.

As discussed above, graphs can be used to find the forward or backward references. Graphs can also be used to discover the user behavior instead of identifying the patterns. Identifying the user behavior helps us find other people in the same community. This can be a special application in case of the social media.

9.5.6   Temporal Applications

Temporal refers to time and temporal applications are those where data is represented either as continuous time series or discrete sequences. The data in continuous time series can be converted into discrete sequences using techniques, e.g., SAX. This method converts average values in a particular time window into discrete values. Now, frequent pattern mining algorithms can be applied directly to discrete values instead of continuous time series values of the data. Methods applied to biological data for mining patterns can also be used for temporal data mining with slight changes. Biological data also has a small number of very long rows like temporal data. Therefore, row-enumeration techniques for biological data can also be applied to temporal data.

In temporal applications, event detection can also be done by applying pattern mining algorithms. Event detection is closely related to classification problems where labels are linked with timestamps. Therefore, classification rules in terms of time can be used to detect events. In these cases, data consists of base feature events and class events. Pattern mining algorithms can be used to predict particular sequences in the data. We construct the temporal classification rules for event detection which classify feature events on the left-hand side and class events on the right-hand side. In the context of the temporal application, these rules have a numerical lag value associated with each event. Frequent pattern mining algorithms for event detection can also be used for intrusion detection. Similarly, this model can be successfully used to predict events other than intrusion scenarios. To predict the behavior of customers in telecommunications, sequence mining is a very useful technique.

The characteristics from temporal and sequential data are mined to understand the important trends of data for classification, e.g., pattern-based rule mining can be used to categorize the sequences in different classes. Sequences are decomposed into multi-granularity representation and on the basis of this representation, rules can be formulated. This is different from the event detection problem; here labels are linked with time series instead of a specific event.

Periodic pattern mining is another temporal problem where patterns related to seasonality in time series sequences are identified. These patterns are useful for clinical diagnosis, e.g., ECG, etc. Another method called MARBLES is used to find association rules between episodes. Such methods can discover the relation between different episodes.

9.5.7 Spatial and Spatiotemporal Applications

Social sensing is an emerging phenomenon in today's world that is based on mobile sensing technology. Mobile companies are continuously tracing mobile phones using GPS technology. Furthermore, such data then can be used to build trajectories on the basis of which the clusters and frequent patterns can be generated. Here data is collected across space and time; such data is referred to as spatiotemporal data. To identify clusters from spatiotemporal data, frequent pattern mining methods have been widely used. One such method is swarm where snapshots of data are used to form clusters on the basis of the discrete value of clusters. Objects which have the same discrete value in different snapshots belong to swarm. In this way, frequently occurring discrete values help us discover sequences of the objects which move together.

There are many methods that can be used to perform trajectory classification based on frequent pattern mining techniques. For example, to predict a rare class, pattern mining can be used to detect important patterns belonging to that rare class. Therefore, we are detecting anomalies in the data using rare class identification. Similarly, outliers can be detected by using pattern-based classification methods.

The methods based on spatial association rules are also an interesting application. Spatial association rules define the relationship between objects having spatial and nonspatial attributes. Any object which is spatial may have spatial and nonspatial attributes, e.g., the temperature of the sea surface has values of spatial location along with temperature value, and the spatial association rule can be defined as "The temperature in the southern region is normally high." However, spatial association rules are quite complex and may cause challenges due to computational inefficiencies. To handle these challenges, the concept of rough spatial approximations has been proposed. Spatial association rules are first determined for low resolution and the rules that qualify are then explored for finer spatial resolution. In another method, spatially close objects are discovered by the pattern mining process.

Images and multimedia data may also be considered as spatial data as locations are mostly linked with pixels of an image; similarly, color may also be related to different locations. All methods mentioned above can also be used for this kind of spatial data, e.g., classification methods for spatial data may be enhanced to accommodate images or multimedia data by considering pixels as properties of objects.

9.5.8 Software Bug Detection

If we consider software programs as structured objects, graphs can be used to represent them. These graphs can be termed as software behavior graphs. In the case of a traditional software program, a graph may have a traditional shape. In a typical software program, different bugs may present, e.g., memory breaches, core dumps, logical errors, etc. Logical errors are difficult to identify during execution. These errors affect the final result of the program but do not generate any warning while running. Errors in the structure of the program may cause logical bugs which are normally non-crashing bugs.

Many techniques based on classifiers have been designed to locate logical errors using a graph of the software program. Frequent graph mining and many classification methods can be employed to detect errors in different regions of the program using software behavior graphs. CP-Miner is used to detect copy-paste code and then find the regions where logical errors exist. Rule-based methods can also be used to formulate rules from the data and then detect logical errors by using these rules, e.g., the PR-Miner approach. Similarly, C-Miner, a software bug detection approach, implemented frequent pattern mining to detect bugs from blocks of the program.

9.5.9 Chemical and Biological Applications

To represent biological and chemical data, the graph structure is mostly used. Similarly, chemical data can also be represented by graphs, e.g., a chemical compound can correspond to a graph, where nodes of the graph represent the atoms. Similarly, biological data, e.g., complex molecules, microarrays, protein interaction networks, etc., can also be represented as graphs. From the underlying graphs, frequent pattern discovery techniques can be used to identify valuable information from the data. The discovered information can then be used to analyze data. Patterns identified from chemical and biological data can also be termed as motifs. Sequence mining and structural mining both are important for biological data analysis.

Pattern mining can relate to classification techniques. Many methods, based on frequent pattern mining, have been designed to perform predictions. The properties like activity, toxicity, absorption, distribution, metabolism, excretion, etc., of compounds, can be identified by applying frequent pattern mining techniques to chemical data. Frequent subgraph mining algorithms can be used for this purpose. All the subgraphs are considered to be frequent subgraphs that are present in a defined minimum number of compounds. This can be called a minimum support requirement. In chemical data, we may have many different descriptors which may have different support values, but the minimum support values may not be present. So, frequent subgraph mining algorithms can be used for setting the value of minimum support. Properties identified during this process are very beneficial for the classification of chemical compounds.

In the context of biological applications, the data may be structured data or in the form of graphs. Frequent pattern mining can mine useful patterns from both types of data. Most of the biological data can be represented in the form of sequences, and many algorithms have been proposed to identify frequent patterns from such data. In biological data, each row may have many attributes, so the scalability issue also needs to be addressed as each sequence is too long. The pattern mining algorithm used for biclustering can also be applied to biological data.

To mine graphs using frequent pattern mining, a community detection mechanism can be used to extract useful patterns. For tree structures, e.g., RNA, glycans, etc., subgraph mining is used. In the case of phylogenies, many different trees are generated for a given set of genes. Then it is often advised to find the broader patterns in these trees which can be named as frequent subtree mining. These trees are called consensus trees or supertrees. For glycan databases, the subtree mining concept has been proved to be very helpful. Similarly, a rule-based mining method can be used to identify whether a glycan belongs to a certain class or not.

9.5.10   Frequent Pattern Mining in Indexing

To implement indexing algorithms, data must be in concise form and for that classification methods are used. For example, transaction data is divided into different groups based on broad patterns. These groups then prove to be helpful in query processing tasks. Frequent pattern mining techniques can also be implemented to design such concise representations, although it is related to clustering problems.

In the context of graphs, frequent mining methods prove more useful. In the gIndex method, discriminative frequent patterns are identified from the indexing structure. All those indexing structures which have similar discriminative patterns are considered to be similar. With this method, infrequent patterns may sometimes also be considered relevant depending on the size of the pattern. There are many similar methods available, e.g., SeqIndex, Graft, PIS, etc.

Summary
In this chapter, we discussed rule association mining and frequent itemset mining. Starting with the basic example of market basket analysis, we discussed different algorithms used at various stages of association rule mining and frequent itemset mining.

We then discussed different frequent itemset mining methods including the Apriori algorithm and the fuzzy Apriori method. Finally, we discussed the applications of frequent itemset mining in different domains.

Further Reading

- *Frequent Pattern Mining* by Charu C. Aggarwal and Jiawei Han, Springer & Co, 2014. The book consists of 18 chapters written by some of the most prominent researchers in the domain. Each chapter provides in-depth details of one of the related aspects of frequent pattern mining. The authors have focused on simplifying the concepts to provide ease for readers of the book. A few of the topics in the book include big data, clustering, graph patterns, etc. The book is intended both for beginners and practitioners from the industry.

- *Frequent Pattern Mining in Transactional and Structured Databases: Different aspects of itemset, sequence and subtree discovery* by Renáta Iváncsy, LAP LAMBERT Academic Publishing, 2010. This book provides details about the extraction of hidden relationships from large datasets. Frequent pattern mining is one of the important topics of data mining as it finds the patterns from the data. The book discusses the three important topics related to frequent pattern mining. These topics include frequent itemsets, frequent sequence, and frequent subtree discovery. The book discusses the detailed literature on these topics.

- *Spatiotemporal Frequent Pattern Mining from Evolving Region Trajectories* by Berkay Aydin and Rafal A. Angryk, Springer & Co, 2018. The book provides an overview of the spatiotemporal frequent pattern mining approach. Along with a detailed discussion on spatiotemporal data mining and related concepts, the authors have discussed some advanced topics including spatiotemporal trajectories, etc. The book is intended both for the students and the practitioners working in the domain of spatiotemporal data mining.

- *High-Utility Pattern Mining Theory, Algorithms and Applications* by Philippe Fournier-Viger, Jerry Chun-Wei Lin, Roger Nkambou, Bay Vo, and Vincent S. Tseng, Springer & Co, 2019. The book discusses different pattern extraction techniques from the data. The book provides details about different pattern mining algorithms along with the theory and the implementation. The book also discusses the concepts related to different data types including discrete data and sequential data. Overall the book is a good resource for anyone interested in pattern mining.

- *Frequent Pattern Mining: Techniques and Applications* by Ong Kok-Leong, VDM Verlag Dr. Müller, 2010. Modern business organizations keep each and every bit of their business in their data store in order to process further for decision-making. This requires solutions that could process the data and extract the information from it. Therefore, the analysts require a tool that could help them perform the analysis process. This book presents advanced topics in the domain of frequent pattern mining. The book provides good help to organizations that need to explore data mining for their business needs.

Exercises

Exercise 9.1: Consider the baskets of the customers in the form of a table given below. TID is representing a basket and items against each TID are items in each basket. You have to use the given data and find the lift, support, and confidence of any two items.

TID	Items
1	Milk, Butter, Eggs, Cheese, Yoghurt
2	Butter, Eggs, Chicken, Beef, Rice
3	Chicken, Beef, Rice, Milk
4	Butter, Chicken, Jam, Honey, Ketchup
5	Oil, Olives, Milk, Eggs, Honey
6	Ketchup, Oil, Olives, Eggs, Rice
7	Yogurt, Chicken, Oil, Olives, Eggs
8	Cheese, Yoghurt, Chicken, Milk, Rice
9	Cheese, Yoghurt, Beef, Rice, Butter
10	Milk, Cheese, Eggs, Chicken, Butter

Exercise 9.2: What is meant by the term null transactions? Explain it with the help of an example.

Exercise 9.3: What are the negative association rules? Explain them with the help of an example.

Exercise 9.4: What are multilevel association rules? Explain them with the help of an example.

Exercise 9.5: What is the difference between the following two concepts in terms of multilevel association rules? Explain with the help of examples.

- Uniform minimum support
- Reduced minimum Support

Exercise 9.6: Consider the following search techniques that are based on a reduced minimum support approach and explain each of them.

- Level by level independent
- Level cross-filtering by a single item
- Level cross-filtering by k-itemset

Exercise 9.7: To normalize the data, normally binning techniques are used. Consider three binning techniques given below and explain their working.

- Equiwidth binning
- Equidepth binning
- Homogeneity-based binning

Exercise 9.8: Explain the following two concepts related to the frequent itemset generation process:

- Pruning
- Joining

Exercise 9.9: Consider the transaction table given below and perform the following process on the given table graphically.

1. Find frequent items for transactional data.
2. Prune the frequent items.
3. Join the pruned items.
4. Repeat steps 2 and 3 until there are no more items.
5. Calculate support, confidence, and lift.

TID	Items
T1	I2, I8, I1, I4
T2	I5, I1, I7
T3	I4, I10, I9
T4	I7, I10, I8
T5	I2, I6, I7, I10
T6	I2, I3
T7	I9, I2, I1, I9
T8	I1, I9, I6, I4, I8
T9	I5, I8, I2, I3
T10	I10, I1, I2

Exercise 9.10: Consider the sample data given below and convert it to categorical data. For categories use low, medium, and high. You can use any range for the low, medium, and high categories.

No.	Humidity (percentage)	Temperature (degree Celsius)	Rainfall (percentage)
1	83	21	45
2	57	9	82
3	62	16	43
4	70	18	54
5	75	44	27
6	88	43	84
7	15	59	74
8	90	45	8
9	88	58	79
10	14	34	45
11	79	0	91
12	97	1	34

(continued)

No.	Humidity (percentage)	Temperature (degree Celsius)	Rainfall (percentage)
13	92	23	50
14	87	5	7
15	93	29	42
16	75	4	48

Exercise 9.11: Consider the table given below and construct a frequent pattern tree.

TID	Items
T1	I2, I8, I1, I4
T2	I5, I1, I7
T3	I4, I10, I9
T4	I7, I10, I8
T5	I2, I6, I7, I10
T6	I2, I3
T7	I9, I2, I1, I9
T8	I1, I9, I6, I4, I8
T9	I5, I8, I2, I3
T10	I10, I1, I2

Exercise 9.12: The "database" below has five transactions. What association rules can be found in this set if the minimum support is 50% and the minimum confidence is 80%?

T-I {K, A, D, B}
T-II {D, A, C, E, B}
T-III {C, A, B, E}
T-IV {B, A, D}
T-V {K, A, D, E}

Exercise 9.13: In identifying frequent itemsets in a transactional database, consider the following four-itemsets:

- {A, B, D, E}   • {A, C, D, E}
- {B, C, E, G}   • {A, B, C, D}
- {A, B, C, D}   • {A, C, D, G}
- {A, B, E, F}   • {C, D, E, F}
- {C, D, E, F}   • {B, C, D, G}
- {A, C, E, F}   • {A, C, D, G}

Which among the following five-itemsets can possibly be frequent?

- {A, B, C, D, E}
- {A, B, D, E, G}
- {A, C, E, F, G}

Exercise 9.14: Consider the following transaction database:

1: {a, d, e}   6: {a, c, d}
2: {b, c, d}   7: {b, c}
3: {a, c, e}   8: {a, c, d, e}
4: {a, c, d, e}   9: {b, c, e}
5: {a, e}   10: {a, d, e}

For the minimum support to be 3, fill in the following table:

1 Itemset	2 Itemset	3 Itemset	4 Itemset

Exercise 9.15: Consider the following dataset:

Serial no	Items
1	Bread, Butter
2	Bread, Butter, Milk
3	Bread, Butter, Eggs
4	Butter, Eggs, Milk
5	Butter, Eggs, Milk, Cola
6	Butter, Eggs, Milk, Beer
7	Butter, Eggs, Milk, Beer, Cola

For the above dataset, calculate the following:
Support and confidence for the following rules:

- If the customer buys {Milk and Butter}, he/she also buys {Beer}.
- If the customer buys {Cola}, he/she also buys {Milk and Butter}.
- If the customer buys {Bread}, he/she also buys {Eggs and Butter}.
- If the customer buys {Eggs}, he/she also buys {Cola and Butter}.
- If the customer buys {Bread and Beer}, he/she also buys {Eggs, Butter, Milk, Cola}.
- If the customer buys {Eggs and Milk}, he/she also buys {Cola and Beer}.

Exercise 9.16: For the dataset shown below, find all frequent itemsets using Apriori where the min_sup_count is 3.

Serial no	Items
1	Bread, Butter
2	Bread, Butter, Milk
3	Bread, Butter, Eggs
4	Butter, Eggs, Milk
5	Butter, Eggs, Milk, Cola
6	Butter, Eggs, Milk, Beer
7	Butter, Eggs, Milk, Beer, Cola

List all of the strong association rules (with support s and confidence c) matching the given criteria where min_sup = 60% and min_conf = 100%.

Exercise 9.17: What is the difference between the Apriori and the ECLAT algorithms?

Exercise 9.18: Discuss different steps of the ECLAT algorithm in detail with examples.

Chapter 10
Regression Analysis

10.1 Regression for Data Science

Regression is one of the important concepts used in data science applications. Here we will provide brief details of the concept.

10.1.1 Basic Concepts

Let's start with some of the basic concepts of regression.

10.1.1.1 Linear Regression (Introduction)

Linear regression is a supervised machine learning technique that is used to predict the relation between two variables. For linear regression two variables are used. One is called the dependent variable and the other is called the independent variable. Normally, linear regression is used for prediction analysis. In prediction analysis, we give the value of one (independent) variable and predict the value of the other (dependent) variable.

10.1.1.2 Example of Linear Regression

The price of a house depends on its size. If the size of the house will be big, then its price will be high and if the size of the house will be small, then the price will be low. So, we can say that there is a relationship between the size of the house and its price. We can build a model on this relationship such that if we enter the size of the house, the model will be able to predict the price of that house. For this, we can use linear

© The Author(s), under exclusive license to Springer Nature Switzerland AG 2023
U. Qamar, M. S. Raza, *Data Science Concepts and Techniques with Applications*,
https://doi.org/10.1007/978-3-031-17442-1_10

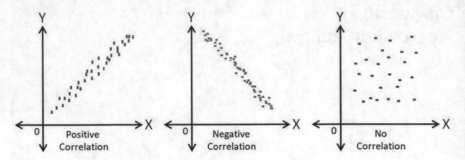

Fig. 10.1 Sample plot between two variables

regression. Linear regression uses the equation of the straight line that can be called a prediction line as well. We use a scatter plot to find the relationship between both variables. If there is some pattern between the values of both variables, we can say that both variables have a correlation, and when there is no pattern, then it will not be possible to use linear regression. For such type of data, we can use any other machine learning technique.

Figure 10.1 presents a sample scatter plot that is showing a correlation between two variables. Linear regression can be used for datasets shown in the first two scatter plots because there are a positive correlation and a negative correlation between two variables. For the third scatter plot, linear regression cannot be used.

We can see in the above diagram that the relationship between two variables can be shown through the correlation coefficient. The range for the coefficient lies between negative and positive one. The coefficient relation of each data point of X and Y shows the strength of association that is observed between X and Y.

The following is the equation of linear regression. As you can see, it is the equation of a straight line in XY-plane.

$$Y = a + bX \tag{10.1}$$

In this equation:

"X" is the independent variable along the X-axis.

"Y" is the dependent variable along the Y-axis.

"b" is the slope of the line that is calculated by the change in "y" divided by the change in "x" between two points.

"a" is the y-intercept (it is the value of y when x is zero or the point where the plotted straight line touches the Y-axis line).

As linear regression is a relationship between two variables, similarly, the equation of a straight line that uses the slope of the line also gives a relationship between two variables. So the formula of linear regression is similar to the formula of a straight line.

$$Y = a + bX$$

The values of "*a*" and "*b*" can be found using the formula given below.

$$b = \frac{m\left(\sum_{i=1}^{m} x_i y_i\right) - \left(\sum_{i=1}^{m} x_i\right)\left(\sum_{i=1}^{m} y_i\right)}{m\left(\sum_{i=1}^{m} x_i^2\right) - \left(\sum_{i=1}^{m} x_i\right)^2} \tag{10.2}$$

$$a = \frac{1}{m}\left(\sum_{i=1}^{m} y_i - b \sum_{i=1}^{m} x_i\right) \tag{10.3}$$

To fit a regression line in the X and Y plot, we can use the least-squares method. This method is used to find the best fit line by reducing the error between the regression line and actual data points. We take the difference between the actual point and prediction at the same value of X. Then we square it so that all positive and negative results do not cancel each other. Next, all the squared differences are added to find the error between actual and predicted values. If the difference between an actual and predicted point is zero, it means the error or deviation of that point is zero. The linear regression process generates a straight line which is also known as the least-square regression line.

Suppose we are given a dependent variable Y and independent variable X; then the following equation can be used to find a regression line:

$$\hat{y} = b_0 + b_1 x \tag{10.4}$$

Here:

"b_0" is a constant.
"b_1" is the regression coefficient.
"x" is the independent variable.
"$\hat{y}$" is the predicted value of the dependent variable.

Let's build a regression line using a single independent and a single dependent variable. Consider the data given in Table 10.1.

In the given data, X is an independent variable and Y is a dependent variable. To find the regression line, we will use the following formula given in Eq. 10.4. First, we will plot the given data using a scatter plot to see the strength of relation of data points as shown in Fig. 10.2.

Now, we need to find the values of b_0 and b_1 using the formulas given in Eqs. 10.2 and 10.3. Now we need to find the values of $\sum X$, $\sum Y$, $\sum XY$, and $\sum X^2$ as shown in Table 10.2.

Table 10.1 Sample data points

X	1	2	3	4	5
Y	1	3	3	2	5

Fig. 10.2 Scatter plot between variables X and Y

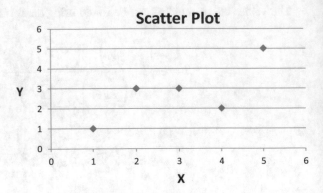

Table 10.2 $\sum X$, $\sum Y$, $\sum XY$, and $\sum X^2$ values

X	Y	XY	X²
1	1	1	1
2	3	6	2
3	3	9	9
4	2	8	16
5	5	25	25
$\sum X = 15$	$\sum Y = 14$	$\sum XY = 49$	$\sum X^2 = 53$

As we have five values, so $n = 5$. Next, we will find $\sum X = 15$, $\sum Y = 14$, $\sum XY = 49$, and $\sum X^2 = 53$.

Let's put them in the formulas to find the values of b_0 and b_1.

$$b_1 = \frac{5 * 49 - 15 * 14}{5 * 53 - (15)^2} = \frac{245 - 210}{265 - 225} = \frac{35}{40} = 0.875$$

$$b_0 = \frac{1}{5}(14 - 0.875 * 15) = \frac{1}{5}(14 - 13.125) = \frac{0.875}{5} = 0.175$$

After putting the values in the formula, the equation of the regression line becomes:

$$\hat{y} = 0.175 + 0.875x$$

Now if we put the values of X in the equation, then we will get a regression line as shown in Fig. 10.3.

In Fig. 10.4, the vertical lines represent the error of each data point. For the first data point, the actual and the predicted values have the same result, so, at this point, there is no error.

Fig. 10.3 Regression line between variables X and Y

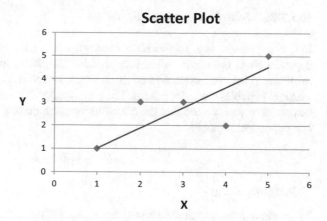

Fig. 10.4 Error representation

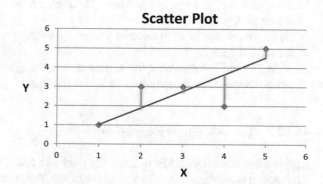

10.1.2 Multiple Regression

Multiple (multivariate) regression is another variation of linear regression that is used for the relationship of multiple independent variables with the dependent variable. Multiple regression is used for the prediction analysis where there is more than one variable. The goal of multiple regression is to model a linear relationship between single dependent and multiple independent variables.

10.1.2.1 Example of Multiple Regression

The price of a house not only depends on its size but some other factors as well, for example, the number of rooms, washrooms, floors, etc. All these factors are collectively used to predict the price of the house. Note that the single-variant linear regression gives a general idea about the price of the house for the provided size but the multivariate regression line gives a more accurate prediction.

10.1.2.2   Multiple Regression Equation

Linear regression is a relationship between two variables where one variable is dependent and the other one is independent. Similarly, multiple regression is the relationship between multiple variables, where one variable is dependent and other multiple variables are independent. The equation for the multiple regression is given below. Here you can see that the equation involves three independent variables and one dependent variable.

$$Y = a + b_1 X_1 + b_2 X_2 + b_3 X_3 \qquad (10.5)$$

In this equation:

"Y" is the dependent variable along the Y-axis of the scatter plot.

"X_1, X_2, and X_3" are the independent variables plotted along the X-axis of the scatter plot.

"a" is the y-intercept (it's the value of y when the value of x is zero or the point where the plotted straight line touches the Y-axis).

"b_1," "b_2," and "b_3" represent the slopes or unit change in the dependent variable when the values of independent variables change.

10.1.2.3   Use of Multiple Regression

Suppose that an analyst wants to know the price of a house by using different factors. The analyst is provided with the size of the house, the number of floors of the house, the number of rooms, and the number of washrooms. For this analysis, multiple linear regression will be used to get a model such that the value of the house can be predicted on the basis of provided factors. To build a multiple linear regression model, the formula that will be used is given below:

The formula of multiple linear regression (MLR) is:

$$\hat{y} = b_0 + b_1 x_1 + b_2 x_2 + b_3 x_3$$

where:

$\hat{y}$ is the predicted value

x_1 is the size of the house

x_2 is the number of rooms in the house

x_3 is the number of washrooms in the house

b_1 is the slope coefficient of the size of the house

b_2 is the slope coefficient of the number of rooms

b_3 is the slope coefficient of the number of washrooms

b_0 is the value of the y-intercept

Table 10.3 Sample dataset for multivariate regression

Y	141	156	160	180	193	201	213	216
X_1	61	63	68	71	72	73	76	79
X_2	23	26	25	21	16	15	15	12

Table 10.4 Sum and mean of Y, X_1, and X_2

	Y	X_1	X_2
	141	61	23
	156	63	26
	160	68	25
	180	71	21
	193	72	16
	201	73	15
	213	76	15
	216	79	12
Sum	1460	563	153
Mean	182.5	70.375	19.125

Table 10.5 Regression sum

	X_1^2	X_2^2	$X_1 X_2$	$X_1 Y$	$X_2 Y$
	3721	529	1403	8601	3243
	3969	676	1638	9828	4056
	4624	625	1700	10,880	4000
	5041	441	1491	12,780	3780
	5184	256	1152	13,896	3088
	5329	225	1095	14,673	3015
	5776	225	1140	16,188	3195
	6241	144	948	17,064	2592
Sum	39,885	3121	10,567	103,910	26,969
Regression sum	263.875	194.875	-200.38	1162.5	-953.5

Let's build a multivariate regression line using two independent variables and a single dependent variable. Consider the data given in Table 10.3.

In the given data, X_1 and X_2 are independent variables and Y is the dependent variable. To find the multiple regression lines, we will use the following formula given in Eq. 10.5.

Now we need to find the values used in the above formulas as shown in Table 10.4:

Now we need to calculate the regression sum. Table 10.5 shows the regression sum.

The regression sum is calculated as follows:

$$\Sigma x_1{}^2 = \Sigma X_1{}^2 - (\Sigma X_1)^2/n$$
$$= 39{,}885 - (563)^2/8$$
$$= \mathbf{263.9}$$

$$\Sigma x_2{}^2 = \Sigma X_2{}^2 - (\Sigma X_2)^2/n$$
$$= 3121 - (153)^2/8$$
$$= \mathbf{194.9}$$

$$\Sigma x_1 y = \Sigma X_1 y - (\Sigma X_1 \Sigma y)/n$$
$$= 103{,}910 - (563^* \ 1460)/8$$
$$= \mathbf{1162.5}$$

$$\Sigma x_2 y = \Sigma X_2 y - (\Sigma X_2 \Sigma y)/n$$
$$= 26{,}969 - (153^* \ 1460)/8$$
$$= -\mathbf{953.5}$$

$$\Sigma x_1 x_2 = \Sigma X_1 X_2 - (\Sigma X_1 \Sigma X_2)/n$$
$$= 9859 - (563^* \ 153)/8$$
$$= -\mathbf{200.4}$$

Now we need to find the values of b_0, b_1, and b_2 using the formulas given below.

$$b_1 = \frac{\left(\sum X_2^2\right)\left(\sum X_1 Y\right) - \left(\sum X_1 X_2\right)\left(\sum X_2 Y\right)}{\left(\sum X_1^2\right)\left(\sum X_2^2\right) - \left(\sum X_1 X_2\right)^2} \qquad (10.6)$$

$$b_2 = \frac{\left(\sum X_1^2\right)\left(\sum X_2 Y\right) - \left(\sum X_1 X_2\right)\left(\sum X_1 Y\right)}{\left(\sum X_1^2\right)\left(\sum X_2^2\right) - \left(\sum X_1 X_2\right)^2} \qquad (10.7)$$

$$b_0 = \bar{Y} - b_1 \bar{X}_1 - b_2 \bar{X}_2 \qquad (10.8)$$

Thus,

$$b_1 = \frac{(194.9)(1162.5) - (-200.4)(-953.5)}{(263.9)(194.9) - (-200.4)^2} = 3.148$$

$$b_2 = \frac{(263.9)(-953.5) - (-200.4)(1162.5)}{(263.9)(194.9) - (-200.4)^2} = -1.656$$

$$b_0 = 182.5 - (3.148)(70.375) - (-1.656)(19.125) = -7.37$$

After putting the values, the formula of the multiple regression line becomes:

$$\hat{y} = -7.37 + 3.148x_1 - 1.656x_2$$

The multiple linear regression is based on the assumption that the independent variables do not have any relation with each other. Every independent variable has a coefficient relation only with the dependent variable.

10.1.3 Polynomial Regression

Linear regression is the relationship between one independent variable and one dependent variable. Multiple linear regression is a relationship of more than one independent variables with one dependent variable. However, the independent variables are only a degree of one. If the degree of independent variables is more than one, then it is called polynomial regression. Here we have used multiple regression and converted it into polynomial regression by including more polynomial elements.

10.1.3.1 Types of Polynomial Regression

Here are a few types of polynomial regression:

- Linear Polynomial Regression: This type of regression has variable(s) with a maximum of one degree.
- Quadratic Polynomial Regression: This type of regression has variable(s) with a maximum of two degrees.
- Cubic Polynomial Regression: This type of regression has variable(s) with a maximum of three degrees.

Table 10.6 shows the comparison of mentioned three types of polynomial regressions.

Polynomial regression cannot be used for every dataset for better decision-making. There are some specific constraints that should be put on the dataset; then

Table 10.6 Comparison of polynomial regression methods

Type	Form	Degree	Example
Linear polynomial	$p(x)$: $mx + n$, $m \neq 0$	De	$2x + 1$
Quadratic polynomial	$p(x)$: $mx^2 + nx + p$, $a \neq 0$	Two-degree polynomial	$5x^2 + 2x + 5$
Cubic polynomial	$p(x)$: $mx^3 + nx^2 + px + q$, $m \neq 0$	Three-degree polynomial	$4x^3 + x^2 + 3x + 7$

polynomial regression can be used for that dataset to make better decision-making. The linear, curved, and additive relationship between the dependent and the independent variables can be used to predict the behavior of the dependent variables. Here again, the relationship should be between the dependent and the independent variables and not between the independent variables themselves.

10.1.3.2   Use Case for Polynomial Regression

We have already examined the linear regression and found there can be big gaps or errors between actual and predicted values. It is also observed that actual values have curved nature on the graph and the best fit line of linear regression is nowhere near the mean points. In this scenario, polynomial regression comes into action because the polynomial regression best fit line is in the form of a curve due to the high degree of independent variables.

Like linear regression, polynomial regression does not need the linear relationship between the dependent and independent variables of the dataset. When linear regression models fail to find an optimum linear relation between the dependent and independent variables, then the polynomial regression comes into action to find the optimum conclusion.

Suppose we have a dataset that is represented by the scatter plot shown in Fig. 10.5.

The straight line of linear regression cannot capture the pattern of the dataset as shown in Fig. 10.6.

As the polynomial regression line is curved in nature, so, it will be suitable to use polynomial regression for this dataset. Figure 10.7 is showing polynomial regression line has fewer gaps between the actual points and predicted points as compared to the gaps of linear regression line.

10.1.3.3   Overfitting vs Under-fitting

If we test the above dataset linearly, we can face the problem of under-fitting. So we can try to resolve this problem by increasing the degree of parameters so that we get the better results. But it will increase the degree too high; for best results then, we may get into the problem called overfitting. So, we must be careful in making the

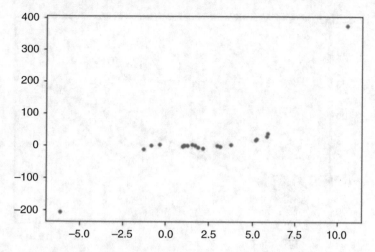

Fig. 10.5 A sample scatter plot showing the dataset

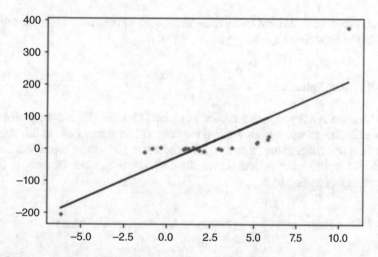

Fig. 10.6 Representation of linear regression using straight line

selection of degree of the variables so that we could not get caught in the problems of overfitting and under-fitting.

10.1.3.4 Choice of Right Degree

To avoid the problem of overfitting and under-fitting, we need to choose right degree of polynomial parameters. Forward selection method can be used that increases the degree of parameter until we reach the optimal conclusions. Similarly, backward

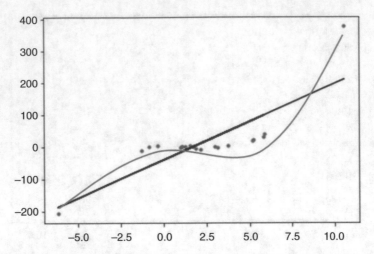

Fig. 10.7 Polynomial regression line

selection method can be used to decrease the degree of parameter unless we reach at the optimum conclusion again.

10.1.3.5   Loss Function

A loss function is a function that is used in machine learning to find the performance of a model. This function is also called the cost and error function. In this function, we subtract the actual value from the predicted value and square the result. Later all the squared differences are added and divided by the number of records or data points as shown in Eq. 10.9.

$$\text{Cost Function} = \frac{1}{n} \sum_{i=1}^{n} \left(\text{Predicted Value} - \text{Actual Value}\right)^2 \qquad (10.9)$$

10.2   Logistic Regression

Linear regression works on continuous data but if the dataset contains discrete values, the linear regression model will not work. For the discrete types of data, we need to use another variation of regression called logistic regression. Logistic regression is a model that uses discrete or binary relations and predicts the outcome on the basis of the observations in the training data.

The logistic regression model takes one or more independent variables as input, analyzes the relationship between independent and dependent variables, and predicts the value of a dependent variable.

For example, logistic regression can take independent variables (symptoms of a disease) as input and predict whether a patient has cancer or not.

Logistic regression can take single or multiple variables as input for decision. For example, with a single variable logistic regression, we can predict that a person will get a job or not. Here the qualification of the person can be an independent variable on the basis of which we can predict the job. Logistic regression may not give best results because there can be other factors which can decide that a person will get the job or not. For example, not only qualification but experience, age, and some other factors can also be used to decide that the person will get the job or not. Because these factors also have high impact, using multiple variables will give better result than using only a single variable.

In machine learning, the logistic regression models have become a very important tool that is used to perform classification on the basis of historical data. When additional relevant factors are included in the data, the model predicts better results. Logistic regression is also used to prepare and clean the data by putting the data in predefined classes using the ETL data cleaning techniques.

10.2.1   Logistic Regression Importance

The logistic regression algorithm is considered to be very important because it can transform complex probabilistic calculations into simple mathematical problems. This transformation is complex, but some modern applications of statistical analysis can automate most of this transformation.

10.2.2   Logistic Regression Assumptions

While deciding to use logistic regression for the classification of a dataset, a few assumptions must be made. The independent variables should be independent of each other and should not have any relationship between them. For example, the age and zip code of a person have no relationship, so the dataset with these two independent variables can be used in logistic regression, but if the variables have any relation, for example, zip code and state have a relation, then using these two independent variables can create the problem in result.

When using logistic regression on a dataset, some less transparent variables can get lost in the noise. For example, we may put a considerable effort to avoid the variables associated with discrimination like gender and civilization, but sometimes these variables can be included indirectly via some other variables that are not correlated such as hobbies and zip code, etc.

Another assumption is that the raw data should not have redundant information. For example, if we conduct a survey of customer satisfaction to get the opinion of different customers and reward the customers through a draw, the result can be slanted because a customer can fill the survey multiple times using different email accounts to increase the chances to get the reward.

Another assumption is that each variable should be represented in the form of discrete values such as yes/no, male/female, young/old, etc. If we will transform a variable into three categories, then our algorithm may not work. For example, the age variable can be transformed into three categories of kid, young, and old.

10.2.3   Use Cases of Logistic Regression

Logistic regression has become popular in digital marketing. It can help the marketer to predict the likelihood of a visitor to click on an advertisement or not.

Logistic regression can also be used in the following scenarios:

- In healthcare logistic regression can be used to identify the risk factors of a specific disease and plan measures to prevent it.
- Finding the effectiveness of a medicine on a person of different ages and gender can help the researchers to adjust and improve the medicine.
- It can be used in weather prediction applications to predict snowfall and rain.
- It can be used in political polls to find whether a voter will give a vote to a particular candidate or not.
- Logistic regression can be used to predict whether a bank customer will default or not, and that is calculated on the basis of annual income and past default records.

10.2.4   Difference Between Linear and Logistic Regression

- Linear regression provides constant output, while logistic regression provides the output in constant form.
- Linear regression can have infinite values of the dependent variable, while logistic regression will give only specific (finite) values to the dependent variable.
- Linear regression is used for the variables like age, height, and distance, while logistic regression is used when we have variables like gender or any other variable which can have categorical values, e.g., yes/no, true/false, and fail/pass.

10.2.5   Probability-Based Approach

Now we will discuss the probability-based regression approach.

10.2.5.1   General Principle

Logistic regression uses one or multiple variables and predicts the odds of the output of a categorical variable. The categorical variables have limited values, levels, or categories such as yes/no and valid/invalid. The major advantage of logistic regression is that the predictions are always in the range of 0 to 1. For example, the logistic model can predict the likelihood of a man visiting the beach by taking the temperature as an input function.

The probability-based logistic regression is also called the logit model or logistic model. It is used to predict the probability of a specific event happening by providing the data to a logistic function.

The logistic function used in logistic regression is represented by a sigmoid or S curve. The logistic curve shows an S-shaped diagram of the growth of some population P. In the first stage, the growth is exponential, then growth starts decreasing and at the end, the growth stops. Mathematically, the logistic function can be defined by the formula given in Eq. 10.10.

$$P(t) = \frac{1}{1 + e^{-t}} \tag{10.10}$$

Here,

"P" is used to represent a population.
"t" is used for the time.
"e" is Euler's number.

In Fig. 10.8, we have used the value range $[-6, 6]$ for t to get S curve or logistic/sigmoid function.

The logistic function can be used to predict the probability of some specific events. In such cases, multiple variables or many factors can be used to check whether an event will happen or not. So, the general formula of logistic regression is shown in Eq. 10.11:

$$P = \frac{1}{1 + e^{-(B_0 + B_1 X_1 + B_2 X_2 + B_3 X_3 + \ldots)}} \tag{10.11}$$

where:

P is the probability of the event
$B_0, B_1, B_2 \ldots$ are regression coefficients
$X_1, X_2, X_3 \ldots$ are independent variables

Fig. 10.8 S curve using the
range $[-6, 6]$

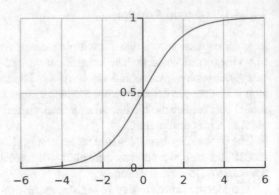

10.2.5.2   Logistic Regression Odds Ratio

The odds of an event happening are defined as the probability of case and non-case
of the event given the value of the independent variable. The probability of a case is
divided by the probability of a non-case given value of the independent variable. The
odds ratio (OR) is used as a primary measure of change in logistic regression that is
computed by comparing the odds of membership in one group with the odds of
membership in another group. So, in simple words, the OR is calculated simply by
dividing the chances of being a case in one group by the chances of being a case in
the other group. This helps us find how much change in the value of an independent
variable affects the change in the value of a dependent variable. Let's take an
example.

Suppose we want to build a model that will predict the gender of the person if we
provide height as the independent variable. Here we are talking about the probability
of being male or female, or we are talking about the odds of being male or female. So
here our OR formula will come into action.

Suppose the probability of being male is 90%; so if we put values in the formula,
we will get the odds of male.

$$\text{odds} = \frac{P}{1-P} = \frac{0.9}{1-0.9} = \frac{0.9}{0.1} = \mathbf{9}$$

Logistic regression takes odds natural log which is also known as logit or
log-odds to create an incessant criterion. The curve of the natural log function is
given in Fig. 10.9.

Then probability-based linear regression analysis is used for the predictors that
use logit of success. The results of logit may not be instinctive. So, the exponential
function is used and the logit is converted back to odds. Therefore, the observed
variables in logistic regression are discrete but the result is in the form of a
continuous variable.

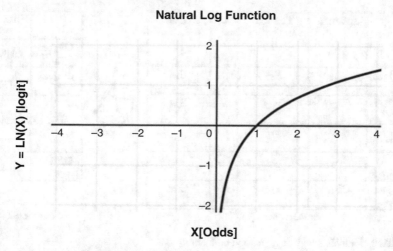

Fig. 10.9 Curve of the natural log

10.2.6 Multiclass Logistic Regression

Normally, the logistic regression predicts two classes, e.g., yes/no, true/false, male/female, etc. But if we have to predict more classes using logistic regression, then multiclass logistic regression is used. It is also called "softmax" or multinomial logistic regression. Some analysts always use multiclass logistic regression because they do not have a complete understanding of how multiclass logistic regression works.

Let's assume we have N number of objects and each object has M number of features and they all belong to C classes. This problem statement can be represented by the following diagram given in Fig. 10.10:

The object-matrix E has M number of rows and N number of columns where E_{ij} represents an object i with feature j.

The matrix Z (actually a vector) has M number of rows with c classes where Z_i represents that an object i belongs to a specific class c.

10.2.6.1 Workflow of Multiclass Logistic Regression

We want to figure out the weight matrix W and use W to predict the class membership of any given observation in E.

The weight matrix W has N features represented as the number of rows and C classes represented as the number of columns. W_{jk} represents the weights for feature j and class k.

Now we know E and W where initially W is filled with all zeros. Figure 10.11 shows the workflow of the multiclass regression using the forward path.

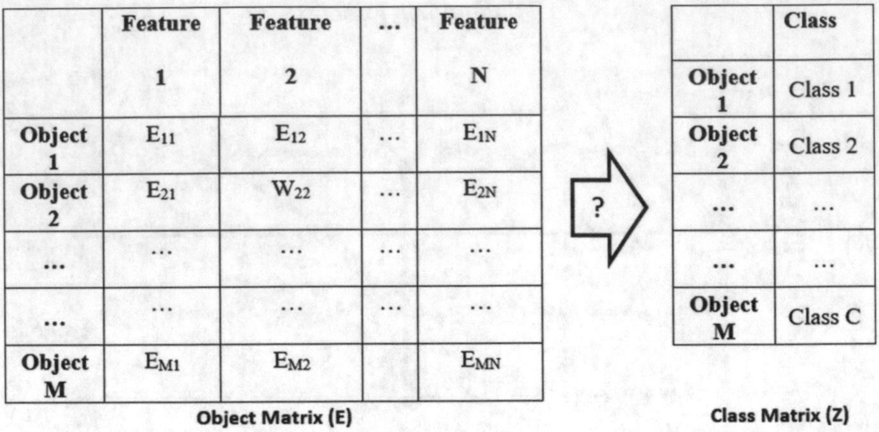

	Feature 1	Feature 2	...	Feature N
Object 1	E_{11}	E_{12}	...	E_{1N}
Object 2	E_{21}	W_{22}	...	E_{2N}
...	...	...	...	...
...	...	...	...	...
Object M	E_{M1}	E_{M2}		E_{MN}

Object Matrix (E)

	Class
Object 1	Class 1
Object 2	Class 2
...	...
...	...
Object M	Class C

Class Matrix (Z)

Fig. 10.10 Multiclass logistic regression problem

In the first step, we will calculate the product of E with W so $K = -EW$.

A negative sign with the product of E and W is not mandatory. Some people use it, while some others use positive signs.

In the second step, we take the softmax of every row of K_i using the formula given in Eq. 10.12.

$$M_i = \text{softmax}(K_i) = exp(K_i) / \sum exp(K_{ik}). \tag{10.12}$$

Every row of k_i is the product of every row of E with the whole matrix W. Now every row of matrix Q adds up to 1.

In the third step, we take the argmax of every row and find the class with maximum probability.

In Fig. 10.12, we have presented another view of multiclass logistic regression using the forward path method with the help of a single object.

In the first step, we will calculate the product of E_i with W, so $K = -E_iW$.

In the second step, we take the softmax of the row of Z_i.

$$M_i = \text{softmax}(K_i) = exp(K_i) / \sum exp(K_i k).$$

In the third step, we take the argmax of row M_i and find the index with the highest probability as Z_i.

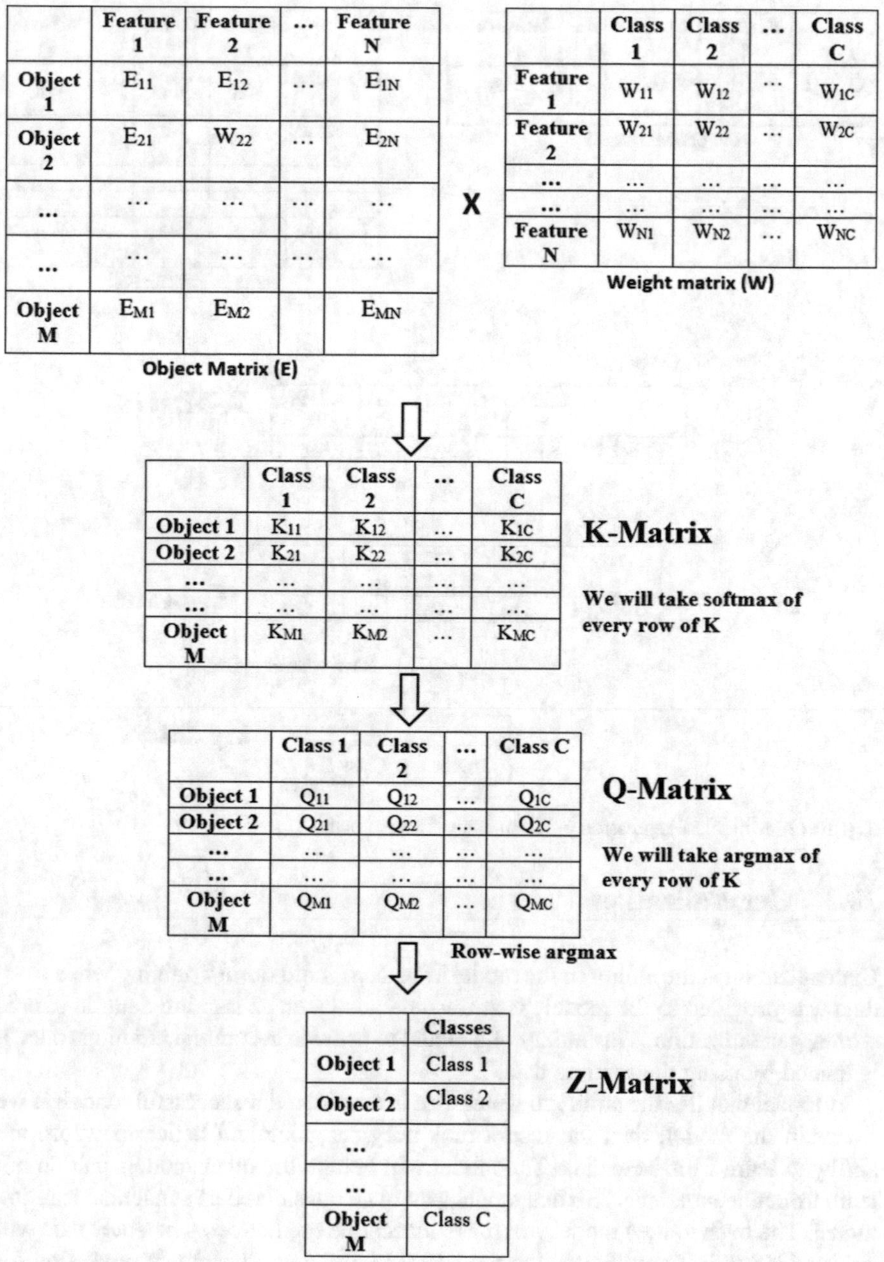

	Feature 1	Feature 2	...	Feature N
Object 1	E_{11}	E_{12}	...	E_{1N}
Object 2	E_{21}	W_{22}	...	E_{2N}
...	...	...	...	...
...	...	...	...	...
Object M	E_{M1}	E_{M2}		E_{MN}

Object Matrix (E)

X

	Class 1	Class 2	...	Class C
Feature 1	W_{11}	W_{12}	...	W_{1C}
Feature 2	W_{21}	W_{22}	...	W_{2C}
...	...	...	...	...
...	...	...	...	...
Feature N	W_{N1}	W_{N2}	...	W_{NC}

Weight matrix (W)

	Class 1	Class 2	...	Class C
Object 1	K_{11}	K_{12}	...	K_{1C}
Object 2	K_{21}	K_{22}	...	K_{2C}
...	...	...	...	...
...	...	...	...	...
Object M	K_{M1}	K_{M2}	...	K_{MC}

K-Matrix

We will take softmax of every row of K

	Class 1	Class 2	...	Class C
Object 1	Q_{11}	Q_{12}	...	Q_{1C}
Object 2	Q_{21}	Q_{22}	...	Q_{2C}
...	...	...	...	...
...	...	...	...	...
Object M	Q_{M1}	Q_{M2}	...	Q_{MC}

Q-Matrix

We will take argmax of every row of K

Row-wise argmax

	Classes
Object 1	Class 1
Object 2	Class 2
...	
...	
Object M	Class C

Z-Matrix

Fig. 10.11 Workflow of the multiclass regression forward path

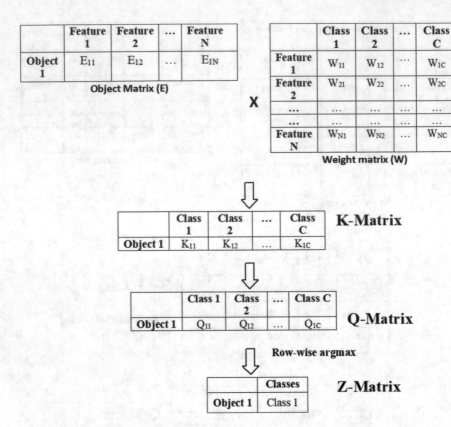

	Feature 1	Feature 2	...	Feature N
Object 1	E_{11}	E_{12}	...	E_{1N}

Object Matrix (E)

X

	Class 1	Class 2	...	Class C
Feature 1	W_{11}	W_{12}	...	W_{1C}
Feature 2	W_{21}	W_{22}	...	W_{2C}
...	...	...	...	...
...	...	...	...	...
Feature N	W_{N1}	W_{N2}	...	W_{NC}

Weight matrix (W)

	Class 1	Class 2	...	Class C
Object 1	K_{11}	K_{12}	...	K_{1C}

K-Matrix

	Class 1	Class 2	...	Class C
Object 1	Q_{11}	Q_{12}	...	Q_{1C}

Q-Matrix

Row-wise argmax

	Classes
Object 1	Class 1

Z-Matrix

Fig. 10.12 Multiclass regression problem using forward path

10.3   Generalization

Generalization is the ability of the model to react as it did during training when fresh data was provided to the model. That fresh data is taken as training data. In simple words, generalization is the ability of a model to provide accurate predictions once it is trained by using the training data.

A model that has the ability to generalize is considered a successful model. If we overtrain the model, then the model may not generalize and hence may lose the ability to learn from new data. The model will be like the other models that do not learn from the experience. So this model cannot be considered as a machine learning model. The over-trained model will make inaccurate predictions when new data will be provided to it. Even though the model is making a good accurate prediction for the training dataset, new data can be different from the training data, and that is the reason for inaccurate predictions.

Similarly, the model can have the issue of under-fitting. If the model will have an under-fitting problem, then again model will not generalize. If we will train the model with less data, then the model may be unable to do good predictions when new data is provided to the model. Because the model is trained with less data, it will not cover the maximum possible samples of data, so when the model will find new data different from training data, its results will not be accurate.

So, to generalize a model, we need to take care of overfitting and under-fitting. There are some other problems that can stop a model from being generalized. Those will be discussed later in the current topic.

Suppose we have a model using natural language processing (NLP) and in input "the plane ___" is given, for instance; then we expect that our model will predict "departs" instead of "depart." Similarly, "leaves" should have high probability than "leave." In this case, the model learned a pattern in a more generalized form but still does not understand the meaning of words that are not given for training. For example, "theee plaaaane ___" is not used in the training set, so the model is not trained for such words, and thus the model will screw up and make incorrect predictions.

10.3.1 Has the Model Learned All?

How would we know that the model has learned all and it is a generalized model now? It is a good idea to plot the learning process of the model and note the loss or error of the model. Our target is that the prediction error should be zero or at least very close to zero. We will keep training the model and testing it from time to time with fresh data and note the error. When the error will get close to zero and at any stage when more training will start increasing, the error will stop the further training of the model. Now, the model is supposed to be generalized. If we will train the model more, the overfitting issue will start arising. Figure 10.13 is showing the relationship between training with loss or error.

Fig. 10.13 Relationship between training and loss

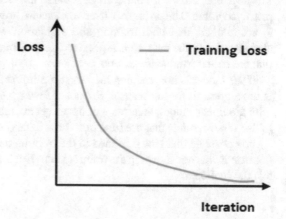

Fig. 10.14 Validation loss
vs training loss

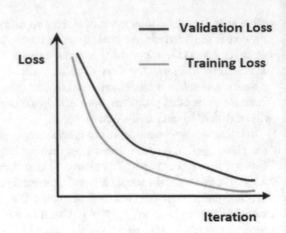

10.3.2 Validation Loss

Decreasing the value of training loss is a good technique to minimize the error during the prediction, but what if the model is not reaching at target minimum error? If the training process has reached its peak, more training will increase the loss. But what if we want to decrease the loss here? At this point, validation comes into action.

The model will become more generalized if we will decrease validation loss too. It will be good if validation loss decreases with training loss. If validation loss decreases, it means that the model has learned unseen patterns of the same words. So by learning unseen patterns, the model will be more generalized. Generally, validation loss is always higher than training loss as shown in Fig. 10.14.

10.3.3 Bias

The term bias is used for the average squared difference between the actual value and predicted value. This is the tool used to measure how well your model fits the data. It is not realistic that both training and validation will get to zero because there is always noise in the data, so it is good to use the "bias" technique to reduce the errors that are called irreducible errors.

If the model is not reaching the target minimum loss, it means the selected model is not a good fit for our dataset. Suppose if we try to fit the exponential relationship with the linear model, then we will never get accurate predictions. So we should try to use other models that are best suited for the exponential relationship of data.

It is possible that bias is hidden in the training data. So we may overlook the bias because it was not shown in the training data. But bias can be identified when we will test the real data.

10.3.4 Variance

If the predictions are very sensitive to small changes in the input data, it is said that the model has high variance. It means the structure of data is not smooth; it is very wiggly, or data points have spread too much which means that both the independent and dependent variables' values have high spread. High variance means that model is over-fitted because our model is capturing outlier and noisy data. Like bias and under-fitting, variance and overfitting are also related but they are not equal.

10.3.5 Overfitting

At the training stage, it is possible that a point arrives where the validation loss may start increasing again but training loss is still decreasing. It means that model has reached an overfitting stage. It also means that the model is still learning but it does not generalize beyond the training dataset as shown in Fig. 10.15. Overfitting normally occurs when we have a very large number of parameters.

The overfitting problem will happen after a certain number of training iterations. A big gap between training loss and validation loss signals that the model is not well generalized. You can see the big gap shown in Fig. 10.16. So the solution is to stop the training early and check validation loss, and if loss starts increasing, then stop the further training. Under-fitting can also happen if we stop training too early. The solution to the problem is a regularization technique to avoid such problems.

Fig. 10.15 Model overfitting

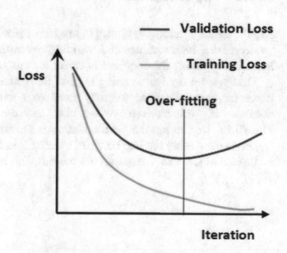

Fig. 10.16 Gap between training loss and validation loss

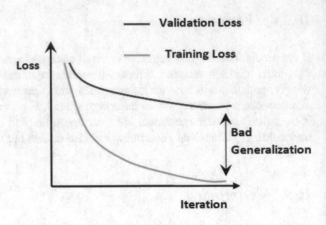

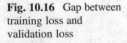

Fig. 10.17 Regularization

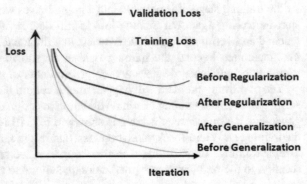

10.3.6   *Regularization*

Regularization is a technique used in machine learning to increase the generalization by decreasing high variance and avoiding overfitting. The ultimate target of regularization is getting the coefficients closer to zero.

Thus we can say that learning is good but generalization is our target. So to get maximum generalization, we may need to avoid overfitting and under-fitting, decrease the variance, remove the bias, and use the regularization method. In Fig. 10.17, we can see that regularization is the solution to all problems. Regularization is decreasing the gap between training loss and validation loss.

Instead of any tool, the graphs are created with hands to give the idea of concept.

10.4   Advanced Regression Methods

In this section we will see the following advance regression methods:

- Bayesian regression
- Regression tree
- Bagging and boosting

10.4.1   Bayesian Regression

In Bayesian regression, we perform linear regression with the help of probability distribution instead of point estimation. The output (point y) is not estimated as a single value, but it is calculated through a probability distribution. The following formula is used to model the Bayesian linear distribution:

$$y \sim N\left(\beta^T X, \sigma^2 I\right) \tag{10.13}$$

The value of y is generated using Gaussian distribution that uses mean and variance. The multiplication of the predictor matrix with the transposed weight matrix is used to calculate the "mean" in linear regression, while variance is calculated by multiplying the square of the standard deviation with the identity matrix. Both of these are provided to Bayesian linear distribution as shown in the above equation.

The main purpose of Bayesian linear regression is not to find the best single prediction, so it does not find the best single value of the model but the subsequent distribution of the model. In this technique, both the response and the input parameters come from a distribution. The resultant subsequent probability is dependent on the training inputs and outputs. Its equation is shown below.

$$P(\beta|y, X) = \frac{P(y|\beta, X) * P(\beta|X)}{P(y|X)} \tag{10.14}$$

In the formula:

$P(\beta|y, X)$ = posterior probability distribution
$P(y|\beta, X)$ = likelihood of the data
$P(\beta|X)$ = prior probability of the parameters
$P(y|X)$ = the normalization constant

The simplest form of the Bayesian theorem is given below.

$$\text{Posterior} = \frac{(\text{Likelihood} * \text{Prior})}{\text{Normalization}} \tag{10.15}$$

The following are the details of posterior and priors that we get from Bayesian linear regression:

Priors

If we have any guess or domain knowledge about the parameters of the model, then we should include it in the model, while the actual approach is that the parameters are taken from the data. If we do not have any guess or domain knowledge, then non-informative priors like normal distribution can be used.

Posterior

The result of Bayesian line regression which is calculated from prior probability, the likelihood of data, and normalization constant is called posterior probability.

The use of data and prior helps us to find the model uncertainty (if any). For small datasets, the posterior probability will increase, whereas, for large data, the likelihood will decrease the effect of priors.

10.4.2   Regression Trees

A regression tree is a machine learning model that is used to partition the data into a set of smaller groups and then solve each smaller group recursively. The single tree model gives a poor and unstable prediction. To overcome this problem, bagging technique is used. This technique provides a fundamental basis for complex tree-based models like gradient boosting and random forest. The tree shown in Fig. 10.18 has two types of nodes. The nodes having dotted outlines are parent nodes, while the nodes having solid outlines are children or leaves.

There are many methods for the construction of a regression tree; however, the conventional one is called the CART.

The mechanism is that the data is partitioned into subgroups where each group forms a subtree. To partition, the dataset recursive partitioning techniques are used. Then the prediction is performed on the bases of the average response values of each subgroup.

Fig. 10.18 Regression tree

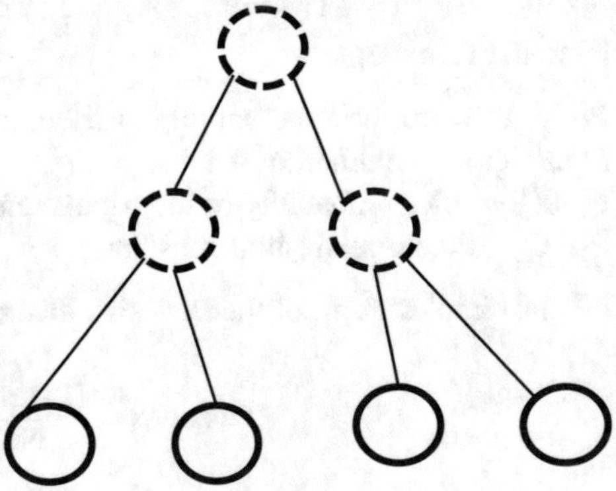

Now we will present some strengths of regression trees.

- We can easily interpret the regression trees.
- The prediction is fast as it does not require any complex calculation.
- It is easy to find the important variables, for example, those that split the data.
- Even if the dataset has missing values, we can still make the predictions to some extent.
- A number of algorithms are available which can be used to get the regression trees.

There are some weaknesses of regression trees.

- Single regression trees may result in unstable predictions due to high variance.
- Due to the high variance, the accuracy of single regression trees may be affected.

10.4.3 Bagging and Boosting

The ensemble is one of the important methods used in machine learning for data analysis. In this method, we train multiple models or weak learners to fix a problem. We then integrate their result to get the final result. By properly combining the result of each learner, we can get the desired results.

We first develop the base models to set up the ensemble learning method that is clustered later. We use a single base learning model in bagging and boosting. This is because we have homogeneous learners that are trained in different ways.

These types of ensemble models are called homogeneous models. Apart from homogeneous models, we may use other methods in which different types of base learning algorithms are used along with heterogeneous weak learners. Such types of models are called heterogeneous ensemble models.

Here we will only discuss the first model along with the two most common ensemble methods. Bagging is one of the homogeneous weak learners' models in which the weak learners learn independently in parallel and finally the result is combined to determine the model average.

Boosting is also a homogeneous weak learners' model; however it works differently from bagging. In boosting, the weak learners learn sequentially and adaptively to improve the predictions.

Now we will look at both of these models in detail about how both methods work. There are multiple factors that affect the accuracy of machine learning models, e.g., noise, variance, bias, etc. The ensemble methods enable us to reduce the effect of these factors in order to get higher accuracy in results.

10.4.3.1   Bagging

Bagging stands for "bootstrap aggregation." We use this method to decrease the variance in prediction models. In bagging, weak learners are implemented in parallel that learn independently from each other. Thus we can train them simultaneously. For bagging, we need some additional data for training. We perform the random sampling (with replacement) on the original dataset. There may be a problem with such random sampling, i.e., an observation may appear in multiple samples. Each observation, however, has an equal probability to appear in the new sample dataset.

The generated multiple samples are used to train the multiple learners that get training in parallel. Then we calculate the average of all the weak learners to get the calculation of the final result. Here we can consider the mechanism called majority voting to perform the classification. Bagging is an effective method to decrease the variance and increase the accuracy of the results.

10.4.3.2   Example of Bagging

The random forest model is an example of bagging where different decision trees are used in parallel to learn from different data samples. Each tree performs its own classification and we finally combine the results.

10.4.3.3   Boosting

As discussed above, boosting is a sequential ensemble learning method in which the weights are adjusted iteratively according to the last classification results. The main idea is to convert weak learners to strong ones. This helps in building strong prediction models that can make more accurate predictions.

We identify the data points that are misclassified in each iteration and increase their weights. So in boosting, we assign the weights to each learner participating in classification. The learner that gives higher results gets the higher weight during the training process.

10.4.3.4   Example of Boosting

AdaBoost uses boosting technique. We use the threshold of error, i.e., less than 50%, to maintain the model. On the basis of the error rate, we either decide to keep or discard the learner. Here, boosting can keep or discard a single learner. The same process is repeated until we get the final results.

10.4.3.5 Comparison of Bagging and Boosting

Both bagging and boosting are commonly used methods. Both are ensemble-based techniques. Now we will discuss some of the similarities and differences between both of them. First, we will mention some similarities.

1. Both bagging and boosting are ensemble-based methods that use N learners for getting the final results.
2. Both use random sampling of the input dataset to get the new samples for each learner.
3. Both generate the final results by averaging the results of N learners or by using the majority voting mechanism.
4. Both techniques attempt to reduce the variance and provide higher accuracy by reducing the error.

Now we will discuss some of the differences between both of these techniques.

1. In bagging, we merge the same types of predictions, whereas in boosting we merge different types of predictions.
2. In bagging, the variance is decreased, whereas in boosting bias is decreased and not the variance.
3. In bagging, each model is assigned the same (equal) weights, whereas in boosting each model gets different weights based on its performance.
4. In bagging, we have independent models which have no impact on each other, whereas in boosting, models are affected by the performance of the previously built model.

In bagging, we randomly build the subsets from the original dataset during the training process. However, in boosting, the new data subsets comprise the elements that were misclassified in the previous attempt. Training data subsets are drawn randomly with a replacement for the training dataset. In boosting, every new subset comprises the elements that were misclassified by previous models.

As variance is reduced in bagging, so it is applied where we have higher variance in results. Boosting, on the other hand, is applied when we have a high bias.

So far, we have discussed both the concepts of bagging and boosting in detail. Both are important methods that are used in various machine learning applications. The application of each model depends on the nature of the data available and the current requirements. So, in the scenario where we need bagging, we can use the random forest approach, and in the scenarios where we need boosting, we can use algorithms like AdaBoost.

It should be noted that the performance of a machine learning algorithm is measured by comparing the training accuracy with its validation accuracy.

10.5   Real-World Applications for Regression Models

Now we will discuss some of the applications of regression models.

10.5.1   Imbalanced Classification

In some real-world applications, e.g., in fraud detection or disease prediction problems, it is important to correctly identify the minority classes. So, it is important that the models should not be biased to detect the majority classes only. So, equal weights should be assigned to minority classes as well. There are different techniques to handle this problem. One of the methods is called resampling.

Using this technique we can up-sample or down-sample a majority or minority class. While using the imbalanced dataset, the minority class can be oversampled by using the replacement technique. Similarly, some rows corresponding to the majority class can be removed from the dataset in order to match the count with those of the minority class. This is called under-sampling. Once the data is sampled, we can say that there is an equal number of majority and minority classes. So the algorithm can be assumed to give equal weights and importance to both classes.

However, it should be noted that the selection of appropriate evaluation metrics is much important. The accuracy of the classifier can be obtained by dividing the number of correct classifications by the total number of classifications. For this we have created the confusion matrix as shown in Table 10.7:

The approach may work well for balanced datasets but may create inaccurate results for datasets with imbalanced classes. There is another metric that can be used, i.e., precision. The precision specifies how accurate the prediction of a specific class of a classifier is. There is another measure called recall which shows the classifier's ability to identify a class.

For a dataset with an imbalanced class problem, the F_1 score is the more appropriate metric. It is the harmonic mean of precision and recall as shown in Eq. 10.16:

$$F_1 = 2 * \frac{(\text{precision} * \text{recall})}{(\text{precision} + \text{recall})} \tag{10.16}$$

Thus in a model that predicts the minority class, if the number of false positive increases, the precision value will be low, and F_1 measure will be low as well.

Table 10.7 Confusion matrix

	Original	
Prediction	Positive	Negative
Positive	True positive	False positive
Negative	False negative	True negative

Similarly, if the number of minority classes is wrongly identified, then false negatives will increase, so the value of recall and F_1 score will remain low. So, it should be noted that the F_1 measure only increases in case both the number and quality of prediction increase.

F_1 score keeps the balance between precision and recall and improves the score only if the classifier identifies more of a certain class correctly.

10.5.2 Ranking Problem

Ranking means finding out the documents according to their relevance to the given query. This is an important problem in the domain of "information retrieval."

Some of the other applications of "ranking" are:

- Search Engines: Search engines need to accurately identify the contents that the user requested through their queries.
- Recommender Systems: Recommender systems also need to find the profiles or recommendations on the basis of the requested information. So, the profiles that rank high according to the query should be selected.
- Travel Agencies: Find out the rooms according to the relevance.

In the ranking problem, we have a ranking model that takes input $X=(Q,D)$ where Q represents the user query and D represents the document. The model outputs a relevance score $s=f(x)$ for each input. Once all the documents are ranked, we can rank the documents according to their score as shown in Fig. 10.19.

Ranking models rely on a scoring function.

The scoring model can be implemented using various approaches.

- Vector Space Models – As the name implies, they produce a vector embedding for each (query, document) and then score each embedding.
- Learning to Rank – They work similar to other traditional machine learning algorithms. They learn to rank on the basis of some ranking data.

10.5.3 Time Series Problem

In general, the analysis task is performed to get information about what has happened in the past and what can be expected in the future by seeing and analyzing the currently available data.

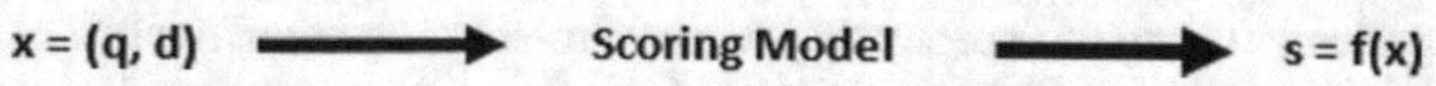

$$x = (q, d) \longrightarrow \text{Scoring Model} \longrightarrow s = f(x)$$

Fig. 10.19 Scoring model

In time series data, we have ordered data where each record has a timestamp. So, in simple words, when we have observations in order of their occurrence of time, the data is called the time series data.

Suppose the people want to know how the rate of inflation increased or decreased with the passage of time. For this, they will take periodic data, e.g., at the end of each year, about the rate of inflation. Once they have sufficient data, it can be analyzed to know how inflation increased and decreased with the passage of time.

A time series data is a set of observations that are taken after a constant time interval. The time here acts as an independent variable that is used to study the change in behavior of the data. The objective is to study the change in behavior of a dependent variable on the other hand.

For example, one can measure:

- Change in temperature per day
- Monthly sales
- Annual change in inflation
- Daily increase or decrease in shares of a company

The time series data can be of three types:

- *Time series data:* a set of observations taken after different time intervals
- *Cross-sectional data:* data values of different variables taken at the same point in time
- *Pooled data:* a combination of both time series and cross-sectional data

We can use different visualization techniques to represent all types of time series data in order to find the hidden patterns in the data. Some of the visual representations are given in Fig. 10.20:

In time series analysis, time acts as a reference point. So, in time series analysis, there is always a relationship between two variables. One of these variables is the time and the other variable is the quantitative variable, the value of which is under consideration.

It should be noted that the values of the variables involved in time series analysis do not always increase. The values may decrease as well, e.g., in a certain period of time in a year, the temperature may keep on decreasing on daily basis.

10.5.3.1   Time Series Analysis

Time series analysis is the process of processing the time series data and extracting the patterns from this data.

In time series data, we have sequential data points mapped to sequential time durations. So, in time series analysis, we have the procedures that summarize the data in order to understand the different patterns and to make predictions on the basis of these patterns. Using time series data to make forecasting about the future requires the development of models that take input from the past time series data and make conclusions on the basis of this past history.

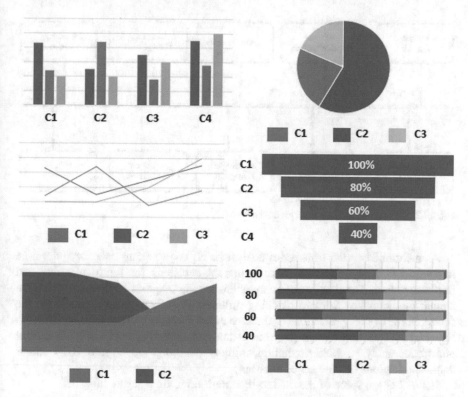

Fig. 10.20 Time series analysis through various modes of data visualization

- One of the objectives of the time series analysis is to explore and understand the patterns in the data according to which the changes occur with the passage of time. A pattern specifies a recurrent behavior according to which the data changes over time.
- When such behaviors exist in data, the model should be developed in order to successfully and accurately extract such existing behaviors because these existing behaviors help predict future behaviors and trends.

For example, suppose that we want to predict the sales of a certain item. We can develop the models on the basis of the existing time series data that will predict the sale of that product by analyzing the previous historical data of that product.

Similarly, the time series data can be used for multiple investigations to predict the future in terms of the expected trends, behaviors, changes, etc. We can use these predictions to answer certain questions related to the interests of the business. For example, what are the reasons behind the certain drop in the sale of a product, and what will be the profit rate at a certain period of a product at certain locations? What steps should be taken in order to keep in competition with the competitors in the market?

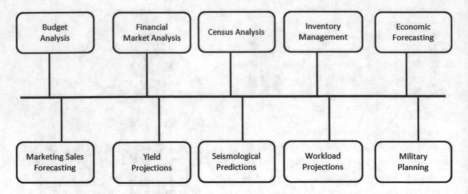

Fig. 10.21 Applications of time series analysis

As mentioned earlier time series analysis is all about taking data at regular time intervals and analyzing it to make important decisions for future planning and implementation. The time series analysis has a wide range of applications from the simple prediction of sales to planning millions of dollars of investments for the future. Some of the applications of the time series analysis include sales forecasting, financial analysis and forecasting, census analysis, inventory analysis, stock market and trends analysis, yield prediction, military planning, etc. Figure 10.21 shows some applications of time series analysis.

Following are some of the models that are used in time series analysis:

- Integrated (I) models
- Autoregressive (AR) models
- Moving average (MA) models
- Autoregressive moving average (ARMA) models
- Autoregressive integrated moving average (ARIMA) models

These models show that the measurements taken after small time intervals show more accuracy in prediction as compared to those measures that are taken after long time intervals in time series analysis.

10.5.3.2 Examples of Time Series Analysis

Now we will consider an example of time series analysis in the domain of finance. The time series analysis can be used to find the trends, identify the seasonal behavior, and figure out the correlation between different variables using time series analysis techniques.

The main objectives may be to:

1. *Predict different utilities:* For a successful business and to compete with the other business competitors, it is essential to accurately predict the future such as future prices of the assets, the expected sale of products, the future demand, etc. All this can be done using different statistical techniques by analyzing time series data.
2. *Simulate series:* Once we have analyzed the time series data, we can use it to perform different types of event simulations. We can use it to find the volumes of trade, costs of the products, investment guidelines for different business campaigns, risk identification and mitigation plans, expected product returns, etc.
3. *Presume relationship:* By finding the relationships between the time and different quantitative values, we can get the guidelines for improving the existing business processes and techniques and can make the right decisions. For example, by predicting the foreign exchange rates, we can predict the volume of the expected trade for a certain period of time. Similarly, we can find out a more effective campaign policy to attract more customers for a certain product in a certain season.

Similarly, we can use the time series analysis to adjust the train schedules. We can use the time series data regarding the different train journeys to find out:

1. The expected delay in arrival of a train at a certain location given certain circumstances
2. Expected revenue that may be generated by a train once a certain campaign has been launched

Social media is one of the most important sources of the generation of time series data. We can analyze the social media data to find out the expected number of responses for a certain hashtag, the expected number of traffic a certain event on social media will generate, etc.

Table 10.8 is showing the particular field and use of time series data in several of the analysis tasks performed in these fields.

Table 10.8 Examples of the use of time series data in various fields

Sr. no	Field	Example topics
1	Finance and economics	Prediction of Consumer Price Index (CPI), prediction of gross domestic product (GDP), employment rates
2	Social sciences	Increase in population, prediction of birth rates, feedback on a certain government policy
3	Epidemiology	Use of a certain medicine by doctors, disease rates, effect of a medicine, etc.
4	Medicine	Tracking heartbeats, prediction of blood pressure in certain circumstances, number of patients in a certain hospital during the heat wave
5	Physical sciences	Fog prediction, temperature prediction, pollution levels, traffic loads, etc.

10.5.3.3   Machine Learning and Time Series Analysis

Machine learning has become a prominent domain these days in processing images, speech, natural languages, etc. It is one of the most followed domains for processing huge data volumes these days.

- In problems related to time series data analysis, the time series data normally does not have well-defined features. There are so many variations in these features as the data is collected from various sources. There may be differences in dataset values in terms of scale, attributes, features, values, etc.
- In time series analysis, we require such machine learning algorithms that could learn from time-dependent data and find out the time-dependent patterns.
- Different machine learning tools require real-time data with proper timestamps like the algorithms for classification, clustering, forecasting, and anomaly detection.

Now we are going to discuss time series forecasting because it is a very important domain of machine learning. There are many problems in machine learning where time is involved in making efficient predictions. For time series forecasting, there are multiple models and methods available to use. Let's understand them more clearly.

10.5.3.4   Time Series Forecasting Using Machine Learning Methods

The univariate time series forecasting methods involve only two variables, one of which is the time and the other one represents the quantity about which we want to forecast. For example, suppose we want to predict the temperature of the city during the next month. Here there will be two variables. One is the timestamp that may be the date of day of the month and the other variable will be the temperature value. Similarly, suppose we want to predict the heartbeat of a person per minute using the previous data. Here again we will have two variables. The one is the timestamp (in minutes) and the other one will be the heartbeat rate.

In contrast to univariate time series forecasting where we have only two variables, in multivariable time series analysis, we may have number of variables to analysis. Definitely one of the variables will be the timestamp. For example, in forecasting the temperature of the city for the next month, we may have multiple variables like humidity, wind speed, etc.; however, the one variable, i.e., temperature, will have the main effect. Following are some of the variables that we may have in our dataset to predict the next month's temperature of the city:

1. Rainfall and the duration (time) of rain
2. Speed of winds
3. Humidity level
4. Atmospheric pressure
5. Precipitation

Summary
Regression is the process of learning relationships between inputs and continuous outputs from data. In this chapter, we provided in-depth details of regression analysis. Starting from the basic concept of what a regression is, we discussed different types of regression.

We also discussed the advance regression methods including Bayesian regression, bagging and boosting, etc. Finally, we also discussed the real-world applications of the regression models.

Further Reading

- *Regression Analysis: An Intuitive Guide for Using and Interpreting Linear Models* by Jim Frost, Statistics By Jim Publishing, 2020. In this book, you'll analyze many topics related to regression analysis including the following: how regression works, evaluation of regression algorithms, and examples of different regression algorithms and their implementations.

- *Regression Analysis: A Practical Introduction* by Jeremy Arkes, Routledge, 2019. This book provides details about regression analysis. After reading the book, the reader will be able to conduct, interpret, and assess the regression analyses. The book provides the concepts in a simple and easy-to-learn way. The book addresses the tools used to conduct regressions along with providing the skills to design the optimized regression models.

- *Handbook of Regression Analysis with Applications in R*, 2nd Edition by Samprit Chatterjee and Jeffrey S. Simonoff, Wiley, 2020. The book provides state-of-the-art methods for conducting complex regression analyses using the R programming language. The book discusses the advance regression topics including time-to-event survival data and longitudinal and clustered data. The book provides in-depth details about the prominent regression methods including the regularization, smoothing, and tree-based methods. The book also presents details of the data analysts' toolkit. Concepts are explained with the help of real-life examples. The book also includes the source code provided through the mentioned website.

- *Linear Regression Analysis: Theory and Computing* by Xin Yan and Xiao Gang Su, by World Scientific Publishing Company, 2009. The book provides the basics of linear regression and its evaluation methods along with the relevant statistical computing strategies. Concepts are explained with the help of examples in a simple and easy way. This book is suitable both for students and practitioners in order to implement the algorithms related to regression analysis. Overall the book is a good resource for anyone interested in regression analysis and the computing concepts related to it.

- *Applied Regression Analysis in Econometrics* (*Statistics: A Series of Textbooks and Monographs*) by Howard E. Doran, CRC Press, 1989. The book is intended to be a one-semester course for graduate and undergraduate economics students. The book covers the basics of the regression techniques in the context of the single-equation econometric models, featuring Minitab and Shazam software examples for attacking real-world problems. Annotation copyright Book News, Inc.

Exercises

Exercise 10.1: What is the difference between the following types of correlation? Explain each of them with the help of a real-life example. Also, mention which type of correlation linear regression analysis can be used.

- Positive correlation
- Negative correlation
- No correlation

Exercise 10.2: Consider the sample data given below in the table that is showing the relationship between petrol price and its demand. Plot the given data using a scatter plot, and also mention what type of correlation it shows.

Demand	Price
1390	84
1421	85
1511	94
1538	104
1552	110
1556	115
1582	118
1634	126
1641	135
1706	142
1708	151
1756	158
1781	160

Exercise 10.3: The sample data given below has one dependent variable Y and one independent variable X. Fit the linear regression line for the given data; also predict what will be the value of Y when the value of X is 90.

Y	X
120	65
126	75
140	76
147	81
162	83
174	87
179	97
189	98
203	102
213	111
223	115
236	125
247	126

Exercise 10.4: The sample data given below has one dependent variable Y and two independent variables X_1 and X_2. Using multivariate linear regression, predict what will be the value of Y when the value of X_1 is 85 and the value of X_2 is 25.

Y	X_1	X_2
126	58	23
133	62	18
138	66	22
144	67	23
159	77	21
166	80	18
174	81	19
183	88	24
194	91	26
200	95	22
210	103	22
225	108	18
235	113	14

Exercise 10.5: Consider the sample dataset given below showing the temperature and yield value measured from experiments, and fit the polynomial regression line using the given data.

Temperature	Yield
55	3.6
60	3.1
62	3.4
65	2.6
69	2.8
73	2.3
77	2.0
79	2.7
82	2.8
83	2.7
87	2.7
92	3.0
94	3.2
95	3.5
100	2.7

Exercise 10.6: What are the terms overfitting and under-fitting in the context of polynomial regression analysis?

Exercise 10.7: While analyzing the polynomial regression line, it is recommended to choose the right degree for the independent variable. What problems can arise if we do not select the right degree? Explain with the help of an example.

Exercise 10.8: What is the loss function and what relation does it have with predicted value and actual value?

Exercise 10.9: What are the similarities and differences between linear regression and logistic regression?

Exercise 10.10: What are validation loss and training loss? How are both related to each other in the context of the generalization of a model?

Exercise 10.11: What is the difference between bias, variance, and overfitting in the process of generalization of models?

Exercise 10.12: What is the relation of regularization with generalization?

Exercise 10.13: What are the similarities and differences in bagging and boosting?

Chapter 11
Data Science Programming Language

11.1 Python

Python is one of the most common programming languages for data science projects. A number of libraries and tools are available for coding in Python. Python source code is available under GNU General Public License.

To understand and test the programs that we will learn in this chapter, we need the following resources and tools:

Python: https://www.python.org/downloads/
Python documentation: https://www.python.org/doc/
PyCharm an IDE for Python: https://www.jetbrains.com/pycharm/download/

Installations of these tools are very simple; you just need to follow the instructions given on the screen during installation.

After installation, we are ready to make out the first program of Python.

Listing 11.1 First Python program

```
1. print ("Hello World!");
```

Output: Listing 11.1

Hello World!

11.1.1 Python Reserved Words

Like other programming languages Python also has some reserved words which cannot be used for user-defined variables, functions, arrays, or classes. Some Python reserved words are given in Table 11.1.

© The Author(s), under exclusive license to Springer Nature Switzerland AG 2023 353
U. Qamar, M. S. Raza, *Data Science Concepts and Techniques with Applications*,
https://doi.org/10.1007/978-3-031-17442-1_11

Table 11.1 Reserved words of Python

And	exec	not
Assert	finally	or
Break	for	pass
Class	from	print
Continue	global	raise
Def	if	return
Del	import	try
Elif	in	while
Else	is	with
Except	lambda	yield

11.1.2   Lines and Indentation

Like other programming languages, Python doesn't use braces "{" and "}" for decision-making, looping statements, function definition, and classes. The blocks of code are denoted by a line indentation. The number of spaces in indentation can vary but all statements in a block must have the same number of spaces.

11.1.3   Multi-line Statements

Normally a statement in Python ends on a new line but when we want to keep a statement on multiple lines, then Python allows us to use the line continuation character "\" with the combination of the "+" sign. The following example will demonstrate to you the use of line continuation character.

Listing 11.2 Line continuation in Python

```
1. name = "Testing " + \
2.     "Line " + \
3.     "Continuation"
4. print (name);
```

Output: Listing 11.2

Testing Line Continuation

If the statements enclosed in brackets like (), {}, and [] are spread into multiple lines, then we do not need to use line continuation character.

Listing 11.3 Statements enclosed in brackets

```
1. friends = ['Amber', 'Baron', 'Christian',
2.    'Crash', 'Deuce',
3.    'Evan', 'Hunter', 'Justice', 'Knight']
4. print (friends[0]+" and " + friends[5] + " are friends")
```

Output: Listing 11.3

Amber and Evan are friends

11.1.4   Quotations in Python

Python uses different types of quotations for strings. Single (') and double (") can be
used for the same strings, but if we have one type of quotation in the string, then we
should enclose the string with another type of quotation. Triple quotations of single
(') and double (") types allow the string to contain line breaks or span the string to
multiple lines.

Listing 11.4 Different types of quotations in Python

```
1.  word = "Don't"
2.  sentence = 'Testing double(") quotation'
3.  paragraph1 = """Testing paragraph
3.  with triple single (') quotations"""
4.  paragraph2 = '''Testing paragraph
5.  with triple double(") quotations'''
6.  print (word)
7.  print (sentence)
8.  print (paragraph1)
9.  print (paragraph2)
```

Output: Listing 11.4

Don't
Testing double(") quotation
Testing paragraph
with triple single (') quotations
Testing paragraph
with triple double(") quotations

11.1.5   Comments in Python

Any line in Python starting with the "#" sign is considered as a comment. If the "#" sign is used inside a string little, then it will not be used as comments. Python interpreter ignores all comments.

Listing 11.5 Comments in Python

```
1. # this is comments
2. print ("Comments in Python") # another comments
```

Output: Listing 11.5

Comments in Python

11.1.6   Multi-line Comments

For multi-line comments in Python, we need to use three single (') quotations at the start and end of comments.

Listing 11.6 Multi-line comments in Python

```
1. '''
2. These are multi-line comments
3. span on
4. multiple lines
5. '''
6. print ("Testing multi-line comments")
```

Output: Listing 11.6

Testing multi-line comments

11.1.7   Variables in Python

In Python when we declare a variable and assign a value, then memory is reserved for it. Variables in Python are loosely typed, which means that we do not need to specify the type of variable while declaring it. The type of variable is decided by assigning the type of value. For the assignment of value, we need to use the equal (=) sign. The operand on the left of the equal (=) sign is the name of the variable and the operand on the right is the value of the variable.

Listing 11.7 Variables in Python

```
1. age   = 25        # Declaring an integer variable
2. radius =10.5        # Declaring a floating point variable
3. name  = "Khalid" # Declaring a string
4.
5. print (age)
6. print (radius)
7. print (name)
```

Output: Listing 11.7

25
10.5
Khalid

11.1.8   Standard Data Types in Python

The data saved in a variable or memory location can be of different types. We can save the age, height, and name of a person. The type of age is an integer, the type of height is float, and the name is of string type.

Python has five standard data types:

- Numbers
- String
- List
- Tuple
- Dictionary

11.1.9   Python Numbers

In Python when we assign a numeric value to a variable, then numeric objects are created. Numeric objects can have a numeric value which can be an integer, octal, binary, hexadecimal, floating point, or complex number. Python supports four different numerical types.

- int
- long
- float
- complex

Listing 11.8 Numbers in Python

```
1. intNum     = 5        # Output: 107
2. binaryNum  = 0b1101011 # Output: 107
3. octalNum   = 0o15      # Output: 13
4. hexNum     = 0xFB      # Output: 251
5.
6. print(intNum)
7. print(binaryNum)
8. print(octalNum)
9. print(hexNum)
```

Output: Listing 11.8

5
107
13
251

11.1.10 Python Strings

Strings in Python are a set of contiguous characters enclosed in single or double quotations. A subset of string can be extracted with the help of slice operators [ ] and [:]. The first character of each string is located at 0 index which we provide inside slice operators. In strings plus (+) sign is used for concatenation of two strings and the asterisk (*) sign is used for repetition of strings.

Listing 11.9 String in Python

```
1. test = "Testing strings in Python"
2.
3. print(test)  # Displays complete string
4. print(test[0])    # Displays the first character of the string
5. print(test[2:5])   # Displays all characters starting from index 2 to
5
6. print(test[2:])   # Displays all characters from index 2 to end of the
string
7. print(test * 2)    # Displays the string two times
8. print(test + " concatenated string") # Display concatenation of two
strings
```

Output: Listing 11.9

Testing strings in Python
T
sti

sting strings in Python
Testing strings in PythonTesting strings in Python
Testing strings in Python concatenated string

11.1.11 Python Lists

Lists are flexible and compound data types in Python. The items are enclosed into square brackets ([ ]) and separated with commas. The concept of a list is similar to an array in C or C++ with some differences. The array in C is also a list, but it can hold only one type of element, but a list in Python can hold different types of elements in one list. Similar to strings, the elements stored in the list are accessed with the help of the slice operator ([ ] and [:]) and starting index starts from zero. In the list plus (+) sign is used for concatenation and the asterisk (*) sign is used for repetition.

Listing 11.10 List in Python

```
1. # This list contains information about a student in order of Name, Age,
CGPA, State, marks
2. student = ['John', 20, 2.23, 85, 'New York']
3.
4. print (student)          # prints the complete list
5. print (student[0])       # prints first element of list
6. print (student[2:5])     # prints all elements from index 2 to 5
7. print (student[3:])      # prints all elements from index 2 to the end of
the list
8. print (student * 2)      # prints the list two times
9. print (student + student) # Concatenate list with itself
```

Output: Listing 11.10

['John', 20, 2.23, 85, 'New York']
John
[2.23, 85, 'New York']
[85, 'New York']
['John', 20, 2.23, 85, 'New York', 'John', 20, 2.23, 85, 'New York']
['John', 20, 2.23, 85, 'New York', 'John', 20, 2.23, 85, 'New York']

11.1.12 Python Tuples

Another sequence-related data type similar to the list is called a tuple. Similar to a list, a tuple is a combination of elements separated with commas, but instead of square brackets, the elements of the tuple are enclosed into parentheses "(( ))". As modification is not allowed in tuple, so we can say tuple is a read-only list.

Listing 11.11 Tuples in Python

```
1. # This tuple contains information of student in order of Name, Age,
CGPA, State, marks
2. tuple = ('John', 20 ,2.23, 85, 'New York')
3.
4. print (tuple)        # prints complete tuple
5. print (tuple [0])   # prints the first element of the tuple
6. print (tuple [2:5])  # prints all elements from index 2 to 5
7. print (tuple [3:])   # prints all elements from index 2 to the end of the
tuple
8. print (tuple * 2)      # prints the tuple two times
9. print (tuple + tuple) # Concatenate tuple with tuple
```

Output: Listing 11.11

['John', 20, 2.23, 85, 'New York']
John
[2.23, 85, 'New York']
[85, 'New York']
['John', 20, 2.23, 85, 'New York', 'John', 20, 2.23, 85, 'New York']
['John', 20, 2.23, 85, 'New York', 'John', 20, 2.23, 85, 'New York']

11.1.13 Python Dictionary

Python has a data type that is similar to a hash table data structure. It works like an associative array or key-value pair. The key of a dictionary can be any data type of Python but most of the time, only a number or string is used for the key. As compared with other data types, numbers and strings are more meaningful for keys. The value part of the dictionary can be any Python object. Dictionary is enclosed by curly braces ({ }), while the value of the dictionary can be accessed using square braces ([ ]).

Listing 11.12 Dictionary in Python

```
1. testdict = { 'Name': 'Mike','Age':20, 'CGPA': 3.23, 'Marks':85}
2. testdict ['State'] = 'New York'
3.print (testdict)  # Prints complete dictionary
4.print (testdict ['Age'])  # Prints value of dictionary have key Age
5.print (testdict.keys())   # Prints all the keys of dictionary
6.print (testdict.values())  # Prints all the values of dictionary
```

Output: Listing 11.12

```
{'Name': 'Mike', 'Age': 20, 'CGPA': 3.23, 'Marks': 85, 'State': 'New York'}
20
dict_keys(['Name', 'Age', 'CGPA', 'Marks', 'State'])
dict_values(['Mike', 20, 3.23, 85, 'New York'])
```

11.1.14 If Statement

If the Boolean expression in the "if" part will output *true*, then statements followed by expression will execute. In the case of *false* output, statements will not execute.

Syntax
```
if expression:
  statement(s)
```

Listing 11.13 If statement in Python

```
1.booleanTrueTest = True
2.if booleanTrueTest: # Testing true expression using boolean true
3.print ("Expression test of if using Boolean true. If the expression is
true then this message will show.")
4.
5.numericTrueTest = 1
6.if numericTrueTest: # Testing true expression using numeric true
7.print ("Expression test of if using numeric true. If the expression is
true then this message will show.")
8.
9.print ("Testing finished")
```

Output: Listing 11.13

Expression test of if using Boolean true. If the expression is true then this message will show.

Expression test of if using numeric true. If the expression is true then this message will show.

Testing finished

11.1.15 If...Else Statement

If the output of Boolean expressions in the "if" part will return *true*, then statements in the "if" part will be executed. If the output will be *false*, then statements in the "else" part will be executed.

Syntax

```
if expression:
  statement(s)
else:
  statement(s)
```

Listing 11.14 If...else in Python

```
booleanTrueTest = False
if booleanTrueTest: # Testing true expression using Boolean true
print ("Expression test of if using Boolean true. If the expression is
true then this message will show.")
else: # It will execute if the expression output is false
print ("Showing this message means the output of expression is false.")

print ("Testing finished")
```

   Output: Listing 11.14

Showing this message means the output of expression is false.
Testing finished

11.1.16   elif *Statement*

In Python "elif" statement is used to check multiple expressions and execute one or
more statements. Similar to "else," the "elif" statement is also optional. The expression of first "elif" will be tested if expression of "if" will evaluate as false. If any
expression of any "elif" is false, then the next "elif" will be tested. The process will
continue until we reach the "else" statement at the end. If the expression of any "if"
or "elif" will be *true*, then all the remaining "elif" and "else" will be skipped.

Syntax

```
if expression1:
  statement(s)
elif expression2:
  statement(s)
elif expression3:
  statement(s)
else:
  statement(s)
```

Listing 11.15 elif in Python

```
1.marks = 95
2.
3.  if marks == 100:
4.print ("Your marks are 100")
5.  elifmarks >= 90 and marks <100:
6.  print ("Your marks are 90 or more but less than 100")
7.  elifmarks == 80:
8.  print ("Your marks are 80")
9.  elifmarks >0:
10.  print ("Your marks are more than 0")
11.else:
12.  print ("Your marks are negative")
13.
14.print ("Testing Finished")
```

   Output: Listing 11.15

Your marks are 90 or more but less than 100
Testing Finished

11.1.17   Iterations or Loops in Python

In Python sometimes we need to execute a statement or a block of statements several times. Repetitive execution of a statement or block of statements or code over and over is called iteration or loop.

   Loops in Python can be divided into two categories on the basis of definite and indefinite iterations. If we know the exact number of iterations, the loop is a definite loop. For example, iterate the block of code ten times.

   If we do not know the exact number of iterations, the number of iterations is decided on the basis of any condition. This type of loop is called an indefinite loop, for example, the execution of a loop to read characters from a provided text file.

11.1.18   While Loop

While loop statements in Python language repeat execution till the provided expression or condition remains *true*.

Syntax
```
while expression:
  statement(s)
```

While loop, just like in other programming languages, executes a statement or set of statements until the specified condition remains true. There is a possibility that while loop may not run even for one time. This can happen if the expression will return *false* on the testing for the first time.

Listing 11.16 While in Python

```
1. character = input ("Enter 'y' or 'Y' to iterate loop, any other key
to exit: ")
2. counter = 1;
3. while character == 'y' or character == 'Y':
4.    print ('Loop iteration :', counter)
5.    counter += 1
6. character = input ("Enter 'y' or 'Y' to iterate loop again, any other
key to exit: ")
7.
8. print ("Testing Finished")
```

Output: Listing 11.16

Enter 'y' or 'Y' to iterate loop, any other key to exit: y
Loop iteration : 1
Enter 'y' or 'Y' to iterate loop again, any other key to exit: Y
Loop iteration : 2
Enter 'y' or 'Y' to iterate loop again, any other key to exit: n
Testing Finished

In the "while" loop program given above, we are taking an input character from the user. If the character entered by the user is "y" or "Y," then we are executing the statements of the "while" loop. When the user enters any other character, then the loop stops iterating.

11.1.19 For Loop

For loop in Python is similar to for loop in other programming languages. It is used to execute a statement or set of statements for a fixed number of times. We can use for loop to iterate through the arrays, strings, items in tuples and dictionaries, etc.

Syntax
```
for iteration_variable in sequence:
   statements(s)
```

Here "iterating variable" is an item of sequence (list, tuple, diction, string).

When we start the "for" loop, the first item from the sequence is assigned to the "iteration_variable" and loop statement(s) execute(s) for the first time. In the next iteration, the next item from the sequence is assigned to the "iterating variable" and loop statement(s) execute(s) again. This process continues till we reach the end of the sequence (list) or the expression in the sequence returns *false*.

Listing 11.17 Iterating list through elements using for loop in Python

```
Colors = ['Red', 'White', 'Black', 'Pink'] # A list of some colors

1. print('Characters of the word "Colors" are:')
2. # this loop is printing each letter of string Colors
3. for letter in 'Colors':
4. print (letter)
5. print("The First loop's iterations finished \n")
6.
7. # this loop is printing each element of a list
8. for color in Colors:
9.print (color)
10.
11. print("The Second loop's iterations finished")
```

Output: Listing 11.17

Characters of the word "Colors" are:
C
o
l
o
r
s
The First loop's iterations finished

Red
White
Black
Pink
The Second loop's iterations finished

In Python, there is another way to iterate through each item of the list by index offset. In the following example, we have performed iteration by using the index offset method.

Listing 11.18 Iterating list through index using for loop in Python

```
1. Colors = [Blue', 'Green', 'White', Black'] # A list of some colors
2. # this loop is printing each element of a list
3. for index in range(len(Colors)):
4.print (Colors[index] + " is at index " + str(index) + " in the list")
5.
6. print("Loop iterations finished")
```

Output: Listing 11.18

Blue is at index 0 in the list
Green is at index 1 in the list
White is at index 2 in the list
Black is at index 3 in the list
Loop iterations finished

In the code given in Listing 11.18, we are using three built-in functions.

The range( ) function is used for the number of iterations. If we want to iterate the loop ten times, then we should provide 10 into the argument of the range( ) function. For this configuration, the loop will iterate from index 0 to 10. We can also use range ( ) for other specific ranges. For example, if we want to iterate the loop from index 5 to 10, then we will provide two arguments to the range ( ) function. The first argument will be 5 and the second will be 10.

The second built-in function that we used is len( ); this function counts the number of elements in the list.

Later we used the str( ) function, which is a data type conversion function that is used to convert a number into a string data type.

11.1.20   Using Else with Loop

Python provides a unique feature of loops that is not available in many other famous programming languages and that is the use of the else statement with loops. In other languages, we can use the else statement only with the if statement.

We can use the else statement with for loop but it will execute only when the loop has exhausted the iteration list. It means if the loop will iterate through the complete list, then else will be executed. But if there will be a break statement to stop the loop iterations, then else will not be executed.

We can use the else statement with the while loop too, but similar to the for loop, the execution of the else statement with the while loop will depend on the loop execution status. If the condition will become false and the loop will stop, then else statement will be executed. But if the loop will stop due to the use of a break statement, then else statement will not be executed.

The following two programs are demonstrating both scenarios of executing and not executing else statement with for and while loop.

Listing 11.19 Else statement with for loop in Python

```
1. for iin range(1, 5):
2.    print(i)
3. else:  # it will execute because the break is not used
4.print("Else statement executed")
5. print("Loop iterations stopped with all iterations finished\n")
6.
```

```
7. for j in range(1, 5):
8. print(j)
9. if j ==3:
10. print("using break statement")
11. break
12. else: # it will not execute because the break is used
13. print("No Break")
14. print("Loop iterations stopped with the use of break statement")
```

Output: Listing 11.19

1
2
3
4
Else statement executed
Loop iterations stopped with all iterations finished

1
2
3
using break statement
Loop iterations stopped with the use of break statement

Listing 11.20 Else statement with while loop in Python

```
1. character = input("Enter 'y' or 'Y' to iterate the loop, any other key
to exit: ")
2. counter = 1;
3. while character == 'y' or character == 'Y':
4.    print ('Loop iteration :', counter)
5.    counter += 1
6. character = input("Enter 'y' or 'Y' to iterate the loop again, any
other key to exit: ")
7. else: # it will execute because the break is not used
8. print('Else statement executed')
9. print('Loop iteration stopped because the condition became false
\n')
10.
11. character = input("Enter 'y' or 'Y' to iterate the loop, any other key
to exit: ")
12. counter = 1;
13. while character == 'y' or character == 'Y':
14.    print ('Loop iteration :', counter)
15.    counter += 1
16. if counter == 3:
17. print('using break statement')
```

```
18.break;
19.    character = input("Enter 'y' or 'Y' to iterate the loop again, any
other key to exit: ")
20. else: # it will execute because the break is used
21.print ('Else statement executed')
22. print ('Loop iteration stopped because the break is used')
```

Output: Listing 11.20

Enter 'y' or 'Y' to iterate the loop, any other key to exit: y
Loop iteration : 1
Enter 'y' or 'Y' to iterate the loop again, any other key to exit: Y
Loop iteration : 2
Enter 'y' or 'Y' to iterate the loop again, any other key to exit: n
Else statement executed
Loop iteration stopped because the condition became false

Enter 'y' or 'Y' to iterate the loop, any other key to exit: y
Loop iteration : 1
Enter 'y' or 'Y' to iterate the loop again, any other key to exit: Y
Loop iteration : 2
using break statement
Loop iteration stopped because the break is used

11.1.21   Nested Loop

Python allows us to put a loop inside another loop. If the outer loop will execute, then the inner loop may also execute. It depends on the truth value of the expression of the inner loop. Normally nested loops are used when we need to extract the smallest item from data. Suppose we have a list that contains different color names and we want to print each character of a color name. In this scenario, we need to use two loops: one outer loop will be used to extract each color name from the list, and the other inner loop will be used to get every character from a color name.

Syntax of nested for loop
```
for iterating_var in sequence:
  for iterating_var in sequence:
    statements(s)
  statements(s)
```

Syntax of nested while loop
```
while expression:
  while expression:
    statement(s)
  statement(s)
```

Listing 11.21 Nested for loop to print every character from the list of colors

```
Colors = [Blue', 'Green', 'White', Black'] # A list of some colors
# this loop is printing each element of a list
for index in range(len(Colors)):
print(Colors[index])
for letter in Colors[index]:
print(letter, end=' ')
print()
print("Loop iterations finished")
```

Output: Listing 11.21

```
Blue
B   l   u   e
Green
G   r   e   e   n
White
W   h   i   t   e
Black
B   l   a   c   k
Loop iterations finished
```

Listing 11.22 Nested while loop to print asterisk in grid

```
1. # this nested loop will create a grid of 3 rows and 4 columns
2. rows = 0
3. while rows <3:
4.    cols = 0
5.    while cols <4:
6.       print("*", end=' ')
7.       cols = cols + 1
8.    print()
9.    rows = rows + 1
```

Output: Listing 11.22

```
* * * *
* * * *
* * * *
```

11.1.22   Function in Python

Sometimes we need to execute some statements multiple times. If we need to execute the statements at the same place, then we can use the loops, but if we need to execute them at different places, then we should create a function. A function is a block of statements that can be used to perform some specific task.

In the coding examples given above, we have already covered some functions like print( ), range( ), len( ), etc., but these functions are built-in functions. The functions provided by language in the form of libraries are called built-in functions. We can design our custom functions which are called user-defined functions.

11.1.23   User-Defined Function

To define a user-defined function, we need to provide some details which are given below.

- To define a function, first we need to use the "def" keyword.
- Give the name of the function; it is always recommended to use a meaningful name. Meaningful means that by looking at the name of a function, anyone could judge what the function is doing.
- Provide the arguments of the function. All functions do not need arguments; the number of arguments depends on our requirements.
- The next step is to provide a string which is called a documentation string or docstring. This string provides information about the functionality of the function.
- Then we will write the statement(s) which we want to execute on the call of this function. These statements are also called function suites.

The last statement can be the return value. All function does not have a return value; in this case, we will use only the return keyword.

Syntax
```
def functionname ( parameters ) :
  "function_docstring"
function_suite
  return [expression]
```

Listing 11.23 User-defined function in Python

```
1. def myfunc(str):
2.    """This function prints any passed in arguments"""
3.    print (str)
4.    return;
6. # calling our user-defined function by providing an argument
8. myfunc ("Hello World!")
9. myfunc (10)
10. myfunc (5.5)
```

Output: Listing 11.23

Hello World!
10
5.5

11.1.24 Pass by Reference vs Value

"Pass by value" means the argument passed to function is a copy of the original object. If we will change the value of the object inside the function, then the original object will not change.

While "pass by reference" means the argument passed to function is the original object. If we will change the value of the object inside the function, then the original object will change.

In Python, arguments are passed to functions using the "pass by reference" method, so if the function will change them, then the original object will change. Let's test it with the help of an example.

Listing 11.24 Function argument pass by value and reference in Python

```
def myfunc ( score ) :
"""This changes the value of a passed list"""
score.append(40);
return
score = [10,20,30];
print ("Status of the list before calling the function: ", score)
myfunc ( score );
print ("Status of the list after calling the function: ", score)
```

Output: Listing 11.24

Status of the list before calling the function: [10, 20, 30]
Status of the list after calling the function: [10, 20, 30, 40]

11.1.25 Function Arguments

There are some other types of arguments that can be used while calling a user-defined function. The following is the list of these arguments:

- Required arguments
- Keyword arguments
- Default arguments
- Variable-length arguments

11.1.26   Required Arguments

In the "required arguments" type, the number of arguments passed to the function definition and those provided in the function call should be the same. If we will not provide the same number of arguments in the correct position, then we will get a syntax error.

Listing 11.25 Function in Python with required type arguments

```
def myfunc(str):
"""This function print any passed in arguments"""
print (str)
return
# calling our user-defined function by providing the argument
myfunc("Hello World!")
myfunc(10)
myfunc() # it will show a syntax error because we did not provide the
argument
```

Output: Listing 11.25

Traceback (most recent call last):
File "C:/Test.py", line 8, in <module>
myfunc() # it will show a syntax error because we did not provide the argument
TypeError: myfunc() missing 1 required positional argument: 'str'
Hello World!
10

11.1.27   Keyword Arguments

In "keyword arguments," it is not compulsory for function definition and function call to match the order of arguments. In this type, arguments are matched by their names. Look at the code given below; here we changed the order of arguments in the function definition and the function call. The argument in the first place in the function call is in second place in the function definition, and the argument in second place in the function call is in first place in the function definition.

Listing 11.26 Function in Python with keyword type arguments

```
def displayinfo( state, marks ):
"""This function prints passed info"""
print ("State: ", state)
print ("Marks ", marks)
return;
# calling function
displayinfo( marks=50, state="NY" )
```

Output: Listing 11.26

State: NY
Marks 50

11.1.28   Default Arguments

In the "default arguments" type, we provide a default value of an argument in the function definition. In a function call, if we do not provide the argument with the default value, then it will be auto-included. In the following code, when we called the function the second time and did not provide the second argument, the function used the default argument.

Listing 11.27   Function in Python with default type arguments

```python
def displayinfo( marks, location = 'USA'):
"""This prints the information passed to this function."""
print ("State: ", location, end=" ")
print ("Marks ", marks)
return;
# Now you can call printinfo function
displayinfo( marks=50, location="NY")
displayinfo( marks=50)
```

Output: Listing 11.27

State: NY Marks 50
State: USA Marks 50

11.1.29   Variable-Length Arguments

In this type of argument, when we call the function, we provide a different number of arguments. Note that we do not have any argument with the default value. For this scenario, the function call which will have fewer arguments will give a syntax error. To handle this problem, we need to use a variable-length argument type. In variable-length argument type, we provide the first argument as a variable and for the rest of the arguments, we use a tuple. In a function call, the first argument will be assigned to the first argument in the function definition. The rest of the arguments in the function call will be assigned to the tuple.

```
def funcname([formal_args,] *var_args_tuple ):
  "function_docstring"
function_suite
  return [expression]
```

To create a tuple, we just place an asterisk (*) before the variable name that holds the values of all non-keyword variable arguments. In a function call, if we will not provide additional arguments, then the tuple will remain empty.

Listing 11.28 Function in Python with variable-length arguments

```
def showmarks( marks, *moremarks ):
"""This function prints the variables passed through arguments"""
print ("Output is: ")
print (marks)
for varin moremarks:
print (var)
return;
# calling function
showmarks( 20 )
showmarks( 30, 40, 50 )
```

Output: Listing 11.28

Output is:
20
Output is:
30
40
50

11.1.30   The Return Statement

In Python, every user-defined function must have a return statement. The return statement returns a value. If we do not want to return any value from a specific function, then we should use only return keyword. Providing an expression to return the statement is optional.

Listing 11.29 Return statement in Python function

```
def findaverage( num1, num2 ):
"""finds the average of two numbers"""
total = num1 + num2
return total/2
```

```
# calling function
result = findaverage ( 10, 20 )
print ("The Average of the two numbers is: ", result)
```

   Output: Listing 11.29

The Average of the two numbers is: 15.0

11.2   Python IDLE

IDLE stands for "Integrated Development and Learning Environment." Python IDLE lets you write the code more efficiently. We can say that Python IDLE is a basic-level tool for a beginner that is available both for Windows and Mac OS.

   Python IDLE provides you with file editing and an interactive code interpreter. It works as a shell that lets you execute the Python command and provides you with the results. Although a basic-level tool, it provides some very essential features, e.g., highlighting of the syntax, auto indentation of the code, code completion, etc.

11.3   R Programming Language

R is another important programming language used for statistical analysis and data visualization. Similar to Python, it is also a common language for data science-related projects. R was created by Ross Ihaka and Robert Gentleman at the University of Auckland, New Zealand, and is currently developed by the R Development Core Team.

   To use the R language, we need to install two applications.

   First, we need to install the R language precompiled binary to run the code of the R language. R language precompiled binaries are available on the following link:
   https://cran.r-project.org/

   RStudio is an integrated development environment (IDE) intended for the development of R programming projects. Major components of RStudio include a console, syntax-highlighting editor that supports direct code execution, as well as tools for plotting, history, debugging, and workspace management. For more information on RStudio, you can follow the link:
   https://rstudio.com/products/rstudio/

11.3.1   Our First Program

Once everything is set, we are ready to create our first R script program.

Listing 11.30 First R program

```
mystr<- "Hello, World!"

print ( mystr)
```

Output: Listing 11.30

[1] "Hello, World!"

11.3.2 Comments in R

Comments are helping text in our R program which is ignored by the interpreter. Comments give us information about any specific portion or piece of code. In R comments start with the "#" sign.

Listing 11.31 Comments in the R program

```
# this is a comment in the R program on a separate line
mystr<- "Comments in R" # assigning a string to a variable

print ( mystr)
```

Output: Listing 11.31

[1] "Comments in R"

11.3.3 Data Types in R

In any programming language, we need variables to store some information; then later we can use this information in our program. Variables are reserved memory locations where we can store some information or data. The information we store in a variable can be string, number, Boolean values, etc. The size of the memory used by the variable depends on the type of the variable which is reserved through the operating system.

In other programming languages like C++ and Java, we decide the type of a variable when declaring it. However, in R, the type of variable is decided by the data type of the R-object. The value of the variable is called R-object. Similar to a variable in Python, a variable of R is loosely typed. It means we can assign any type of value to a variable. If we assign an integer, then the type of variable will become an integer. If we assign a string to a variable, then the type of variable will become a string.

Listing 11.32 Checking the type of variable after assigning R-object

```
var <- 10
print (paste("Type of variable:", class(var)))

var <- "string"
print (paste("Type of variable:", class(var)))

var <- 50L
print (paste("Type of variable:", class(var)))

var <- 2+5i
print (paste("Type of variable:", class(var)))

var <- TRUE
print (paste("Type of variable:", class(var)))
```

Output: Listing 11.32

[1] "Type of variable: numeric"
[1] "Type of variable: character"
[1] "Type of variable: integer"
[1] "Type of variable: complex"
[1] "Type of variable: logical"

There are many types of R-objects; the most frequently used ones are given below.

- Vectors
- Lists
- Matrices
- Arrays
- Factors
- Data frames

11.3.4   Vectors in R

To create a vector with more than one element, we need to use a built-in function called "c( )". This function is used to combine the elements of a vector.

Listing 11.33 Vectors example in R

```
#declaring a vector
colors <- c('red', 'green', 'yellow')

print(colors)

# getting the data type or class of vector
print(class(colors))
```

Output: Listing 11.33

[1] "red" "green" "yellow"
[1] "character"

11.3.5   Lists in R

A list is an R-object that contains different types of elements. An element of a list can be another list or a vector.

Listing 11.34 Example of list in R

```
#declaring a list
list <- list(c(2,5,3), 21.3, sin)

print(list)

# getting the data type or class of the list
print(class(list))
```

Output: Listing 11.34

[[1]]
[1] 2 5 3

[[2]]
[1] 21.3

[[3]]
function (x). Primitive("sin")

[1] "list"

11.3.6   Matrices in R

In R, a matrix is a collection of dataset arranged in the form of rows and columns. The following is an example of a matrix with three rows and two columns:

Listing 11.35 Matrices example in R

```
# create a matrix
M = matrix (c ('a', 'a', 'b', 'c', 'b', 'a'), nrow = 3, ncol = 2, byrow =
TRUE)

print (M)

# getting the data type or class of M
print (class (M))
```

Output: Listing 11.35

	[, 1]	[, 2]
[1,]	"a"	"a"
[2,]	"b"	"c"
[3,]	"b"	"a"

[1] "matrix"

11.3.7   Arrays in R

An array in R is similar to a matrix; however, a matrix can have only two dimensions but an array can have any number of dimensions. To create an array, we need to call a built-in function called "array( )". It uses an attribute named "dim" which has information about dimensions.

Listing 11.36 Arrays example in R

```
# Creating an array.
A <- array (c ('White', 'Black'), dim = c (3, 3, 2))

print (A)

# getting the data type or class of A
print (class (A))
```

Output: Listing 11.36

, , 1

	[, 1]	[, 2]	[, 3]
[1,]	"White"	"Black"	"White"
[2,]	"Black"	"White"	"Black"
[3,]	"White"	"Black"	"White"

, , 2

	[, 1]	[, 2]	[, 3]
[1,]	"Black"	"White"	"Black"
[2,]	"White"	"Black"	"White"
[3,]	"Black"	"White"	"Black"

[1] "array"

11.3.8   Factors in R

Factors are R-objects that are created with the help of a vector. To create a factor, we need to use the factor( ) function. It uses a vector as an argument and returns a factor object. Numeric and character variables can be converted into factors, but the levels of a factor will always be in the form of characters. To get the levels of a factor, we need to use the nlevels( ) function.

Listing 11.37 Factors example in R

```
# Creating a vector
colors <- c('white', 'green', 'black', 'yellow')

# Creating a factor object
F <- factor(colors)

# Print the factor.
print(F)
print(nlevels(F))

# getting the data type or class of F
print(class(F))
```

Output: Listing 11.37

[1] white green black yellow
Levels: black green white yellow
[1] 4
[1] "factor"

11.3.9   Data Frames in R

Data frames are R-objects in the form of tabular data which is similar to matrices in R, but in data frames, each column can have a different type of data. For example, the first column can be logical, the second can be numeric, and the third can be of character type. To create a data frame, we need to use built-in function data.frame( ).

Listing 11.38 Data frames example in R

```
# Creating a data frame
Frm<-    data.frame(
 gender = c("Male", "Male", "Female"),
 height = c(145, 161, 143.3),
 weight = c(70,65,63),
 Age = c(35,30,25)
)

print(Frm)

# getting the data type or class of Frm
print(class(Frm))
```

Output: Listing 11.38

	gender	height	weight	Age
1	Male	145.0	70	30
2	Male	161.0	65	35
3	Female	143.3	63	25

[1] "data.frame"

11.3.10   Decision-Making in R

Decision-making in R is performed by testing a Boolean expression (one or more conditions) that returns a true or a false value. The execution of a set of statements depends on this true or false value.

11.3.11   If Statement in R

"If" statement specifies the R statements which will be executed once the output of the condition will be true. The syntax of the "if" statement in R is given below.

Syntax
```
if(boolean_expression) {
  // statement(s) will execute if the Boolean expression will return true
}
```

Listing 11.39 If statement in R

```
num <- 50L
if(is.integer(num)) {
  print("num is an integer")
}
```

   Output: Listing 11.39

[1] "num is an integer"

11.3.12   If...Else in R

"If...else" is similar to the "if" statement; however, here we specify the else part as well. It means that we specify the statements that should be executed when the "if" condition becomes false. Following is the syntax of the "if...else" statement in R:

Syntax
```
if(boolean_expression) {
  // statement(s) will execute if the Boolean expression will return true
} else {
  // statement(s) will execute if the Boolean expression will return
false
}
```

Listing 11.40 If...else statement in R

```
num <- "50"
if(is.integer(num)) {
  print("num is an integer")
} else {
  print("num is not an integer")
}
```

Output: Listing 11.40

[1] "num is not an integer"

11.3.13   Nested If...Else in R

In R, "if" and "else if" statements are used as a chain of statements to test various conditions. Nested if, else if, and else statements start from an "if" statement followed by an optional "else" of the "if else" statement. We can use many "else if" statements after the first "if" statement which will end with an optional "else" statement. Execution of these statements starts from the top; if any of "if" or "else if" will return true, then the remaining "else if" and "else" statements will not run. Following is the syntax of the nested if, else if, and else statements:

Syntax
```
if(First Condition) {
  // statement(s) will be executed if First Condition will be true
} else if(Second Condition) {
  // statement(s) will be executed if Second Condition will be true
} else if(Third Condition) {
  // statement(s) will execute if Third Condition will be true
} else {
  // statement(s) will be executed all of the above conditions will be
false
}
```

Listing 11.41 Nested if...else statement in R

```
num <- 50

if(num > 80) {
  print("number is greater than 80")
} else if (num > 60) {
  print("number is greater than 60")
} else if (num > 40) {
  print("number is greater than 40")
} else if (num > 20) {
  print("number is greater than 20")
```

```
} else {
  print ("Number is less than or equal to 20")
}
```

   Output: Listing 11.41

[1] "number is greater than 40"

11.3.14   Loops in R

The R language has three types of loops as discussed below.

11.3.15   Repeat Loop in R

In R, to execute a statement or group of statements again and again, we can use a repeat loop. When we need to execute a block of code multiple times at the same place, then we use a repeat loop. The syntax of the repeat loop is given below.

Syntax
```
repeat {
  statement(s)
  if(boolean_expression) {
    break
  }
}
```

Listing 11.42 Repeat loop in R

```
vector <- c("Loop Test")
counter <- 0

# repeat loop
repeat {
  print(vector)
  counter <- counter + 1

  if(counter > 4) {
    break
  }
}

print(paste(c("Repeat loop iterated:", counter, "times"), collapse = "
"))
```

Output: Listing 11.42

[1] "Loop Test"
[1] "Loop Test"
[1] "Loop Test"
[1] "Loop Test"
[1] "Loop Test"
[1] " Repeat loop iterated: 5 times"

11.3.16 While Loop in R

In R, the repeat loop stops execution with the help of a break statement inside the "if" statement. While loop in R stops execution when the Boolean expression or condition of while loop returns false. Following is the syntax of the while loop in R:

Syntax
```
while (boolean_expression) {
  statement(s)
}
```

Listing 11.43 While loop in R

```
vector <- c("Loop Test")
counter <- 1

# while loop
while(counter < 5) {
  print(vector)
  counter <- counter + 1

}

print(paste(c("While loop iterated:", counter, "times"), collapse = "
"))
```

Output: Listing 11.43

[1] "Loop Test"
[1] "Loop Test"
[1] "Loop Test"
[1] "Loop Test"
[1] "While loop iterated: 5 times"

11.3.17 For Loop in R

In R language when we need to iterate one or more statements for a defined number of times, then we should use for loop. The syntax of using for loop in R language is given below.

Syntax
```
for (value in vector) {
  Body of Loop (Statements)
}
```

R's loops can be used for any data type including integers, strings, lists, etc.

Listing 11.44 For loop in R

```
vector <- c("Loop Test")
counter <- 0

# for loop
for(i in 6:10) {

  print(vector)
  counter <- counter + 1
}

print(paste(c("For loop iterated:", counter, "times"), collapse = "
"))
```

    Output: Listing 11.44

```
[1] "Loop Test"
[1] "Loop Test"
[1] "Loop Test"
[1] "Loop Test"
[1] "Loop Test"
[1] "For loop iterated: 5 times"
```

11.3.18 Break Statement in R

If we want to stop the execution of any loop at any iteration, we can use the break statement. Normally, we use the break statement with an "if" statement. When the "if" condition becomes true during the execution of the loop, the break statement stops the loop iterations. The break statement can be used with repeat, for, and while loops.

Listing 11.45 Checking the type of variable after assigning R-object

```
vector <- c("Loop Test")
counter <- 0

# repeat loop
repeat {
  print(vector)
  counter <- counter + 1

  if(counter > 4) {
    break
  }
}

print("Repeat loop stopped with the help of break statement"))
```

Output: Listing 11.45

[1] "Loop Test"
[1] "Loop Test"
[1] "Loop Test"
[1] "Loop Test"
[1] "Loop Test"
[1] " Repeat loop stopped with the help of break statement "

11.3.19   Functions in R

There are two types of functions in the R language called built-in and user-defined functions. The functions provided by the R language library are called built-in or predefined functions. The functions created and defined by the user are called user-defined functions. The following are components of a user-defined function:

- Function name
- Arguments
- Function body
- Return value

The following is the syntax of the user-defined function:

```
function_name<- function(arg_1, arg_2, ...) {
  Function body
}
```

Listing 11.46 Function in R

```
# creating a user-defined function
my.function<- function(arg) {
    print(arg)
}

# calling function
my.function(10)
my.function(5.8)
my.function("Hello World!")
```

Output: Listing 11.46

[1] 10
[1] 5.8
[1] "Hello World!"

11.3.20   Function Without Arguments

Listing 11.47 Function without argument in R

```
# creating a user-defined function
my.function<- function() {
    print("This function does not have any argument")
}

# calling function
my.function()
```

Output: Listing 11.47

[1] "This function does not have any argument"

11.3.21   Function with Arguments

Similar to Python, functions in R language can take the data in the form of arguments. The argument values are provided when the specific function is called.

Listing 11.48 Function with the argument in R

```
# creating a user-defined function
my.function<- function(arg) {
  print("This function has an argument")
  print(paste(c("The value of the provided argument is", arg), collapse
= " "))
}

# Call the function my.function supplying 30 as an argument.
my.function(30)
```

Output: Listing 11.48

[1] "This function has an argument"
[1] "The value of the provided argument is 30"

Summary

In this chapter, we have discussed two very important and most commonly used tools for developing data science applications. The first one is Python and the second one is the R programming language. We have provided the details of basic concepts along with programming examples. The overall intention was to provide you with a simple tutorial to enable you to write and analyze the data science programs and algorithms using these tools.

Further Reading

The following are some valuable resources for further reading:

- *Programming with Python* (Author: T R Padmanabhan), Springer & Co, 2016. The book provides complete and in-depth details of Python programming. The book uses real-life applications as a case study for teaching Python programming. Each and every practical example is properly explained along with the terminologies and the underlying concepts. By using the real-life applications, the book not only provides Python programming details but also how the language can help in developing these real-life applications.

- *Python Programming Fundamentals* (Author: Kent D. Lee), Springer & Co, 2014. The book explains how to learn and code in the Python programming language. Just reading the programming syntax is not sufficient. The reader should be able to apply the programming concepts to solve different problems. For this purpose, the book explains the concepts with the help of practical examples and exercises. The book also provides details about the use of the debugger tool to debug the programs.

- *Python Programming* by Ryan Turner, Nelly B.L. International Consulting LTD, 2020. The book provides a step-by-step guide to the readers. A few of the topics covered in the book include benefits of the Python, installation, and running of Python, complete coding instructions and their details, etc. One of the main features of the book is that it explains the concepts with real-life examples.

- *R for Data Science* by Garrett Grolemund and Hadley Wickham, O'Reilly Media, 2017. The book provides all the details a programmer may require to develop applications in the R programming language. The book discusses a range of topics including models, communication, tibbles, data imports, data visualizations, etc. Overall the book is an excellent resource that covers the majority of the R-related topics.

Exercises

Exercise 11.1: Consider the table given below and write Python code that will use if...else statements to check the grade of a student based on the marks of that student.

Marks range	Grade
<55	F
55–59	C−
60–62	C
63–65	C+
66–69	B−
70–74	B
75–79	B+
80–84	A−

Exercise 11.2: How many types of loops do we have in Python? Explain each of them with the help of a scenario and Python code.

Exercise 11.3: In Python, there is a built-in mechanism that checks whether the "for" loop has stopped after completing all iterations or due to the "break" statement; explain that mechanism with the help of Python code.

Exercise 11.4: Consider the table given below which contains the age of different people living in a community. Use nested loop in Python and display the ages of all the people.

59	70	44	23	75	66	67	72	60	71
91	71	15	39	46	39	91	49	6	88
78	21	32	2	36	94	46	88	26	5
53	71	93	14	35	37	77	54	54	66
100	72	51	15	83	47	51	66	45	74
18	10	90	54	42	98	66	81	6	37
98	77	28	82	67	54	61	41	100	9
85	59	30	53	19	82	31	69	87	23
58	58	67	87	50	36	2	89	54	42
71	20	9	68	78	72	65	80	13	25

Exercise 11.5: What will be the output of the following Python code?

```python
rows = 0
while rows <5:
 cols = 0
 while cols <3:
  print("*", end=' ')
  cols = cols + 1
 print()
 rows = rows + 1
```

Exercise 11.6: Consider the table shown below and write R code that will use if...else statements to check the grade of the student based on the marks of a student.

Marks range	Grade
<55	F
55–59	C–
60–62	C
63–65	C+
66–69	B–
70–74	B
75–79	B+
80–84	A–
85–89	A
>89	A+

Exercise 11.7: How many types of loops do we have in R? Explain each of them with the help of a scenario and R code.

Exercise 11.8: Consider the table given below which contains the age of different people living in a community. Use nested for loop in R and display the ages of all the people.

71	26	73	19	52	91	95	9	14	74
87	14	2	41	45	56	32	26	94	59
39	67	32	7	21	43	60	22	50	30
72	91	25	67	44	45	39	46	93	35
17	71	35	62	70	79	85	51	47	70
73	16	35	94	79	25	7	12	4	5
75	48	70	37	39	48	77	6	49	40
7	46	42	65	27	17	86	44	16	17
90	5	76	52	11	66	41	31	20	77
79	22	19	8	18	61	1	92	35	56

Chapter 12
Practical Data Science with WEKA

12.1 Installation

WEKA can be downloaded from the following official website:
https://www.cs.waikato.ac.nz/ml/weka/

The installation setup is available for Windows, Linux, and Mac OS. Download the setup file and execute it. Follow the installation steps as mentioned on the installation wizards that appear once the installation process starts. Once the installation is complete, launch the WEKA application just like any other normal applications. The following main window, shown in Fig. 12.1, will appear.

The interface comprises the main menu bar and five control buttons that are used to launch different interfaces of the WEKA. On the left bottom of the main interface, the information about WEKA along with its current version is provided. We will explain and discuss all of these components one by one.

Explorer is the first interface that is provided by the WEKA. Once the Explorer button is clicked, the Explorer interface launches as shown in Fig. 12.2.

In the Explorer interface, you can perform six main tasks related to data mining including preprocessing, classification, clustering, association, attribute selection, and visualization. It should be noted that once the Explorer interface is opened, the first tab that will be enabled will be the preprocessing tab.

The following are the main components of the Explorer interface:

1. These are six tabs that let you perform different tasks.
2. The dataset loading buttons which let you open the datasets from different file formats.

© The Author(s), under exclusive license to Springer Nature Switzerland AG 2023   393
U. Qamar, M. S. Raza, *Data Science Concepts and Techniques with Applications*,
https://doi.org/10.1007/978-3-031-17442-1_12

3. Filters let you choose and apply different filters on your loaded data as preprocessing task.
4. Attributes section shows and lets you select different attributes.
5. When you select each attribute, you can see the details of that attribute including the name, type, unique values, distinct values, etc.
6. You can also see some statistical calculations, e.g., the minimum value in the attribute, the maximum value, the mean and the standard deviation, etc. Similarly, you can see the other details, e.g., the number of counts of a single value in an attribute value set.
7. The graph: Graph represents the same measures in graphical format.

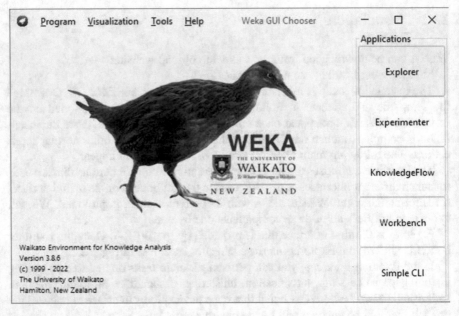

Fig. 12.1 WEKA main interface

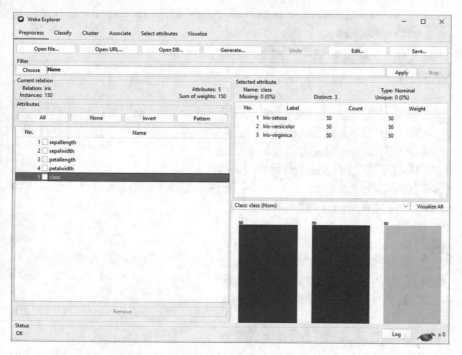

Fig. 12.2   WEKA main interface

12.2   Loading the Data

Now we will see how you can load your datasets in WEKA. You can load data from multiple formats in WEKA. The most common formats include Attribute-Relation File Format (ARFF), comma-separated values (CSV), and JSON file format. You can load the data from other file formats as well.

You can check the documentation about the details of the other file formats.

To open the dataset, you will have to perform the following tasks:

1. Click on the "Open file" button.
2. Select the required dataset. In the file extension filter box, you can select different file formats to open a specific file.
3. Select the dataset file and click "Open."
4. To load the dataset from an online URL, click the "Open URL…" button and then specify the URL path where the dataset resides.
5. Loading the dataset from the database is a bit complicated; you will have to provide the username and password along with the path of the database and query for fetching the data.

Once the dataset is loaded, the Explorer interface will be updated.

It should be noted that the WEKA already comes with many datasets that you can include for your experimentation. We will open the most common "IRIS" dataset.

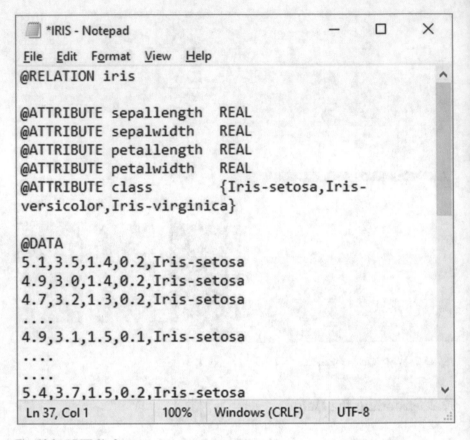

Fig. 12.3 ARFF file format

You can load the dataset from an existing file, from an online website, or from an existing database using ODBC connectivity, or you can generate your own dataset. Open the dataset named "IRIS" that already comes with the WEKA and check the Explorer interface. The dataset comes in ARFF file format. The ARFF file format is given in Fig. 12.3.

The details about the format are as below:

1. The keyword "@RELATION" shows the name of the relation which is IRIS in our case.
2. The keyword "@ATTRIBITE" shows the name of the attributes in the dataset along with their details, e.g., the data type, the values, etc.
3. "@DATA" represents the rows in data. For example, the first row shows that the value of the first attribute is "5.1," the value of the second attribute is "3.5," and so on. Finally, the value of the last attribute, i.e., "class" is "Iris-setosa."

Now when the dataset has been loaded, you can get the complete picture of the dataset. For IRIS:

- There are 150 instances
- There are five attributes
- There are four numeric attributes and one of nominal type
- The "class" attribute contains three values

Similarly, you can check the in-depth details of each individual attribute.

Many times an attribute that is not of a particular interest needs to be removed as part of preprocessing so that it does not contribute to later analysis. You can do this in WEKA by selecting the attribute and then clicking the "Remove" button at the bottom of the list of attributes.

12.3 Applying Filters

Once the data is loaded, the next task is to apply filters on the dataset attributes in order to prepare the data for data analysis tasks. We will convert the numeric attribute "sepallength" from numeric to nominal type. For this purpose you will have to perform the following steps:

1. Select the "sepallength" attribute from the attributes window.
2. Click the "Choose" button from the "Filter" section. A drop-down menu will appear.
3. Select filters, supervised, attribute, and then Discretize.
4. Click on the "Apply" button.
5. The name of the filter will appear in the textbox next to the "Filter" button and the filter will be applied.
6. The "Selected attribute" window will be updated and will show the details of the attribute according to its new type.

Figure 12.4 shows the list of filters available for preprocessing attributes in the case of supervised datasets. Similarly, you can select any filter for unsupervised datasets as well.

As you can see, the previous data has now been defined in terms of the range. The following ranges have been defined on the basis of the value:

1. From −infinity to 5.55 and the total values that fall in this range are 59.
2. From 5.55 to 6.15 and the total values that fall in this range are 36.
3. From 6.15 to infinity and the total values that fall in this range are 55.

Note that the weights are assigned according to the number of counts of a certain value in an attribute.

Note how the representation has changed before and after applying the filter. The new ranges and the total number of values in each range are represented in the form of the bar graph now. In Fig. 12.5, the graph on the left side shows the graphical representation before applying the filter, and the graph on the right side shows the graphical representation after applying the filter.

Now let's apply another filter. Suppose, we want to perform the attribute selection, i.e., we want to remove certain attributes. Last time, we saw that the attributes

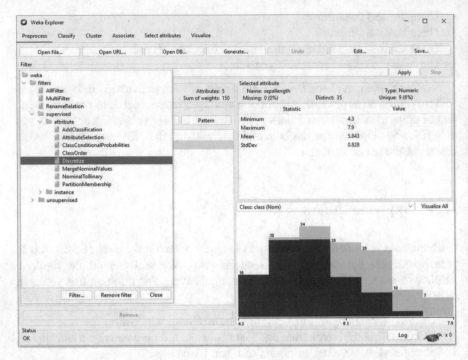

Fig. 12.4 List of filters available for preprocessing attributes in case of supervised datasets

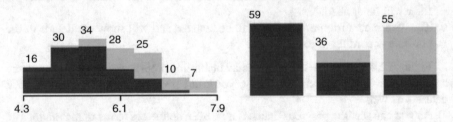

Fig. 12.5 Graphical representation of the attribute values before and after applying the filter

can be removed just by selecting the attribute and clicking the "Remove" button. However, that was the manual method. WEKA provides us with different filters for attribute selection where we can select the attributes according to different criteria.

1. Click the "Filter" button; select filters, supervised, attribute, and AttributeSelection.
2. Click the "Apply" button and the filter will be applied.

Note that now the two attributes "sepallength" and "sepalwidth" have automatically been removed by the attribute selection algorithm.

Once, you have performed all the preprocessing tasks and your data is ready for further processing, click the "Save" button to save the changes.

12.4   Classifier

We have already discussed classification in detail which lets you label the unknown
data by using the model based on your training set. For example, you may need to
classify whether the animal in a picture is a dog or cat. WEKA lets you load your
dataset, train the model, and then perform the classification process. A number of
classification algorithms are already available in WEKA for classification purposes.
We will provide the complete list of these algorithms later on.

As we have already discussed, when you load the dataset first, the "Preprocessing"
tab will be enabled. So, once you have preprocessed the data and saved it, click the
"Classify" tab, and the following interface will appear as shown in Fig. 12.6.

The interface lets you perform the classification. Mainly by using this interface,
you can select the classification algorithm and test options and see the output.

You have the following four test options:

- Use training set
- Supplied test set
- Cross-validation
- Percentage split

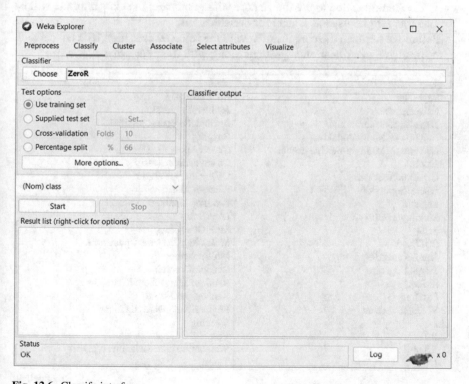

Fig. 12.6 Classify interface

These are the four most commonly used test options; however, you may have a number of others as given below:

- Output model
- Output model for training splits
- Output per-class stat
- Output entropy evaluation measures
- Output confusion matrix
- Store test data and predictions for visualization
- Collect predictions for evaluation based on AUROC, etc.
- Error plot point size proportional to the margin
- Cost-sensitive evaluation
- Preserve order for percentage split
- Output source code

You can check the details from the documentation. If you don't have your own training data, you can use the cross-validation method using n folds provided by the WEKA. Now you will specify the number of folds to split the data for training and testing purposes. You may have another option where you can specify the percentage value to split the data for testing purposes.

Below the "Test Options" sub-window, you can select the class attribute; however, it is recommended to use this feature with great care; the class attribute will be the one that will be used as a label for classification.

Before we perform the classification with the selected classifier, let's have a look at the available classification algorithms along with their categories:

Bayes:	Meta:
BayesNet	AdaBoostM1
NaiveBayes	AdditiveRegression
NaiveBayesMultinomial	AttributeSelectedClassifier
NaiveBayesMultinomialTest	Bagging
NaiveBayesMultionmialUpdateable	ClassificationViaRegression
Function:	CostSensitiveClassifier
GaussianProcesses	CVParameterSelection
LinearRegression	FilteredClassifier
Logistic	IterativeClassifierOptimizer
MultilayerPerceptron	LogitBoost
SGD	MultiClassClassifier
SGDText	MultiClassClassifierUpdateable
SimpleLinearRegression	MultiScheme
SimpleLogistic	RandomCommittee
SMO	RandomizableFilteredClassifier
SMOreg	RandomSubSpace
VotedPerceptron	RegressionByDiscretization
Lazy:	Stacking
IBK	Vote
KStar	WeightedInstancesHandlerWrapper
LWL	Trees:

(continued)

Rules:	DEcisionStemp
DecisionTAble	HoeffdingTree
JRip	J48
M5Rules	LMT
OneR	M5P
PART	RandomForest
ZeroR	RandomTree
	REPTree
	Misc:
	InputMappedClassifier
	SerializedClassifier

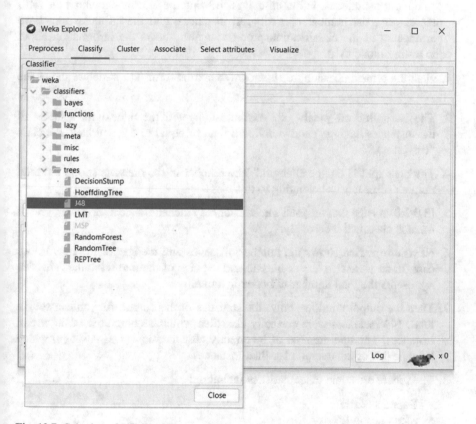

Fig. 12.7 Selection of J48 algorithm

The next step is to select the classifier. Note that we have loaded the IRIS dataset. You will perform the following steps to select the classifier:

1. Click on the Choose button.
2. A dropdown will appear that will show the list of options.
3. Select WEKA, then classifiers, then trees, and then J48 as shown in Fig. 12.7.

4. Select the Cross-validation option with 10 folds.
5. Click the Start button.
6. Classification will be performed and the output window will be updated with the results.

Note that the name of the classifier algorithm appears next to the "Choose" button. Once the algorithm completes the classification process, the output is displayed in the output window as shown in Fig. 12.8.

Now we will see what information is represented by the output windows:

1. First of all it provides run information, i.e., information about the data types and some input parameters. In our case, it is showing the information about the IRIS dataset, e.g., the number of attributes, instances, and the attributes that have participated in the classification process. It also shows the information about the testing mode used.

2. Next the pruned decision tree is shown in textual form. The structure shows the complete details.

3. The first split is on "petalwidth" attribute along with the value of the split. So all the instances having "petalwidth" less than or equal to 0.6 will have the class "Iris-setosa."

4. The next split is on "petallength." The number in the brackets shows the total number of instances belonging to this class.

5. (3.0/1.0) means that the total three instances reached this leaf; however, one of them is classified incorrectly.

6. Next information shows the number of leaves and the size of the tree. In our case, there were five leaves in total and the size of the tree was nine. The size represents the total number of nodes in a tree.

7. Then the output window shows the statistics of the output. As you can see, in total, 144 instances were correctly classified which become 96% of the actual instances. The total number of incorrectly classified instances is six (6) which become 4% of the overall classified instances.

8. Then there are some other statistics including:

   • Kappa statistic
   • Mean absolute error
   • Room mean squared error
   • Relative absolute error
   • Root relative squared error

9. After statistical measures, the detailed accuracy is shown with respect to each class.

10. Finally the confusion matrix is represented.

Note that Fig. 12.8 just shows the upper portion of the output window. Figure 12.9 shows the complete statistics that are represented in the form of results. We will examine all of these statistics one by one.

The following measures are shown for each class:

TP Rate	MCC
FP Rate	ROC Area
Precision	PRC Area
Recall	
F-Measure	

Note that the output window shows the resulting classification tree in textual form. WEKA provides you the facility to visually see the tree in graphical form as well.

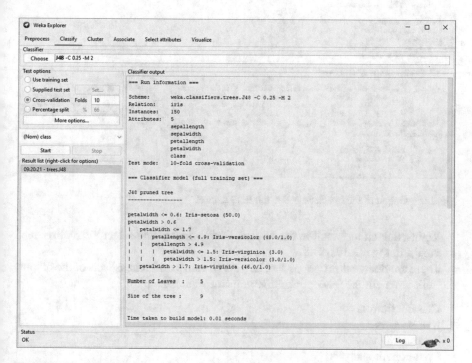

Fig. 12.8 The output window showing the results of the classification

```
=== Run information ===
Scheme:        weka.classifiers.trees.J48 -C 0.25 -M 2
Relation:      iris
Instances:     150
Attributes:    5
               sepallength
               sepalwidth
               petallength
               petalwidth
               class
Test mode:     10-fold cross-validation

=== Classifier model (full training set) ===

J48 pruned tree
------------------

petalwidth <= 0.6: Iris-setosa (50.0)
petalwidth > 0.6
|   petalwidth <= 1.7
|   |   petallength <= 4.9: Iris-versicolor (48.0/1.0)
|   |   petallength > 4.9
|   |   |   petalwidth <= 1.5: Iris-virginica (3.0)
|   |   |   petalwidth > 1.5: Iris-versicolor (3.0/1.0)
|   petalwidth > 1.7: Iris-virginica (46.0/1.0)

Number of Leaves  :    5

Size of the tree :     9

Time taken to build model: 0.01 seconds

=== Stratified cross-validation ===
=== Summary ===

Correctly Classified Instances         144               96     %
Incorrectly Classified Instances         6                4     %
Kappa statistic                          0.94
Mean absolute error                      0.035
Root mean squared error                  0.1586
Relative absolute error                  7.8705 %
Root relative squared error             33.6353 %
Total Number of Instances              150

=== Detailed Accuracy By Class ===

                 TP Rate  FP Rate  Precision  Recall  F-Measure  MCC    ROC Area  PRC Area  Class
                 0.980    0.000    1.000      0.980   0.990      0.985  0.990     0.987     Iris-setosa
                 0.940    0.030    0.940      0.940   0.940      0.910  0.952     0.880     Iris-versicolor
                 0.960    0.030    0.941      0.960   0.950      0.925  0.961     0.905     Iris-virginica
Weighted Avg.    0.960    0.020    0.960      0.960   0.960      0.940  0.968     0.924

=== Confusion Matrix ===

  a  b  c   <-- classified as
 49  1  0 |  a = Iris-setosa
  0 47  3 |  b = Iris-versicolor
  0  2 48 |  c = Iris-virginica
```

Fig. 12.9 Output of the J48 classifier on the IRIS dataset

Right-click on the result in the "Result List" window and select "Visualize tree" as shown in Fig. 12.10:

There are many other options that let you visualize different aspects of the output; the following are the few options that you can visualize:

- Classifier errors
- Margin curve
- Threshold curve
- Cost-benefit analysis
- Cost curve

Fig. 12.10 How to graphically visualize the tree

Once you click the "Visualize tree" option, a new window will appear that will graphically show the decision tree as shown in Fig. 12.11:

The "Visualize Classifier Errors" option lets you check and analyze the classification errors in visual form. Figure 12.12 shows the classification errors in our case.

The cross symbols represent the correct classifications, whereas the square symbols represent incorrect classifications. From the dropdowns showing the X- and Y-planes, you can select the attributes for which you want to see the visualization. For example, Fig. 12.12 shows the visualization between the "class" and the "sepallength" attribute.

Note that you are given some horizontal stripes on the right side of the visualization window. You can achieve the same visualization as shown above by using these strips as well.

Each strip represents an attribute. Clicking on the left side of the strip sets the attribute on the X-axis, whereas clicking the strip on the right side sets the attribute on the Y-axis.

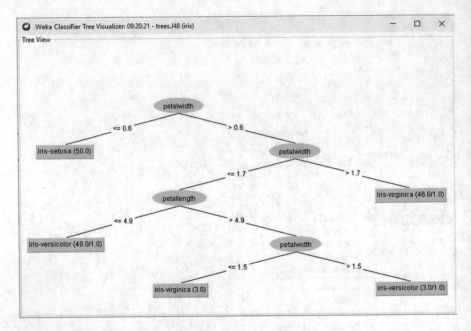

Fig. 12.11 Graphical representation of the visual tree

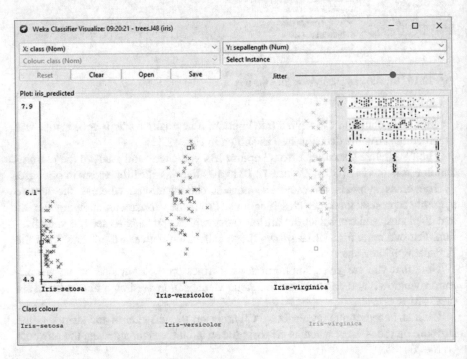

Fig. 12.12 Visualization of classification errors

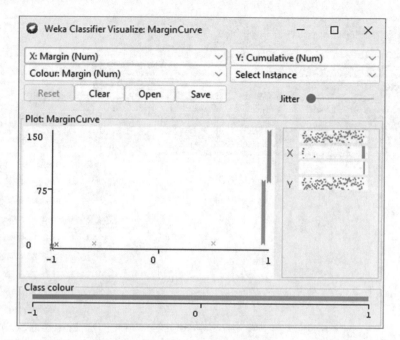

Fig. 12.13 Visualization of margin curve

There are many other visualization aids provided by WEKA. You can use these aids to refine your model. For example, Fig. 12.13 shows the margin curve.

Then we have the threshold curve. The threshold curve shows the tradeoffs in the prediction that can be achieved by varying the value of the threshold. The default value that is used for the threshold is 0.5.

It means that to predict an instance to be positive, the probability of the positive must be greater than 0.5. Figure 12.14 shows the threshold curve for the class "Iris-setosa."

Note that you can visualize the threshold curves for each decision class separately. You can also see the cost curve which provides the probability cost tradeoffs that can be obtained by varying the threshold value between the classes.

Again, you can find the cost curve for each class separately. Figure 12.15 shows the cost curve for class "Iris-setosa" in our IRIS dataset.

These were some of the details about the "Classify" tab. Next, we will discuss the "Cluster" tab in detail.

12.5   Cluster

Clustering is an important data mining task that deals with unsupervised data. We have already discussed the clustering process in detail. In this section, we will provide an in-depth discussion on how the WEKA helps you perform the clustering

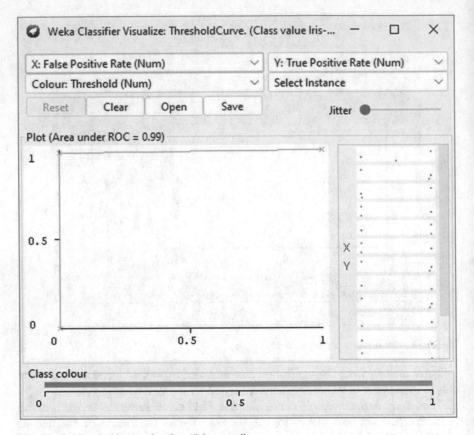

Fig. 12.14 Threshold curve for class "Iris-setosa"

tasks. For this purpose, again you will have to load the dataset and perform the preprocessing tasks if any are required. Then you will have to click the "Cluster" tab next to the "Classify" tab as shown in Fig. 12.16.

The clustering interface shows:

- The Cluster mode option that can be used for the testing purpose
- The Result display options
- The output window

We have already discussed the functionality of each of these windows. Here we will discuss them from the point of view of clustering algorithms. Note that till this point, you have already uploaded the dataset as was in the case of classification. However, because the clustering deals with unsupervised data, you will have to load the unsupervised dataset. To make things consistent, we have used the IRIS dataset; however, we will treat the "class" attribute just like a normal attribute.

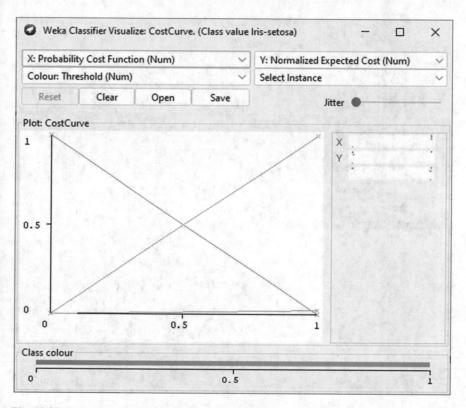

Fig. 12.15 Cost curve for Iris-setosa class in IRIS dataset

Once the dataset is loaded, the next task is to select the clustering algorithm. For this purpose you will have to perform the following steps:

1. Click the "Choose" button.
2. From the dropdown, select the clustering algorithm. We will use "EM" for this purpose.
3. The name of the clustering algorithm will appear next to the "Choose" button.

Once the clustering algorithm has been selected, the next task is to select the evaluation option. The WEKA provides the following evaluation options on the main interface:

- Use training set
- Supplied test set
- Percentage split
- Classes to clusters evaluation

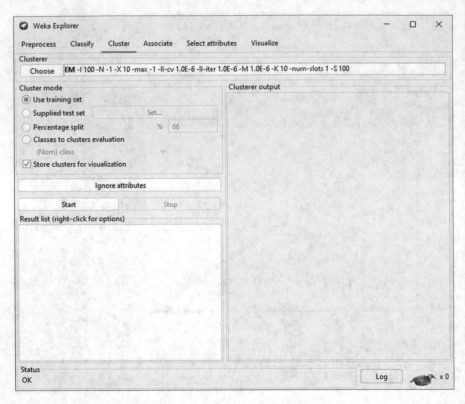

Fig. 12.16 Clustering interface

You can select any appropriate option according to your requirements. We have selected the "Classes to clusters" option as shown in Fig. 12.17:

Note that you have the option to ignore certain attributes from the clustering process. By clicking the "Ignore attributes" option, the following dialog box will appear as shown in Fig. 12.18:

Once the parameters are set, now you will click the "Start" button for the algorithm to execute. The algorithm will be completed and the output will be displayed in the output window just like we had in the case of the "Classify" interface.

The first part of the output window shows the information about the clustering along with the size of the dataset. Then it shows the number of clusters selected by cross-validation and the total number of iterations taken.

Then the standard deviation and the mean value are shown for each attribute in each cluster. The output window then shows the time taken by the WEKA to build the model. It may vary depending on the algorithm and the machine used. The output window then shows the number of instances in each cluster, for example, in the first cluster, there are 28 instances, which are almost 19% of the total clustered instances. Similarly, the second cluster comprises 35 instances in total.

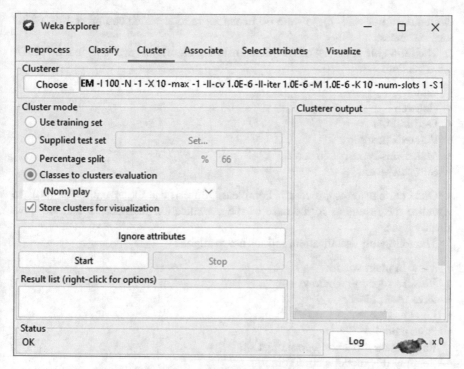

Fig. 12.17 Selection of "Classes to clusters evaluation" option

Fig. 12.18 Ignore attributes
dialog box

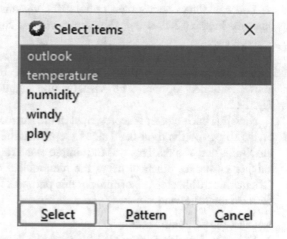

On the basis of the specified class attribute, the clusters have been named as well. The first cluster is named as "Iris-setosa," the second cluster is named as "Iris-virginica," and so on. Note that the last two clusters are not given any name. Finally, the output window displays the total number of instances that have been incorrectly clustered. In our case, these were 60 instances in total which comprise 40% of the total instances.

The following is the list of the other clustering algorithms available in the WEKA 3.8.6 version:

- Canopy
- Cobweb
- FarthestFirst
- FilteredClusterer
- MakeDensityBasedClusterer
- SimpleKMeans

Once the clustering process is complete, you can use the visualization tools to visualize the results as in the case of the classification interface, which we have already done.

The following visualization options are available:

- View in main window
- View in separate window
- Save result buffer
- Load model
- Save model
- Re-evaluate model on current test set
- Reapply this model's configuration
- Visualize cluster assignments

You can select appropriate options for visualization according to the type of analysis. Figure 12.19 shows the output window for the EM algorithm used for the IRIS dataset.

Now we will see how you can pictorially visualize the cluster using the "Visualize cluster assignments" options. Right-click on the result list and select the abovementioned option. The visualization window will appear as shown in Fig. 12.20.

Note that each cluster is represented in the form of a separate color. You can click on the stripes on the right-hand side to represent the attributes that should appear on the X-axis and Y-axis. This was the simple clustering method. Now we will perform another clustering practical using the hierarchical clustering method. We will use "HierarchicalClusterer" algorithm for this purpose. For this purpose you will have to perform the following steps:

1. Click the "Choose" button.
2. From the dropdown, select the "HierarchicalClusterer" as shown in Fig. 12.21.
3. The name of the clustering algorithm will appear in the textbox next to the "Choose" button.
4. Select "Classes to clusters evaluation" as the evaluation method.
5. Click the Start button.

```
=== Run information ===

Scheme:      weka.clusterers.EM -I 100 -N -1 -X 10 -max -1 -ll-cv
Relation:    iris
Instances:   150
Attributes:  5
             sepallength
             sepalwidth
             petallength
             petalwidth
Ignored:
             class
Test mode:   Classes to clusters evaluation on training data

=== Clustering model (full training set) ===

EM
==
Number of clusters selected by cross validation: 5
Number of iterations performed: 16

                    Cluster
Attribute           0       1       2       3       4
                    (0.18)  (0.23)  (0.28)  (0.15)  (0.15)
==========================================================
sepallength
  mean              4.7748  6.8585  6.1613  5.2823  5.5432
  std. dev.         0.2405  0.5228  0.4138  0.2407  0.3159

sepalwidth
  mean              3.1789  3.0862  2.8547  3.7037  2.5786
  std. dev.         0.2599  0.2891  0.2687  0.2857  0.2512

petallength
  mean              1.4194  5.7859  4.7484  1.5173  3.863
  std. dev.         0.1692  0.4745  0.3193  0.1592  0.3516

petalwidth
  mean              0.1948  2.1327  1.5757  0.3028  1.1696
  std. dev.         0.0557  0.2359  0.2196  0.1212  0.1351

Time taken to build model (full training data) : 0.37 seconds
=== Model and evaluation on training set ===
Clustered Instances
0       28 ( 19%)
1       35 ( 23%)
2       42 ( 28%)
3       22 ( 15%)
4       23 ( 15%)


Log likelihood: -1.60803

Class attribute: class
Classes to Clusters:

  0  1  2  3  4  <-- assigned to cluster
 28  0  0 22  0 | Iris-setosa
  0  0 27  0 23 | Iris-versicolor
  0 35 15  0  0 | Iris-virginica

Cluster 0 <-- Iris-setosa
Cluster 1 <-- Iris-virginica
Cluster 2 <-- Iris-versicolor
Cluster 3 <-- No class
Cluster 4 <-- No class

Incorrectly clustered instances :       60.0      40      %
```

Fig. 12.19 The output window of clustering interface

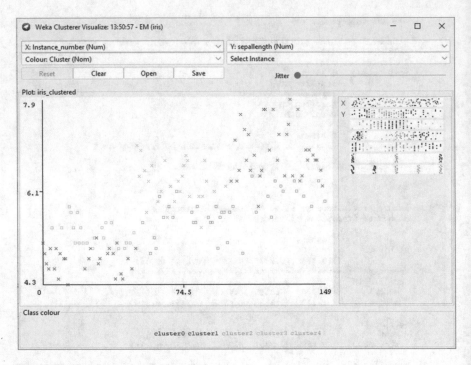

Fig. 12.20   The cluster visualization window

The algorithm will be executed, and the output window will be updated again. Note that the Result window now shows two clustering algorithms. You can click on any clustering algorithm to show its results. This helps you apply multiple algorithms to the same datasets and compare the results.

Now let's see the results of the hierarchical clustering algorithm. The output window first shows the following information:

- The clustering scheme and the information about the dataset used.

- The clusters formed are shown in textual form. It should be noted that two clusters are formed.

- The total number of instances is shown in each cluster. As it can be seen, the first cluster has 50 instances which form 33% of the total instances, and the second cluster has 100 instances which form 67% of total instances.

- The classes are assigned to each cluster. It can be seen that in our case, the class "Iris-setosa" was assigned to 50 instances in Cluster 0. Similarly, the class "Iris-versicolor" is assigned to 50 instances in Cluster 1, and the class "Iris-virginica" was assigned to 50 instances in Cluster 1.

- Finally, the output window shows the total number of instances that were incorrectly clustered. In our case, 50 instances were clustered incorrectly.

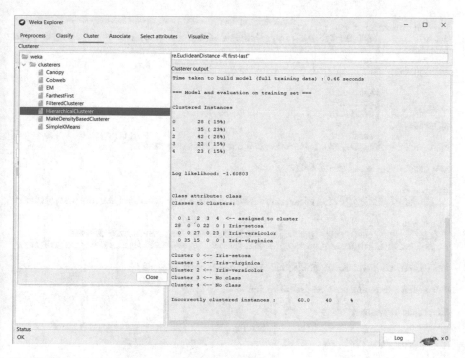

Fig. 12.21 Selection of hierarchical clustering algorithm

Figure 12.22 shows the complete picture of the output window for "HierarchicalClusterer" in the case of the IRIS dataset.

Note that the complete cluster structure has been truncated due to insufficient place on the page. Now you can visualize the result of the clustering algorithm just like we have done before. However, as you have used a hierarchical clustering algorithm now, the WEKA tools let you view the hierarchical structure of the clusters as well.

To view the simple cluster assignments, perform the following tasks:

1. Right-click on the "HierarchicalClusterer" in the "Result List" window.
2. Select "Visualize cluster assignments" from the context menu.
3. The cluster visualization window will appear as shown in Fig. 12.23.

Again the clusters are represented using different coloring schemes. The cross symbols represent the correctly clustered instances, whereas the squares represent the incorrectly clustered instances. To view the hierarchical clustering structure, perform the following tasks:

1. Right-click on the "HierarchicalClusterer" in the "Result List" window.
2. Select "Visualize tree" from the drop-down menu.
3. The cluster visualization window will appear as shown in Fig. 12.24.

```
=== Run information ===
Scheme:        weka.clusterers.HierarchicalClusterer -N 2 -L SINGLE -P -A "weka.core.Euc
Relation:      iris
Instances:     150
Attributes:    5
               sepallength
               sepalwidth
               petallength
               petalwidth
Ignored:
               class
Test mode:     Classes to clusters evaluation on training data

=== Clustering model (full training set) ===

Cluster 0
(((((((((((((((((((((((0.2:0.03254,0.2:0.03254):0.00913,(0.3:0.03254,0.3:0.03254):0.00913

Cluster 1
(((((((((((((((((((((((((((((1.4:0.07344,(((1.5:0.06508,1.5:0.06508):0.00066,
(1.4:0.05008,1.4:0.05008):0.01566):0.00224,1.3:0.06798):0.00546):0.00188,(1.3:0.07137,

Time taken to build model (full training data) : 0.04 seconds

=== Model and evaluation on training set ===

Clustered Instances

0        50 ( 33%)
1       100 ( 67%)

Class attribute: class
Classes to Clusters:

  0  1  <-- assigned to cluster
 50  0 | Iris-setosa
  0 50 | Iris-versicolor
  0 50 | Iris-virginica

Cluster 0 <-- Iris-setosa
Cluster 1 <-- Iris-versicolor
Incorrectly clustered instances :       50.0      33.3333 %
```

Fig. 12.22 The output window of the "HierarchicalClusterer" algorithm

12.6   Association Rule Mining in WEKA

Just like classification and clustering, association rule mining is also an important task that helps data scientists analyze transactional data and find the association rules from this. We have already discussed the association rules in detail in the previous chapter. Here we will discuss how the WEKA can help us find the association rules in data using different algorithms. Although there is a very limited number of algorithms available in WEKA 3.8.6, you can extend as per your own requirements.

One of the important and most common algorithms for association rule mining is the Apriori algorithm. WEKA provides a complete implementation of this algorithm along with setting different of its parameters.

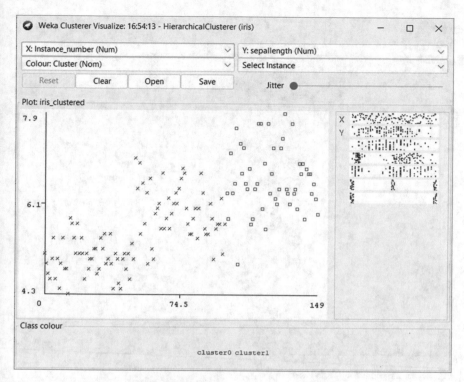

Fig. 12.23 Hierarchical clustering assignments

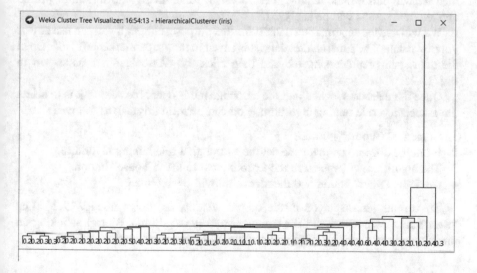

Fig. 12.24 Hierarchical clustering structure

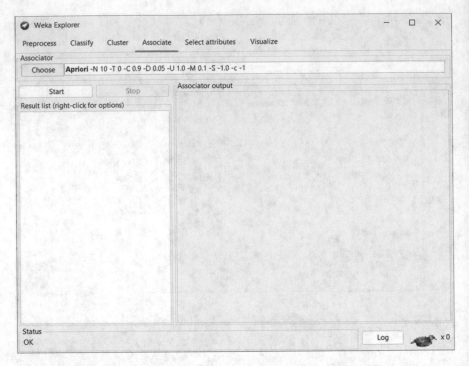

Fig. 12.25 Associate tab in WEKA

You will invoke the association rule mining algorithms in a similar way you have used other clustering and classification algorithms, i.e., first we will load the dataset, then we will perform the required preprocessing, and, finally, the algorithm will be applied, and results will be analyzed.

We will use the supermarket dataset for this purpose, although you can use any of your own data. The supermarket dataset is saved in the "supermarket.arff" file. Open the dataset in preprocessing tab and then click the "Associate" tab as shown in Fig. 12.25:

Once the dataset is loaded and the associate tab is open, the next task is to select the association rule mining algorithm. You can perform this task as follows:

1. Click the "Choose" button.
2. From the drop-down menu, select the association rule mining algorithm.
3. The algorithm will appear in the textbox next to the "Choose" button.
4. Click the "Start" button and the algorithm will be executed.

The results are displayed in the output window as shown in Fig. 12.26. The "Result List" window shows the completion of the algorithm. By right-clicking on the name of the algorithm, you will see the following visualization options:

- View in main window
- View in separate window
- Save result buffer
- Delete result buffer(s)
- Reapply this model configuration

Figure 12.26 shows the output window of the Apriori algorithm.
The output window shows the following information about the results:

1. The association rule mining scheme.
2. Dataset information (number of instances and attributes).
3. Minimum support and confidence level value.
4. Number of cycles taken by the algorithm.
5. The number of best rules found. In our case, 10 rules were found as mentioned in the settings of the algorithm.

By right-clicking on the name of the algorithm in the Explorer window, we can set the following options:

- Metric type
- Mini metric value
- Number of rows
- Output item sets
- Significance level
- Lower bound minimum support
- Upper bound minimum support
- Verbose

```
=== Run information ===

Scheme:       weka.associations.Apriori -N 10 -T 0 -C 0.9 -D 0.05 -U 1.0 -M 0.1 -S -1.0 -c -1
Relation:     supermarket
Instances:    4627
Attributes:   217
              [list of attributes omitted]
=== Associator model (full training set) ===

Apriori
=======

Minimum support: 0.15 (694 instances)
Minimum metric <confidence>: 0.9
Number of cycles performed: 17

Generated sets of large itemsets:

Size of set of large itemsets L(1): 44

Size of set of large itemsets L(2): 380

Size of set of large itemsets L(3): 910

Size of set of large itemsets L(4): 633

Size of set of large itemsets L(5): 105

Size of set of large itemsets L(6): 1

Best rules found:

 1. biscuits=t frozen foods=t fruit=t total=high 788 ==> bread and cake=t 723     <conf:(0.92)> lift:(1.27) lev:(0.03) [155] conv:(3.35)
 2. baking needs=t biscuits=t fruit=t total=high 760 ==> bread and cake=t 696     <conf:(0.92)> lift:(1.27) lev:(0.03) [149] conv:(3.28)
 3. baking needs=t frozen foods=t fruit=t total=high 770 ==> bread and cake=t 705     <conf:(0.92)> lift:(1.27) lev:(0.03) [150] conv:(3.27)
 4. biscuits=t fruit=t vegetables=t total=high 815 ==> bread and cake=t 746     <conf:(0.92)> lift:(1.27) lev:(0.03) [159] conv:(3.26)
 5. party snack foods=t fruit=t total=high 854 ==> bread and cake=t 779     <conf:(0.91)> lift:(1.27) lev:(0.04) [164] conv:(3.15)
 6. biscuits=t frozen foods=t vegetables=t total=high 797 ==> bread and cake=t 725     <conf:(0.91)> lift:(1.26) lev:(0.03) [151] conv:(3.06)
 7. baking needs=t biscuits=t vegetables=t total=high 772 ==> bread and cake=t 701     <conf:(0.91)> lift:(1.26) lev:(0.03) [145] conv:(3.01)
 8. biscuits=t fruit=t total=high 954 ==> bread and cake=t 866     <conf:(0.91)> lift:(1.26) lev:(0.04) [179] conv:(3)
 9. frozen foods=t fruit=t vegetables=t total=high 834 ==> bread and cake=t 757     <conf:(0.91)> lift:(1.26) lev:(0.03) [156] conv:(3)
10. frozen foods=t fruit=t total=high 969 ==> bread and cake=t 877     <conf:(0.91)> lift:(1.26) lev:(0.04) [179] conv:(2.92)
```

Fig. 12.26 Output of the Apriori algorithm

12.7   Attribute Selection

The data in the real world is growing day by day and it is common to have
applications processing data with hundreds and thousands of attributes. Attribute
selection is an important task that lets the data scientists reduce the amount of data
that needs to be processed for performing the analysis tasks.

WEKA provides you with a number of algorithms to select attributes. First, we
open the dataset. You can use any dataset; however, we will use "labor.arff" here.
Figure 12.27 shows the labor dataset loaded in the Explorer window.

You can see the dataset comprises 57 instances and 17 attributes. As we don't
need to preprocess this dataset, so we skip the preprocessing step and move to
attribute selection. For this purpose click on the "Select attributes" tab on the main
Explorer window, and you will get the WEKA interface to perform the attribute
selection.

On this interface, you will have to select the attribute selector algorithm and the
searching algorithm. The attribute selector algorithm performs the task of ranking
the attribute so that after a certain criterion, the selected attributes could be eliminated. The searching algorithms provide the searching scheme for the attributes.
Figure 12.28 specifies the "Select attributes" interface.

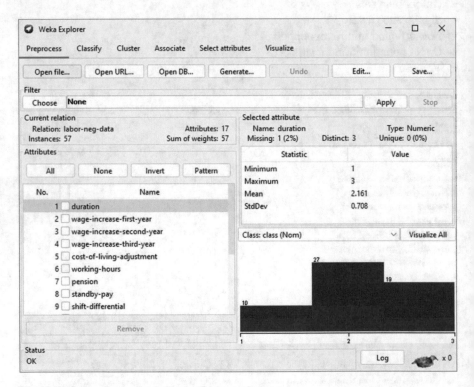

Fig. 12.27 Labor dataset loaded in the Explorer window

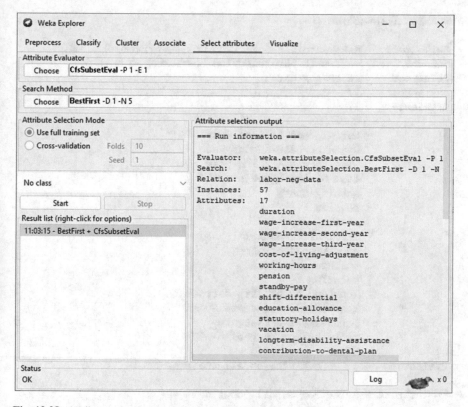

Fig. 12.28 Attribute selector tab on Explorer interface

We have selected the "CfsSubsetEval" algorithm for evaluation of the attributes and the "BestFirst" searching algorithm for search purposes. The following is the list of the algorithms provided by WEKA for attribute evaluation:

- CfsSubsetEval
- ClassfierAttributeEval
- ClassifierSubsetEval
- CorrelationAttributeEval
- GrainRatioAttributeEval
- InformationGainAttributeEval
- OneRAttributeEval
- PrincipalComponents
- ReliefAttributeEval
- SymmetricalUncertAttributeEval
- WrapperSubsetEval

Version 3.8.6 provides the three searching algorithms named BestFirst, GreedyStepwise, and Ranker. Figure 12.29 shows the complete output of the output window.

```
=== Run information ===

Evaluator:      weka.attributeSelection.CfsSubsetEval -P 1 -E 1
Search:         weka.attributeSelection.BestFirst -D 1 -N 5
Relation:       labor-neg-data
Instances:      57
Attributes:     17
                duration
                wage-increase-first-year
                wage-increase-second-year
                wage-increase-third-year
                cost-of-living-adjustment
                working-hours
                pension
                standby-pay
                shift-differential
                education-allowance
                statutory-holidays
                vacation
                longterm-disability-assistance
                contribution-to-dental-plan
                bereavement-assistance
                contribution-to-health-plan
                class
Evaluation mode:    evaluate on all training data

=== Attribute Selection on all input data ===

Search Method:
        Best first.
        Start set: no attributes
        Search direction: forward
        Stale search after 5 node expansions
        Total number of subsets evaluated: 114
        Merit of best subset found:    0.363

Attribute Subset Evaluator (supervised, Class (nominal): 17 class):
        CFS Subset Evaluator
        Including locally predictive attributes

Selected attributes: 2,3,5,11,12,13,14 : 7
                     wage-increase-first-year
                     wage-increase-second-year
                     cost-of-living-adjustment
                     statutory-holidays
                     vacation
                     longterm-disability-assistance
                     contribution-to-dental-plan
```

Fig. 12.29 Complete output of the output window

The output window first shows the attribute selection scheme and dataset information. Then the window mentions the search method and finally the output is displayed. It can be seen that we started with a total of 17 attributes and the final selection resulted in 7 attributes. So, the remaining 10 attributes were eliminated.

You can select the properties of both the attribute evaluation method and the searching algorithm by clicking right on the name of the algorithm and then selecting "Show properties." The following properties can be set for the attribute evaluation method:

- Debug
- donNotCheckCapabilities
- LocallyPredictive
- missingSeparate
- numThreads
- poolSize
- preComputeCorrelationMatrix

Similarly, the following properties can be set for the searching algorithm:

- Direction
- lookupCacheSize
- searchTermination
- startSet

You can visualize the results by clicking right on the results in the "Result List" window and selecting "Visualize Reduced data" as shown in Fig. 12.30.

12.8   The Experimenter Interface

As the name implies, the Experimenter interface lets users experiment on multiple algorithms, compare their results, and make conclusions. You can compare any number of algorithms. For this purpose, first, you will have to select the dataset and then the algorithms that you want to compare. We will execute these algorithms and then compare the results. In WEKA's main window, you will find the button to invoke the Experimenter interface. Click on this button and the Experimenter will be displayed as shown in Fig. 12.31. When the interface is opened, you will see the "New" button on the right side of the window. Click this button to launch a new experiment. Then click on the "Browse" button to specify the path of the file in which the results will be saved. Note that we will load the IRIS dataset and compare two classification algorithms here.

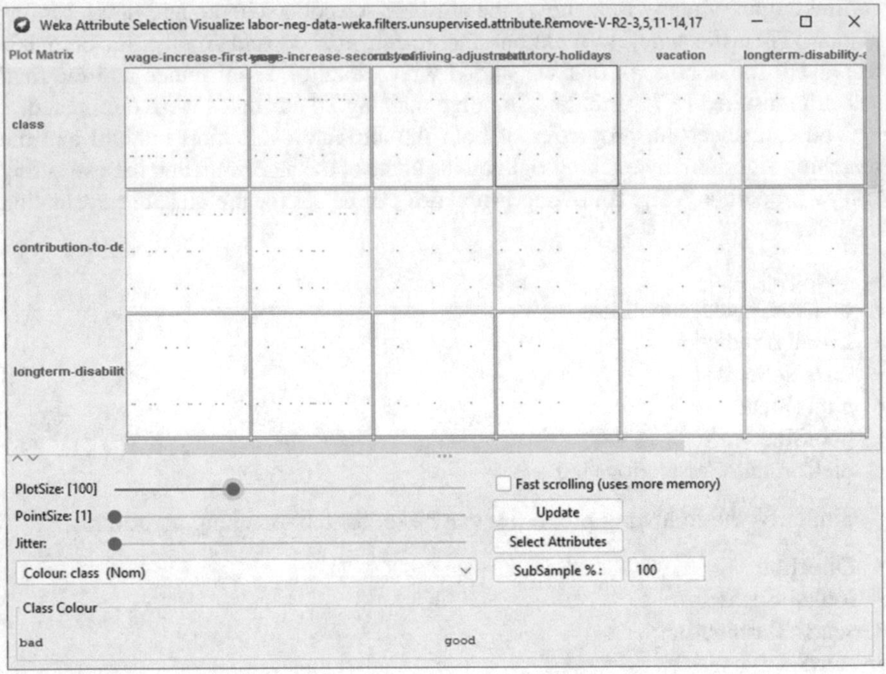

Fig. 12.30 Result visualization window

In the experiment type sub-window, you will specify the validation method. The following validation methods are available:

- Cross Validation
- Test/Train percentage split (data randomized)
- Test/Train percentage split (order preserved)

We will use the cross-validation method with 10 folds. To open the dataset, click on the "Add new" button on the left side of the window and specify the path of the dataset file. Once the path is specified, the path will appear in the textbox below the "Add new" button. So far, you have specified the dataset on which the experiment will be performed. The next step is to select the algorithms. For this, click on "Add new" in the "Algorithm subwindow" and from the algorithm selector, select the first algorithm. We have selected the "ZeroR" algorithm first. Then we have selected the "OneR" algorithm as shown in Fig. 12.32:

You can select any number of algorithms. Similarly, you can specify the greater number of datasets as well. You can edit the values of a particular dataset as follows:

1. Click the dataset file.
2. Click the 'Edit selected' button.
3. A new edit window will appear.
4. Double-click on any value and edit it.

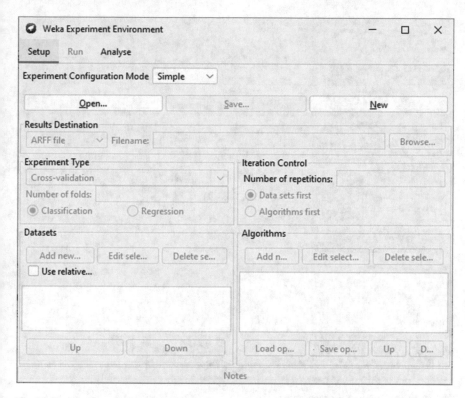

Fig. 12.31 The Experimenter interface

Once all the data and the algorithms are set, you are ready to perform your experimentation. Click on the "Run" tab and click the Start button. The results will be shown in the log window as shown in Fig. 12.33:

You can see that the experiment has been successfully completed without any error. The window also shows the start and the end time of the experiment.

Once the experiment has been completed, you can check the results of the experiment in the "Analyse" tab. This tab helps you perform different types of comparisons and view the results. You can specify different measures for result comparison.

Figure 12.34 shows the comparison of "ZeroR" and "OneR" algorithms using the "Percent_correct" measure on the IRIS dataset. To select a measure, click on the "Comparison field". A dropdown will appear and you will select a measure from the given list.

The output window will provide a comparison of the results. The window first shows the testing scheme and then the measure used for comparison. Finally, the window shows the results of the comparison in the form of a table. Note that you can view the complete output in a separate window by clicking right on the experiment name in the "Result List" window.

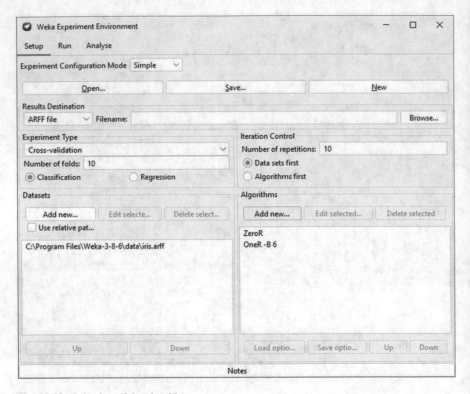

Fig. 12.32 Selection of the algorithms

You can save the output by clicking on the "Save output" button as well. Some important and most common measures that can be used are:

- F1 measure
- Root mean squared error
- Absolute error
- Kappa statistics
- Mean error

Note that we used a simple version of the Experimenter. You can use advance properties to perform advance settings. For this purpose you will have to select "Advance" from the drop-down menu on the Experimenter interface as shown in Fig. 12.35:

Note that now you can set the extra parameters, e.g., the total number of runs, the distribution of experiment, iteration controls, etc.

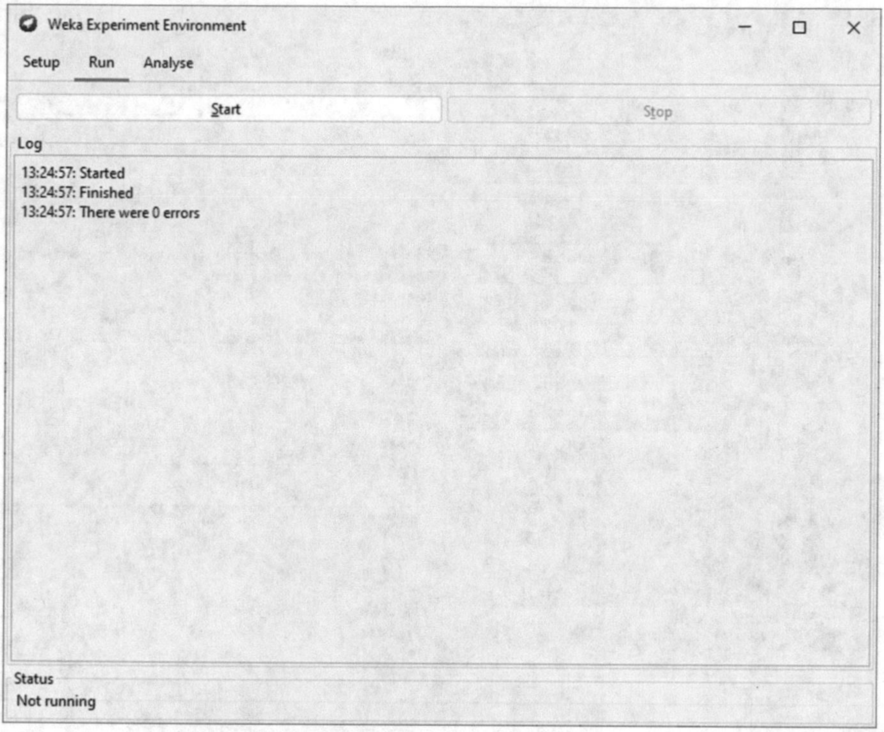

Fig. 12.33 Experiment result log window

12.9   WEKA KnowledgeFlow

So far, all the tasks that have been performed were through the controls provided on the WEKA interface. WEKA provides you the additional usability by giving you a Workbench interface, which you can use to mention your tasks using simple drag-drop controls. You will simply specify your workflow in the form of the flow graph, and the rest of the tasks will be performed by WEKA in the same way you have performed previously. To open the KnowedgeFlow interface, click on the "KnowledgeFlow" button on the main WEKA interface; the KnowedgeFlow interface will be opened as shown in Fig. 12.36:

As you can see on the left-hand side, there are different components of the KnowledgeFlow interface which you can use to perform your analysis tasks. You will have to select the appropriate component and drop it on the working window on the right-hand side. In this way, you can develop your complete model. Once the model is developed, you can then click on the play button in the toolbar to run the model. Once the model is completed, you can see the results.

The WEKA provides you with the complete set of tools right from the loading of data to the visualization of results that can be used. We will discuss this interface

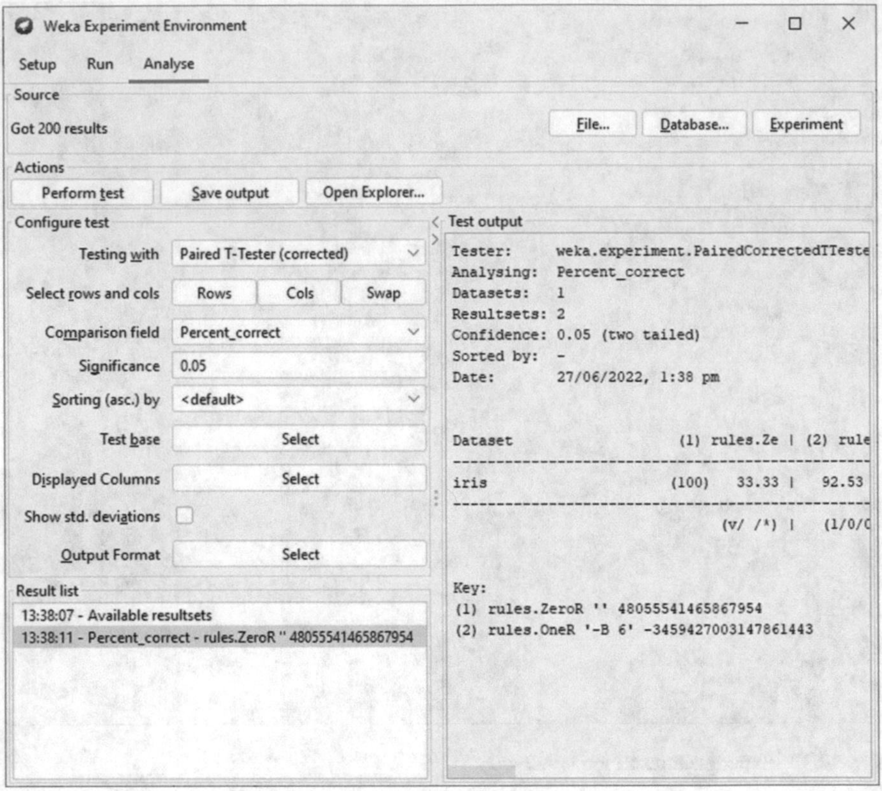

Fig. 12.34 Result comparison using the "Percent_correct" measure

with the help of an example. We will load our old IRIS dataset and then perform the classification task and finally, we will display the results.

To load the dataset, click on the "DataSources" node in the "Design" sub-window. You will have a number of dataset loading options, e.g., ARFFLoader, CSVLoader, JSONLoader, etc. You can select the appropriate loader component depending on the format in which your dataset is stored. As our IRIS dataset is stored in ARFF format, we will use ARFFLoader. Click on it and drop it on the working window as shown in Fig. 12.37:

Once the component is loaded, the next task is to link it with the dataset file. For this purpose perform the following tasks:

1. Right-click on the ArffLoader component.
2. Select the "configure" option from the popup menu.
3. A dialog box will appear as shown in Fig. 12.38.
4. Click on the "Browse" button and specify the dataset file path.
5. Once the ArffLoader component is configured, click on the OK button.

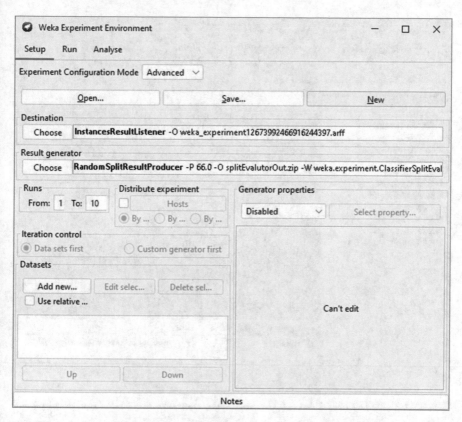

Fig. 12.35 Advance experimental interface

The next task is to assign a classification method to this dataset. For this purpose, we will use the "ClassAssigner" control. You will select this control from the "Evaluation" node as shown in Fig. 12.39:

Once the "ClassAssigner" control is placed on the working board, we need to assign this "ClassAssigner" to our dataset (i.e., ArffLoader control). For this purpose, right-click on the ArffLoader and select the "dataset" option from the popup menu. A line connector will appear that will move with the mouse pointer. Drag your mouse pointer toward the ClassAssigner and click on it. Both ArffLoader and ClassAssigner will be connected as shown in Fig. 12.40:

So far, our dataset is ready as input to the classification algorithm. However, before doing that, we need to split the data between the test set and the training set so that it could be used by the classification algorithm accordingly. For this purpose, we will use the "CrossValidationFoldMaker" component and connect it with the ClassAssigner component. Perform the same steps as performed previously and place the "CrossValidationFoldMaker" component on the working window.

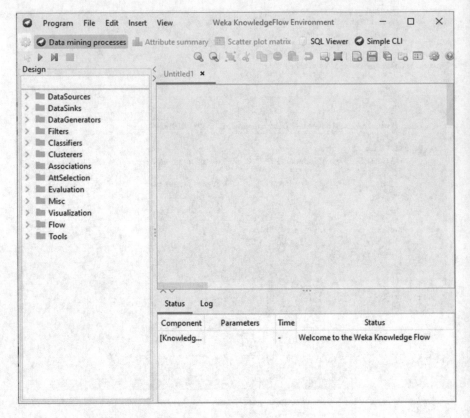

Fig. 12.36   WEKA KnowedgeFlow interface

Next, we need to assign the "ClassAssigner" component with the "CrossValidationFoldMaker." For this purpose right-click on the "ClassAssigner" component and select the "dataset" from the popup window. Connect the "ClassAssigner" component with the "CrossValidationFoldMaker." Note that we need to specify the number of folds for validation. For this purpose, right-click on the "CrossValidationFoldMaker" component and select "configure" from the popup menu. In the configuration dialog box, specify the number of folds. In our case, the value will be ten (10). Now click the "OK" button as shown in Fig. 12.41.

Our next task is to use the classifier. For this purpose, we will use the J48 classifier. So select the J48 classifier component from the "Design" sub-window and place it on the working window. Now we need to connect the "CrossValidationFoldMaker" with the J48 components twice – first for the training set and then for the test set.

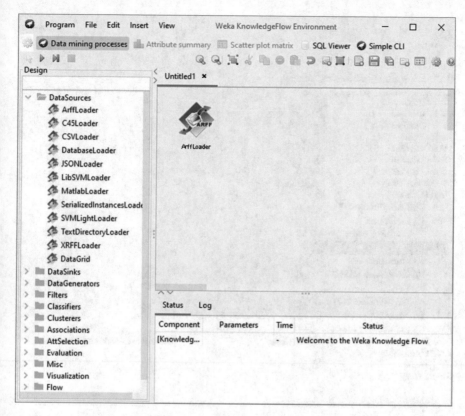

Fig. 12.37 ArffLoader component placed on the working window

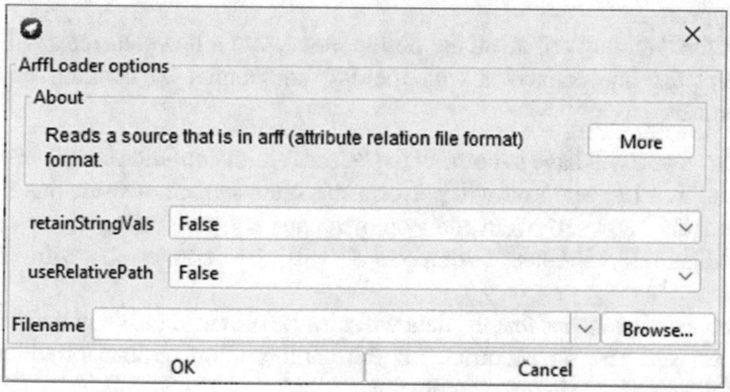

Fig. 12.38 Dataset linking dialog box

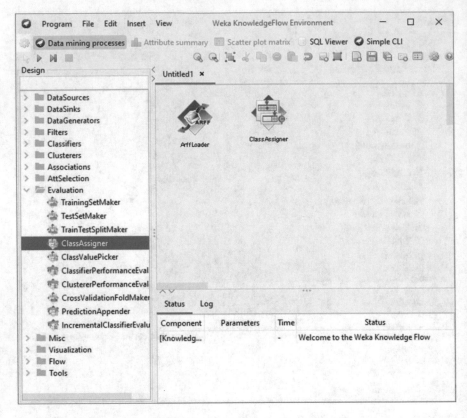

Fig. 12.39 ClassAssigner control

For connecting "CrossValidationFoldMaker" as a training set with the J8 component, perform the following tasks:

1. Right-click on the "CrossValidationFoldMaker" component, and a popup menu will appear.
2. Select the "training set" from the popup menu, and a line connector will appear.
3. Connect the line connector with the J48 component by clicking on the J48 component.

Similarly, you will have to connect the "CrossValidationFoldMaker" with the J48 component as a test set. You will perform the same steps; however, this time, you will select the "test set" from the popup menu. By performing these steps, the "CrossValidationFoldMaker" component will be connected with the J48 component.

This connection means that the data will now be provided to J48 in the form of the training set and test set according to the configuration options provided in the "CrossValidationFoldMaker" component.

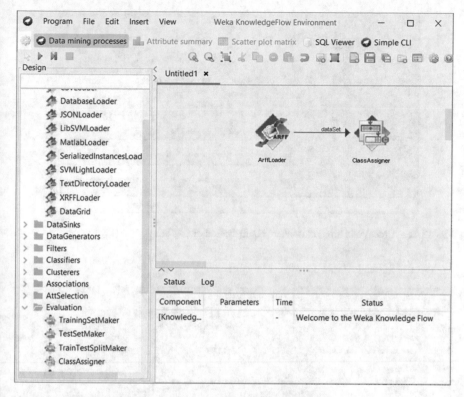

Fig. 12.40 Connector between the "ArffLoader" and "ClassAssigner"

Figure 12.42 shows the connection of "CrossValidationFoldMaker" with J48 components.

So far, we have read our dataset and have assigned it to a classifier. However, we need to insert the performance evaluation component for measuring the performance of our classification algorithm.

For this purpose, we will use the "ClassifierPerformanceEvaluator" component that can be selected from the "Evaluation" node.

Once the performance evaluation component has been inserted, we will connect the J48 classifier component with the performance evaluator component. For this purpose you will perform the following tasks:

1. Right-click on the J48 component and select "Batch classifier" from the popup menu.
2. A line connector will appear; drag the mouse pointer toward the performance evaluator component.
3. Click on the performance evaluator component to connect the classifier component with the performance evaluator component.

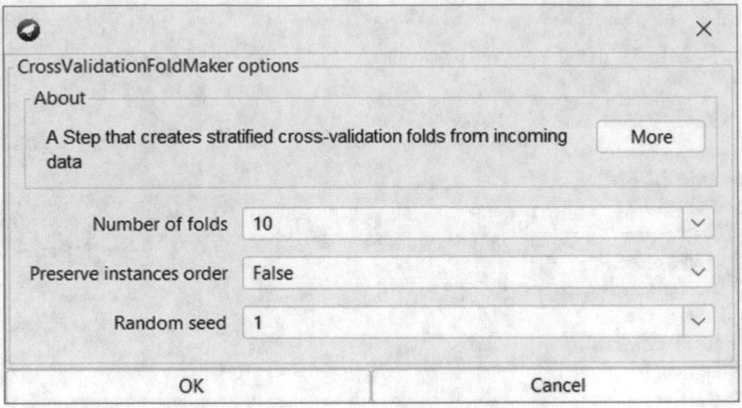

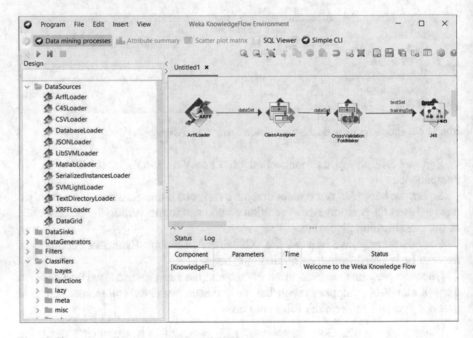

Fig. 12.41 "CrossValidationFoldMaker" configuration dialog box

Fig. 12.42 Cross-validation connection with J48 component

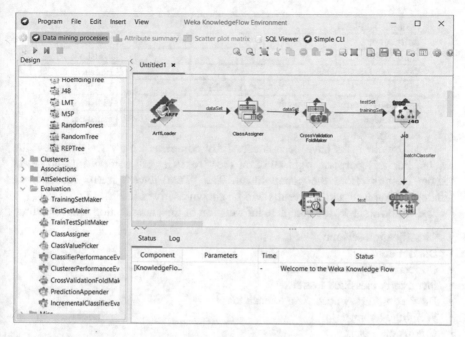

Fig. 12.43 Classification model in WEKA "KnowledgeFlow" Environment

Next, we need to insert the visualization component to display the results. WEKA provides a number of visualization components; however, we will use the simple "TextViewer" component. The component is available under the "Visualization" node in the "Design" sub-window.

Once the "TextViewer" component is inserted, you need to connect the output of the "ClassifierPerformanceEvaluator" component with the "TextViewer" component. For this purpose right-click on the "ClassifierPerformanceEvaluator" component and select "Text" from the popup menu. Now connect the "ClassifierPerformanceEvaluator" with the "TextViewer" component as before.

Modeling of our classification task through WEKA "KnowledgeFlow" component is complete as shown in Fig. 12.43:

Our next task is to execute our model. For this purpose, click on the "Run" button on the left top of the KnowledgeFlow window. Once you click this button, the model will be executed and the results will be calculated. An important point to note here is the information that is shown in the "Status" window below the model window. Note that it provides the status of each task. In our case the following information is provided by the status window:

[KnowledgeFlow]		-	OK.
ArffLoader		-	Finished.
ClassAssigner		-	Finished.
CrossValidationFoldMaker		-	Finished.
J48	-C 0.25 -M 2	-	Finished.
ClassifierPerformanceEvaluator		-	Finished.
TextViewer		-	Finished.

So all of our tasks have been successfully completed. Now we will view the results. For this purpose, right-click on the "TextViewer" component and select "Show Results" from the popup menu. The "TextViewer" result visualization window will appear and the results will be shown as given in Fig. 12.44.

The window shows the complete information of the classification task including:

- Classification scheme
- Dataset name
- Correctly classified instances
- Incorrectly classified instances
- Detailed accuracy according to each class
- Weighted average
- Confusion matrix

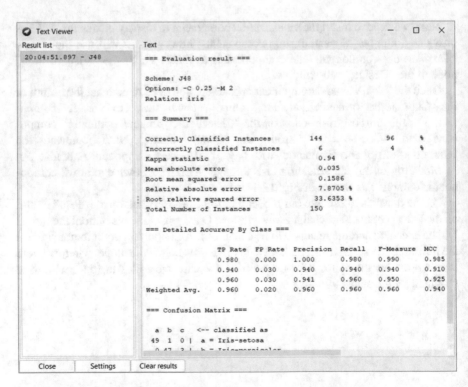

Fig. 12.44 TextViewer result visualization window

12.10   WEKA Workbench

WEKA 3.8.6 comes with another interface called the WEKA Workbench. Apparently, the Workbench seems to integrate all the other interfaces in one place; however, it comes with an important concept of "perspectives," where "perspective" may be a complete application, e.g., an Experimenter, or it may contain components from the other interfaces as separate perspective. Figure 12.45 shows the main Workbench interface in WEKA 3.8.6 version.

The Workbench interface can be customized by selecting/deselecting different perspectives that should be displayed on the interface. The settings can be achieved by clicking on the "Program" menu and then clicking the "Settings." The Workbench settings dialog box will appear which can be used to adjust different settings on the Workbench interface. As shown in Fig. 12.46, the Workbench settings can be used to adjust the look and feel of the interface along with some other settings for each tab that appears on the Explorer interface. For example, you can adjust which classifier algorithm will be displayed when the "Classify" tab is clicked.

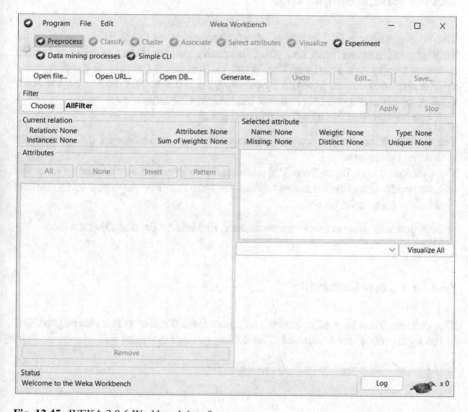

Fig. 12.45 WEKA 3.8.6 Workbench interface

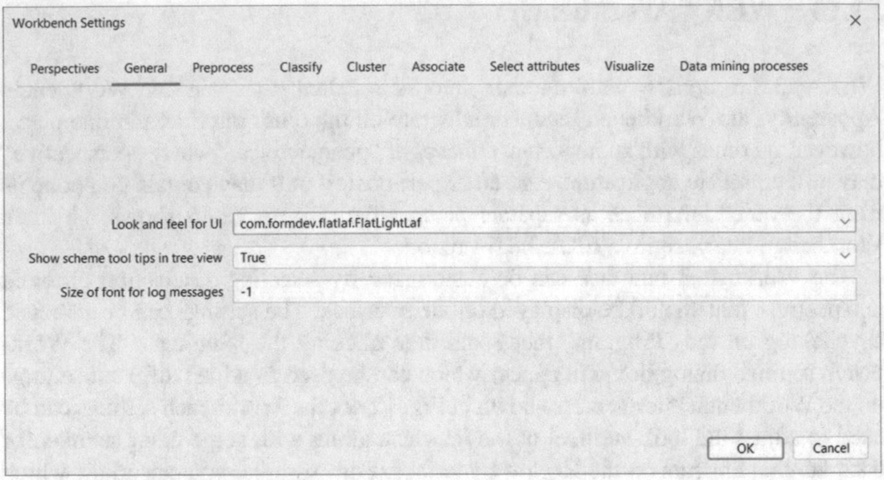

Fig. 12.46 Workbench settings dialog box

12.11 WEKA Simple CLI

The WEKA's Simple CLI provides complete access to all the WEKA classes using the command line interface, i.e., you can specify your tasks in the form of textual commands using this interface. The WEKA's Simple CLI comes in the form of a WEKA shell that can be invoked by clicking the 'Simple CLI' button on the main WEKA interface. Figure 12.47 shows the Simple CLI interface.

It can be seen that the interface has two parts:

The Output Window: The output window displays the results of the commands entered by the user.

The Command Line Input Box: The command line interface box is used to enter the commands. Once the command is entered, the result is displayed in the output window mentioned above.

Now we will discuss different commands that can be used at this interface.

12.11.1 *Java Command*

This command can be used to execute any Java class. The Java class can be provided in the argument of this command. The syntax of this command is:

Java *class_name the_command_line_arguments*

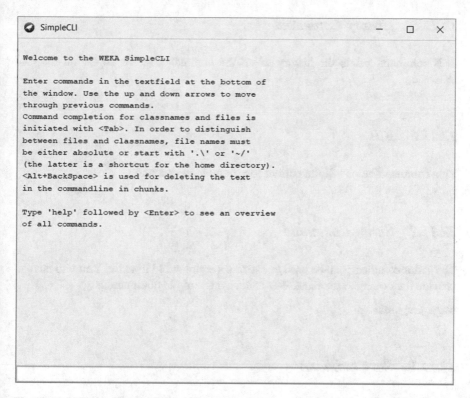

Fig. 12.47 WEKA's Simple CLI interface

12.11.2 Cls Command

The cls command clears the output window. All the previous results are erased from the output window. This command has no arguments.

12.11.3 Echo Command

The echo command is used to display a message in the output window. The command requires a string message as an argument. The syntax of the command is:

Echo "Message string"

12.11.4 Exit Command

Exit command lets you exit the Simple CLI interface. This command has no arguments.

12.11.5   History Command

This command prints the history of all the commands given so far on the CLI interface.

12.11.6   Kill

This command helps kill the current job if any is in execution.

12.11.7   Script Command

The script command can be used to execute a script saved in a file. You will have to provide the complete filename. The exact syntax of the command is:

script *file_name*

12.11.8   Set Command

This command is used to set a variable. The variable and values are provided in the form of key and value pair.

12.11.9   Unset Command

This command is used to unset the previously set variable using the "set" command.

12.11.10   Help Command

This command displays the details of all the commands or the details of the command that is provided in the form of an argument. The exact syntax of the command is:

Help
OR
Help "Command Name"

12.11.11   Capabilities Command

Capabilities command displays the capabilities of the class. The exact syntax of the command is:

Capabilities *class_name arguments*

Note that in WEKA's Simple CLI, you don't need to set any CLASSPATH variable. In fact, it supports the one with which WEKA was started. Figure 12.47 shows the interface of Simple CLI.

Now we will discuss some important components of the WEKA graphical user interface that can help you a lot during the analysis tasks. One such component is the ArffViewer. We will discuss ArffViewer in detail here.

12.12   ArffViewer

ArffViewer lets you open and edit the Arff files. As Arff files do not appear in a simple way just like the other plain files, you need a special editor for this purpose to conveniently view and edit these files.

ArffViewer displays the Arff files in a simple tabular format just like how the files are displayed in MS Excel. This lets you easily read a file's contents and make the changes.

You can open the ArffViewer from the main interface by clicking the "Tools" menu and then selecting "ArffViewer." Figure 12.48 shows the main interface of the ArffViewer tool.

As you can see, the data can be viewed in the tabular format which is a more convenient way to view the data. The attributes in the data are viewed as columns, whereas the instances are displayed in the form of rows.

The attribute names appear at the top of the viewer along with the data type of the attribute. The major functionality of the ArffViewer is available through its menus. Here we will see some important tasks that can be done through the ArffViewer.

The file menu provides simple file-related tasks, e.g., opening, saving, closing, etc. From the file menu, you can see the overall summary of the data as well by clicking the File menu and then selecting the "Properties" submenu. The "Select item" dialog box shows the overall summary of the dataset. The following information is displayed:

- File Name
- Relation Name
- Number of Instances
- Number of Attributes
- Class attribute
- Number of class labels

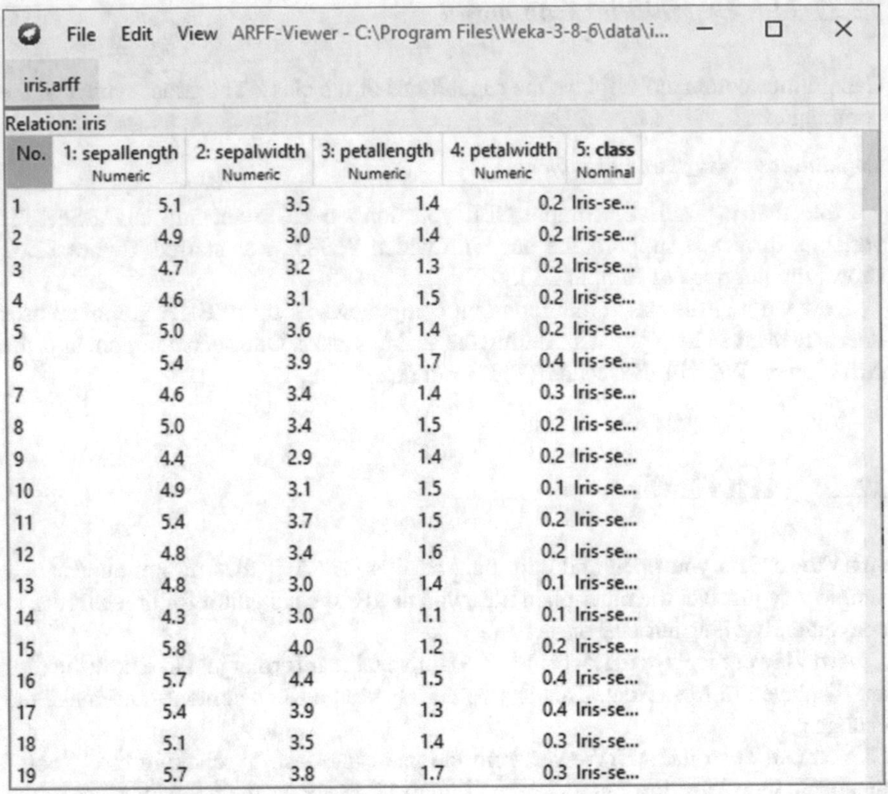

| ○ | File | Edit | View | ARFF-Viewer - C:\Program Files\Weka-3-8-6\data\i... | | — | □ | × |

iris.arff

Relation: iris

No.	1: sepallength Numeric	2: sepalwidth Numeric	3: petallength Numeric	4: petalwidth Numeric	5: class Nominal
1	5.1	3.5	1.4	0.2	Iris-se...
2	4.9	3.0	1.4	0.2	Iris-se...
3	4.7	3.2	1.3	0.2	Iris-se...
4	4.6	3.1	1.5	0.2	Iris-se...
5	5.0	3.6	1.4	0.2	Iris-se...
6	5.4	3.9	1.7	0.4	Iris-se...
7	4.6	3.4	1.4	0.3	Iris-se...
8	5.0	3.4	1.5	0.2	Iris-se...
9	4.4	2.9	1.4	0.2	Iris-se...
10	4.9	3.1	1.5	0.1	Iris-se...
11	5.4	3.7	1.5	0.2	Iris-se...
12	4.8	3.4	1.6	0.2	Iris-se...
13	4.8	3.0	1.4	0.1	Iris-se...
14	4.3	3.0	1.1	0.1	Iris-se...
15	5.8	4.0	1.2	0.2	Iris-se...
16	5.7	4.4	1.5	0.4	Iris-se...
17	5.4	3.9	1.3	0.4	Iris-se...
18	5.1	3.5	1.4	0.3	Iris-se...
19	5.7	3.8	1.7	0.3	Iris-se...

Fig. 12.48   Main ArffViewer interface

Now let us see the editing options available in the ArffViewer.

You can get the editing options from the "Edit" menu or by right-clicking on the title of the attribute. The first thing that you can do is you can rename the attribute. For this click the attribute that you want to rename and then select the "Rename attribute" from the file menu. The "Rename attribute" dialog box will appear. You can enter the new name and then press the "OK" button. The attribute will be renamed.

Next, you can set an attribute as a class. For this just right-click on the attribute header and then select the "Attribute as class." Alternatively, you can click on the attribute and select the same from the "Edit" menu. The attribute will be selected as a class and will be displayed in the last column as shown in Fig. 12.49:

Note that the attribute header is appearing as "bold" to indicate that it is a class attribute now.

Similarly, you can delete an attribute just by clicking on the attribute name and then selecting "Delete attribute" from the "Edit" menu. Alternatively, you can click on the attribute header and then select the same "Delete attribute" from the popup menu that is displayed on the right-click menu.

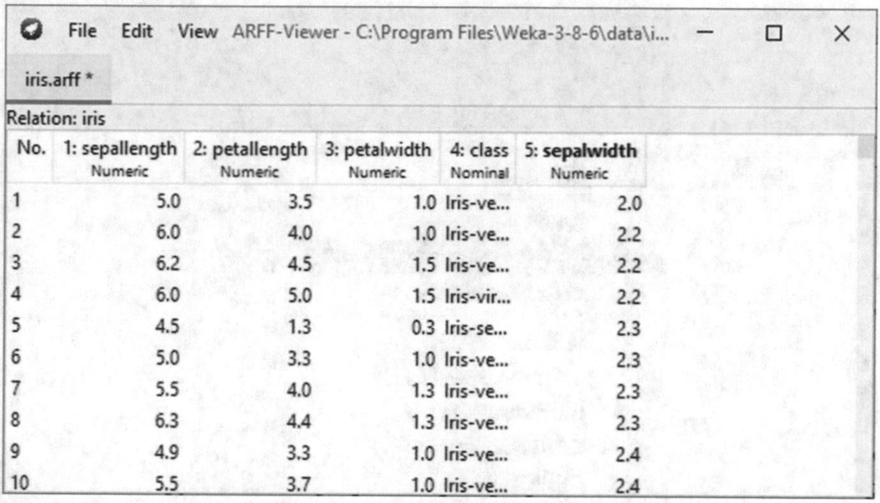

No.	1: sepallength Numeric	2: petallength Numeric	3: petalwidth Numeric	4: class Nominal	5: sepalwidth Numeric
1	5.0	3.5	1.0	Iris-ve...	2.0
2	6.0	4.0	1.0	Iris-ve...	2.2
3	6.2	4.5	1.5	Iris-ve...	2.2
4	6.0	5.0	1.5	Iris-vir...	2.2
5	4.5	1.3	0.3	Iris-se...	2.3
6	5.0	3.3	1.0	Iris-ve...	2.3
7	5.5	4.0	1.3	Iris-ve...	2.3
8	6.3	4.4	1.3	Iris-ve...	2.3
9	4.9	3.3	1.0	Iris-ve...	2.4
10	5.5	3.7	1.0	Iris-ve...	2.4

(File Edit View ARFF-Viewer - C:\Program Files\Weka-3-8-6\data\i... — □ ✕)
iris.arff *
Relation: iris

Fig. 12.49 The attribute "sepalwidth" is set as a class attribute

You can sort the data as well. By clicking on the attribute header and then selecting "Sort data (ascending)," you can sort the data of the attribute.

An important feature of the ArffViewer is that you can manually replace the missing values or an already existing value with the new values. For this, you will perform the following tasks:

1. Right-click the attribute in which you want to replace the missing value as shown in Fig. 12.50.
2. Select "Set missing values to" from the popup menu.
3. A new dialog box will appear. Specify the new value and click the "OK" button. The missing value will be replaced with the new value.

Furthermore, you can view each individual attribute and its values by using the view menu. Click on the attribute and from the "View" menu select "values". The "Select item" dialog box will appear. From this dialog box, select the attribute and click the "Select" button. The values of the selected column will appear in the new dialog box.

12.13 WEKA SqlViewer

WEKA SQL viewer enables you to connect with the database and fetch the data. For this, you will have to provide the database details including the database driver, username, and password. Once you are successfully connected with the database, SqlViewer lets you open and fetch any data from your database. For this purpose, WEKA SqlViewer provides you Query panel which you can use to write any SQL

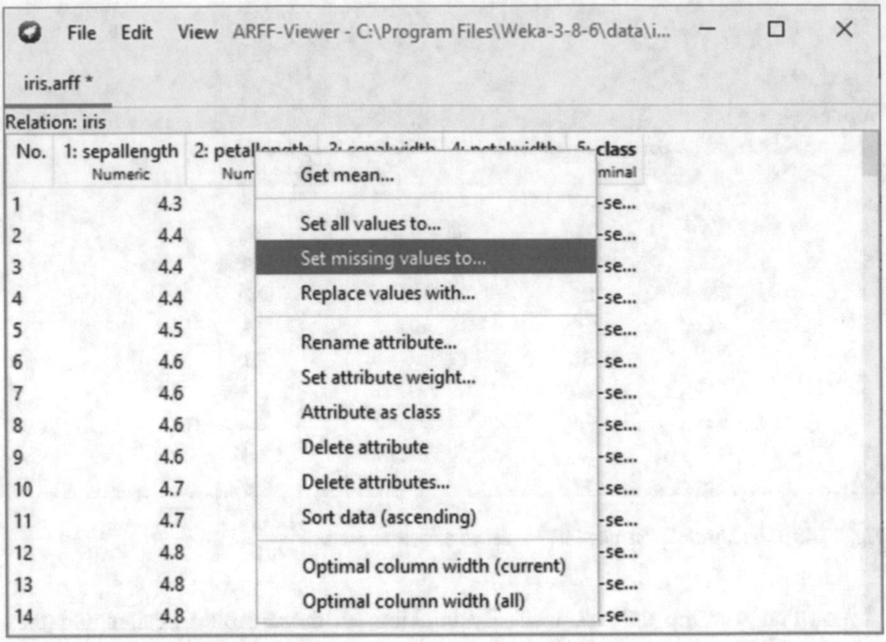

Fig. 12.50 "Set missing values" option on WEKA's ArffViewer

query. You can execute the query by clicking the "Execute" button on the right side of the Query panel.

Once the query is executed and the results are fetched, the results will be displayed in the results panel in a tabular form containing the data fetched by the query. Finally, the "Info" panel at the bottom shows the information about the current data fetching task. Figure 12.51 shows the WEKA's SQL Viewer interface.

Once you have provided all the parameters, you will click the "Connect to database" button on the right side of the Connection panel in order to connect to the database. In case there is some error in connecting with the database, you will get the information in the "Info" panel at the bottom of the SqlViewer's main window.

Note that you can set the maximum number of rows that can be fetched by the query at the time of execution. Similarly, you can clear the query box by clicking the "Clear" button.

12.14   Bayesian Network Editor

WEKA's Bayesian network editor is another important tool that comes with WEKA as a standalone application. You can invoke the editor by clicking the "Bayesian net editor" from the Tools menu on the main WEKA interface.

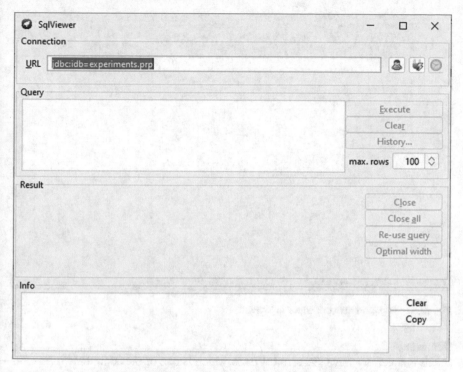

Fig. 12.51 WEKA's SqlViewer interface

The interface lets you open the Bayesian networks and completely edit them with your hands. The following are some major functionalities provided by the Bayesian network editor:

- Learning Bayesian network from data
- Manually editing Bayesian network
- Generation of a new dataset from the existing Bayesian network
- Getting inference of evidence using the junction trce
- View the cliques in the junction tree

Figure 12.52 shows the main interface of WEKA's Bayesian network editor.

Regarding editing of the Bayesian network, the following are some common tasks provided by the Bayesian network editor GUI:

- Add node
- Delete node
- Add Arc
- Delete Arc

Fig. 12.52 Bayesian network editor in WEKA

Summary

In this chapter, we have discussed WEKA which is one of the commonly used tools for working with algorithms related to data science. We have discussed the majority of the functionalities provided by the tool along with step-by-step details of how to use each function. We have discussed different components of the tool and how these components can help in working with the different types of algorithms. We also discussed miscellaneous components provided by WEKA including ArffViewer, SqlViewer, and command line interface. After reading this chapter, one can easily work with WEKA to perform the required tasks.

Further Reading

- *Instant Weka How-to* Paperback – June 21, 2013, by Bostjan Kaluza, Packt Publishing, 2013. WEKA is one of the maximum popular software programs used for machines. This book explains how you can use WEKA for developing your applications. The book explains how you can load data, apply filters, train a classifier, etc. You can also create your own classifiers. It also provides details about how you can test and evaluate your models. It helps you design your own recommender systems. Overall the book is a good resource for anyone interested in WEKA.
- *Machine Learning Mastery With Weka* by Jason Brownlee, Machine Learning Mastery, 2016. The book helps you develop machine learning models with the help of WEKA. In total WEKA is explained with the help of 18 lessons. The book explains the WEKA right from its installation to the final building of the models and making predictions. A few of the topics discussed include a loading of the

datasets, data visualization, selection of the best features, evaluation of the algorithms, improving the model performance, etc.

- *Data Mining: Practical Machine Learning Tools and Techniques* by Mark Hall, Ian Witten, and Eibe Frank, Morgan Kaufmann, 2011. Almost all the details required by a developer to start the use of WEKA are provided in the book. A few of the topics that are covered include the classification interface, clustering interface, association rule mining interface, Experimenter, and visualization tools. The book can be used as a good resource for learning WEKA by anyone interested in the development and execution of various machine learning-related algorithms.

Exercises

Exercise 12.1: What are the steps to load a dataset in WEKA?

Exercise 12.2: Consider the sample dataset given below. Save it in CSV format and name the file as "dataset.csv". Now load it into WEKA.

ID	Degree	Finance	Marks	Registration
1	Post Graduate	Self-Finance	140	Not-Permitted
2	Under Graduate	Scholarship	115	Not-Permitted
3	Under Graduate	Self-Sponsored	55	Not-Permitted
4	Post Graduate	Scholarship	135	Not-Permitted
5	Under Graduate	partial sponsored	110	Permitted
6	Under Graduate	Scholarship	75	Not-Permitted
7	Post Graduate	partial sponsored	205	Not-Permitted
8	Under Graduate	Self-Sponsored	100	Permitted
9	Under Graduate	Scholarship	89	Not-Permitted
10	Under Graduate	Self-Sponsored	105	Permitted

Exercise 12.3: Graphically show all the steps that are needed to apply a classification filter in WEKA.

Exercise 12.4: Consider the following four testing options provided by WEKA to test an algorithm. Explain the purpose of each of these options.

- Use training set
- Supplied test set
- Cross-validation
- Percentage split

Exercise 12.5: Consider the following dataset given in the table below. Open the dataset in WEKA ArffViewer and sort the attributes in ascending order.

- 1, 1, 2, 1, 2   • 3, 2, 2, 1, 1
- 1, 2, 1, 2, 1   • 1, 2, 2, 1, 2
- 2, 2, 2, 1, 2   • 3, 2, 1, 2, 1
- 3, 3, 2, 2, 1   • 1, 4, 0, 2, 1

Exercise 12.6: What is the purpose of WEKA Simple CLI? How is it different from WEKA GUI?

Exercise 12.7: Discuss the purpose of following WEKA commands (available in Simple CLI).

- Java command
- Cls command
- Echo command
- Exit command
- Kill command

Appendix: Mathematical Concepts for Deep Learning

Mathematics and statistics form the core of deep learning. Practitioners on their way to implement the machine learning models have to face them one way or the other. This is the reason that we have separate mathematics for machine learning courses these days. Here we will present some information about the mathematical concepts that may be required as essential for machine learning algorithms.

A.1 Vectors

An ordered n-tuple is called a sequence where n = {1,2, 3...}. So, (n1, n2, n3, ...) is a sequence. The following are some n-tuple sequences:

(3): 1-tuple sequence
(8,9): 2-tuple sequence
(6,7,8): 3-tuple sequence

It should be noted that the sequence is always an ordered arrangement. In the 3-tuple sequence above, "6" will appear before "7" and "7" will appear before "8."

A.1.1 Vector

A vector can be called a sequence comprising n-tuples. However, there is another common definition which states that a vector is an arrangement of elements in the form of a column.

© The Author(s), under exclusive license to Springer Nature Switzerland AG 2023
U. Qamar, M. S. Raza, *Data Science Concepts and Techniques with Applications*,
https://doi.org/10.1007/978-3-031-17442-1

The generic representation of vectors is as follows:

$$a = \begin{bmatrix} a_1 \\ a_2 \\ a_3 \end{bmatrix}$$

The following is the example of a 2-tuple vector:

$$a = \begin{bmatrix} 5 \\ 7 \end{bmatrix} \qquad b = \begin{bmatrix} 9 \\ 8 \end{bmatrix}$$

Vectors can be represented in two-dimensional XY-space. The basic unit length vectors, i.e., i, j, are as follows:

$$i = \begin{bmatrix} 1 \\ 0 \end{bmatrix} \qquad j = \begin{bmatrix} 0 \\ 1 \end{bmatrix}$$

The vector "i" represents the vector from the origin of the XY-space, i.e., (0,0) to the point (1,0) on the X-axis, and "j" represents the vector from the origin (0,0) to the point (0,1) on the Y-axis.

The vectors a and b shown above can be represented in the XY coordinate system as shown in Fig. A.1:

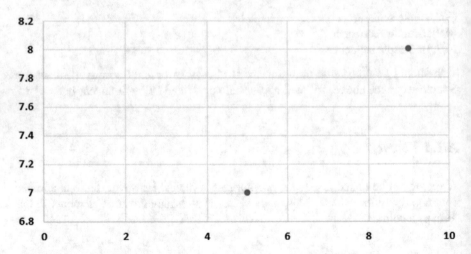

Fig. A.1 Representation of vector 8.2 in the form of XY coordinate system

A.1.2 Vector Addition

Vector addition is performed by adding the corresponding values in each vector with each other using the following method:

$$a = \begin{bmatrix} a_1 \\ a_2 \\ a_3 \end{bmatrix} \qquad b = \begin{bmatrix} b_1 \\ b_2 \\ b_3 \end{bmatrix} \qquad a + b = \begin{bmatrix} a_1 + b_1 \\ a_2 + b_2 \\ a_3 + b_3 \end{bmatrix}$$

For example, consider the following two vectors:

$$a = \begin{bmatrix} 1 \\ 5 \\ 7 \end{bmatrix} \qquad b = \begin{bmatrix} 14 \\ 3 \\ 4 \end{bmatrix} \qquad c = a + b = \begin{bmatrix} 15 \\ 8 \\ 11 \end{bmatrix}$$

A.1.3 Vector Subtraction

Vector subtraction is performed by subtracting the values of the second vector from the corresponding value of the first vector using the following method:

$$a = \begin{bmatrix} a_1 \\ a_2 \\ a_3 \end{bmatrix} \qquad b = \begin{bmatrix} b_1 \\ b_2 \\ b_3 \end{bmatrix} \qquad c = a - b = \begin{bmatrix} a_1 - b_1 \\ a_2 - b_2 \\ a_3 - b_3 \end{bmatrix}$$

For example, consider the following two vectors:

$$a = \begin{bmatrix} 1 \\ 5 \\ 7 \end{bmatrix} \qquad b = \begin{bmatrix} 14 \\ 3 \\ 4 \end{bmatrix} \qquad c = a - b = \begin{bmatrix} -13 \\ 2 \\ 3 \end{bmatrix}$$

A.1.4 Vector Multiplication by Scalar Value

A vector can be multiplied by a scalar value. The scalar value is multiplied by each element of the vector by the following method:

$$a = \begin{bmatrix} a_1 \\ a_2 \\ a_3 \end{bmatrix} \qquad x.a = x * \begin{bmatrix} a_1 \\ a_2 \\ a_3 \end{bmatrix} = \begin{bmatrix} xa_1 \\ xa_2 \\ xa_3 \end{bmatrix}$$

Consider the following example:

$$a = \begin{bmatrix} 1 \\ 5 \\ 7 \end{bmatrix} \qquad 5a = 5 * \begin{bmatrix} 1 \\ 5 \\ 7 \end{bmatrix} = \begin{bmatrix} 5 \\ 25 \\ 35 \end{bmatrix}$$

A.1.5 Zero Vector

A zero vector is the one in which all the elements are zero. Following is the example of a zero vector:

$$a = \begin{bmatrix} 0 \\ 0 \\ 0 \end{bmatrix}$$

A.1.6 Unit Vector

A vector having the unit length in XY or XYZ coordinate system is called the unit vector. The following are the unit vectors:

$$i = \begin{bmatrix} 1 \\ 0 \\ 0 \end{bmatrix} \qquad j = \begin{bmatrix} 0 \\ 1 \\ 0 \end{bmatrix} \qquad k = \begin{bmatrix} 0 \\ 0 \\ 1 \end{bmatrix}$$

A.1.7 Vector Transpose

The transpose of a vector a will be a^T. In simple words the transpose of a vector is obtained by converting the columns of a vector into rows by the following method:

$$a = \begin{bmatrix} a_{11} \\ a_{21} \\ a_{31} \end{bmatrix} \qquad a^T = [a_{11} \ a_{12} \ a_{13}]$$

Consider the following example:

$$a = \begin{bmatrix} 1 \\ 5 \\ 7 \end{bmatrix} \qquad a^T = [1 \ 5 \ 7]$$

Similarly, the transpose of a row vector b will be as follows:

$$b = [1 \ 5 \ 7] \qquad b^T = \begin{bmatrix} 1 \\ 5 \\ 7 \end{bmatrix}$$

A.1.8 Inner Product of Vectors

The inner product is also called a dot product. The dot product of two vectors of the same dimensions is obtained by taking the transpose of the first vector and then taking the sum of the multiplication of the corresponding elements as follows:

$$a = \begin{bmatrix} a_1 \\ a_2 \\ a_3 \end{bmatrix} \qquad b = \begin{bmatrix} b_1 \\ b_2 \\ b_3 \end{bmatrix}$$

Now by taking the transpose of vector a and multiplying it with vector b:

$$a^T b = [a_1 \ a_2 \ a_3] \begin{bmatrix} b_1 \\ b_2 \\ b_3 \end{bmatrix} = a_1 b_1 + a_2 b_2 + a_3 b_3$$

Note that the inner product of two vectors results in a scalar value.

A.1.9 Orthogonal Vectors

If the dot product of two vectors is zero, the vectors will be called orthogonal vectors. For example, the following two vectors a and b are orthogonal as their dot product is zero.

$$a = \begin{bmatrix} 0 \\ 3 \end{bmatrix} \qquad b = \begin{bmatrix} 5 \\ 0 \end{bmatrix}$$

$$a^T b = \begin{bmatrix} 0 & 3 \end{bmatrix} \begin{bmatrix} 5 \\ 0 \end{bmatrix} = 0 * 5 + 3 * 0 = 0$$

A.1.10 Distance Between Vectors

Sometimes we may need to calculate the distance between two vectors. We can calculate the distance between the two vectors a and b as follows:

$$a = \begin{bmatrix} a_1 \\ a_2 \\ a_3 \end{bmatrix} \qquad b = \begin{bmatrix} b_1 \\ b_2 \\ b_3 \end{bmatrix}$$

$$d(a, b) = \sqrt{(a_1 - b_1)^2 + (a_2 - b_2)^2 + (a_3 - b_3)^2} \qquad (A.1)$$

Now consider the following two vectors:

$$a = \begin{bmatrix} 5 \\ 7 \end{bmatrix} \qquad b = \begin{bmatrix} 0 \\ 4 \end{bmatrix}$$

$$d(a, b) = \sqrt{(5 - 0)^2 + (7 - 4)^2}$$

$$d(a, b) = \sqrt{(5)^2 + (3)^2}$$

$$d(a, b) = \sqrt{(5)^2 + (3)^2}$$

$$d(a, b) = \sqrt{25 + 9}$$

$$d(a, b) = \sqrt{34}$$

A.1.11 Outer Product

Suppose we have two vectors a and b. If we take the transpose of vector b and then perform the operation ab^T, the product will be called outer product. The method of outer product is as follows:

$$a = \begin{bmatrix} a_1 \\ a_2 \\ a_3 \end{bmatrix} \qquad b = \begin{bmatrix} b_1 \\ b_2 \\ b_3 \end{bmatrix}$$

$$ab^T = \begin{bmatrix} a_1 \\ a_2 \\ a_3 \end{bmatrix} \begin{bmatrix} b_1 & b_2 & b_3 \end{bmatrix} = \begin{bmatrix} a_1b_1 & a_1b_2 & a_1b_3 \\ a_2b_1 & a_2b_2 & a_2b_3 \\ a_3b_1 & a_3b_2 & a_3b_3 \end{bmatrix}$$

For example, consider the following vectors a and b:

$$a = \begin{bmatrix} 1 \\ 5 \\ 3 \end{bmatrix} \qquad b = \begin{bmatrix} 2 \\ 1 \\ 6 \end{bmatrix}$$

$$ab^T = \begin{bmatrix} 1 \\ 5 \\ 3 \end{bmatrix} \begin{bmatrix} 2 & 1 & 6 \end{bmatrix} = \begin{bmatrix} 2 & 1 & 6 \\ 10 & 5 & 30 \\ 6 & 3 & 18 \end{bmatrix}$$

A.1.12 Angles Between the Vectors

We can find the angles between the vectors in the context of their geometrical representation. Consider the two vectors a and b. The angle between them can be calculated as:

$$\cos \theta = \frac{a^T b}{\|a\| \|b\|} \tag{A.2}$$

For example, consider the following two vectors:

$$a = \begin{bmatrix} 0 \\ 2 \end{bmatrix} \qquad b = \begin{bmatrix} 2 \\ 2 \end{bmatrix}$$

$$\cos\theta = \frac{a^T b}{\|a\|\|b\|}$$

$$\cos\theta = \frac{0*2+2*2}{4*\sqrt{8}}$$

$$\cos\theta = \frac{4}{4*\sqrt{8}}$$

$$\cos\theta = \frac{1}{\sqrt{8}}$$

$$\theta = \cos^{-1}\left(\frac{1}{\sqrt{8}}\right)$$

$$\theta = \cos^{-1}\left(\frac{1}{\sqrt{8}}\right)$$

So the angle between the vectors a and b will be almost 70 degrees.

A.2 Matrix

A matrix is an arrangement of elements in rows and columns. So, depending upon these rows and columns, the dimensions of the matrix are specified. For example, consider the following "3*1" matrix:

$$a = \begin{bmatrix} 6 \\ 9 \\ 4 \end{bmatrix} \tag{A.3}$$

The "3" means that the matrix comprises three rows and the "1" means that the matrix comprises "one" column. Similarly, the following is an example of a "3*3" matrix:

$$a = \begin{bmatrix} 8 & 1 & 0 \\ 4 & 0 & 9 \\ 7 & 3 & 5 \end{bmatrix} \tag{A.4}$$

So the dimension of the matrix, in general terms, is defined as "$m*n$" where "m" represents rows and "n" represents columns. The general notion for a matrix is as follows:

$$a = \begin{bmatrix} a_{11} & a_{12} & a_{13} \\ a_{21} & a_{22} & a_{23} \\ a_{31} & a_{32} & a_{33} \end{bmatrix} \qquad (A.5)$$

A.2.1 Matrix Addition

Matrix addition is performed by adding the corresponding values in each matrix with each other using the following method:

$$a = \begin{bmatrix} a_{11} & a_{12} & a_{13} \\ a_{21} & a_{22} & a_{23} \\ a_{31} & a_{32} & a_{33} \end{bmatrix} \qquad b = \begin{bmatrix} b_{11} & b_{12} & b_{13} \\ b_{21} & b_{22} & b_{23} \\ b_{31} & b_{32} & b_{33} \end{bmatrix}$$

$$a + b = \begin{bmatrix} a_{11} + b_{11} & a_{12} + b_{12} & a_{13} + b_{13} \\ a_{21} + b_{21} & a_{22} + b_{22} & a_{23} + b_{23} \\ a_{31} + b_{31} & a_{32} + b_{32} & a_{33} + b_{33} \end{bmatrix}$$

For example, consider the following two matrices:

$$a = \begin{bmatrix} 8 & 1 & 0 \\ 4 & 0 & 9 \\ 7 & 3 & 5 \end{bmatrix} \qquad b = \begin{bmatrix} 3 & 2 & 4 \\ 5 & 6 & 7 \\ 1 & 3 & 9 \end{bmatrix} \qquad c = a + b = \begin{bmatrix} 11 & 3 & 4 \\ 9 & 6 & 16 \\ 8 & 6 & 14 \end{bmatrix}$$

A.2.2 Matrix Subtraction

Matrix subtraction is performed by subtracting the values of the second matrix from the corresponding value of the first matrix using the following method:

$$a = \begin{bmatrix} a_{11} & a_{12} & a_{13} \\ a_{21} & a_{22} & a_{23} \\ a_{31} & a_{32} & a_{33} \end{bmatrix} \qquad b = \begin{bmatrix} b_{11} & b_{12} & b_{13} \\ b_{21} & b_{22} & b_{23} \\ b_{31} & b_{32} & b_{33} \end{bmatrix}$$

$$a - b = \begin{bmatrix} a_{11} - b_{11} & a_{12} - b_{12} & a_{13} - b_{13} \\ a_{21} - b_{21} & a_{22} - b_{22} & a_{23} - b_{23} \\ a_{31} - b_{31} & a_{32} - b_{32} & a_{33} - b_{33} \end{bmatrix}$$

For example, consider the following two matrices:

$$a = \begin{bmatrix} 8 & 1 & 0 \\ 4 & 0 & 9 \\ 7 & 3 & 5 \end{bmatrix} \qquad b = \begin{bmatrix} 3 & 2 & 4 \\ 5 & 6 & 7 \\ 1 & 3 & 9 \end{bmatrix}$$

$$c = a - b = \begin{bmatrix} 5 & -1 & -4 \\ -1 & -6 & 2 \\ 6 & 0 & -4 \end{bmatrix}$$

A.2.3 Matrix Multiplication by Scalar Value

A matrix can be multiplied by a scalar value. The scalar value is multiplied by each element of the matrix by the following method:

$$a = \begin{bmatrix} a_{11} & a_{12} & a_{13} \\ a_{21} & a_{22} & a_{23} \\ a_{31} & a_{32} & a_{33} \end{bmatrix} \qquad x.a = \begin{bmatrix} x.a_{11} & x.a_{12} & x.a_{13} \\ x.a_{21} & x.a_{22} & x.a_{23} \\ x.a_{31} & x.a_{32} & x.a_{33} \end{bmatrix}$$

Consider the following example:

$$a = \begin{bmatrix} 8 & 1 & 0 \\ 4 & 0 & 9 \\ 7 & 3 & 5 \end{bmatrix}$$

$$5a = 5 * \begin{bmatrix} 8 & 1 & 0 \\ 4 & 0 & 9 \\ 7 & 3 & 5 \end{bmatrix} = \begin{bmatrix} 40 & 5 & 0 \\ 20 & 0 & 45 \\ 35 & 15 & 25 \end{bmatrix}$$

A.2.4 Unit Matrix

The unit matrix is also called an identity matrix. A unit matrix is an $n*n$ square matrix in which all the elements of the main diagonal are "1." For example, the following is the example of the unit matrix:

$$I = \begin{bmatrix} 1 & 0 & 0 \\ 0 & 1 & 0 \\ 0 & 0 & 1 \end{bmatrix}$$

The transpose of a matrix a will be a^t. In simple words the transpose of a matrix is obtained by converting the columns of a matrix into rows by the following method:

$$a = \begin{bmatrix} a_{11} \\ a_{21} \\ a_{31} \end{bmatrix} \qquad a^T = \begin{bmatrix} a_{11} & a_{12} & a_{13} \end{bmatrix}$$

Consider the following example:

$$a = \begin{bmatrix} 1 \\ 5 \\ 7 \end{bmatrix} \qquad a^T = \begin{bmatrix} 1 & 5 & 7 \end{bmatrix}$$

Similarly, the transpose of a 3×3 matrix will be as follows:

$$a = \begin{bmatrix} 8 & 1 & 0 \\ 4 & 0 & 9 \\ 7 & 3 & 5 \end{bmatrix} \qquad a^T = \begin{bmatrix} 8 & 4 & 7 \\ 1 & 0 & 3 \\ 0 & 9 & 5 \end{bmatrix}$$

A.2.5 Matrix Multiplication

Matrices can be multiplied by multiplying all the elements of the first row of matrix A with all elements of the first column of matrix B and then placing the sum as the element c_{11} in the resultant matrix C.

This process is done through the following formula:

$$a = \begin{bmatrix} a_{11} & a_{12} \\ a_{21} & a_{22} \end{bmatrix} \qquad b = \begin{bmatrix} b_{11} & b_{12} \\ b_{21} & b_{22} \end{bmatrix}$$

$$c = a * b = \begin{bmatrix} (a_{11}b_{11} + a_{12}b_{21}) & (a_{11}b_{12} + a_{12}b_{22}) \\ (a_{21}b_{11} + a_{22}b_{21}) & (a_{21}b_{12} + a_{22}b_{22}) \end{bmatrix}$$

Consider the following example:

$$a = \begin{bmatrix} 5 & 7 \\ 4 & 3 \end{bmatrix} \qquad b = \begin{bmatrix} 1 & 3 \\ 0 & 8 \end{bmatrix}$$

$$c = a * b = \begin{bmatrix} (5 * 1 + 7 * 0) & (5 * 3 + 7 * 8) \\ (4 * 1 + 3 * 0) & (4 * 3 + 3 * 8) \end{bmatrix}$$

$$c = a * b = \begin{bmatrix} (5 + 7) & (15 + 56) \\ (4 + 0) & (12 + 24) \end{bmatrix}$$

$$c = a * b = \begin{bmatrix} 12 & 71 \\ 4 & 36 \end{bmatrix}$$

A.2.6 Zero Matrix

A zero matrix is one in which all the elements are zero. For example, the following matrix is a zero matrix:

$$a = \begin{bmatrix} 0 & 0 & 0 \\ 0 & 0 & 0 \\ 0 & 0 & 0 \end{bmatrix}$$

A.2.7 Diagonal Matrix

In a matrix, the entries starting from the left top element and ending down at the right bottom element are called the main diagonal of the matrix. A diagonal matrix is the one in which all the entries across the main diagonal are nonzero and all the other entries are zero. The matrix below is an example of a diagonal matrix.

$$a = \begin{bmatrix} 7 & 0 & 0 \\ 0 & 5 & 0 \\ 0 & 0 & 3 \end{bmatrix}$$

A.2.8 Triangular Matrix

A triangular matrix is a square matrix in which all the elements above the diagonal or below the diagonal are zero as shown below:

$$a = \begin{bmatrix} a_{11} & 0 & 0 \\ a_{21} & a_{22} & 0 \\ a_{31} & a_{32} & a_{33} \end{bmatrix} \quad OR \quad a = \begin{bmatrix} a_{11} & a_{12} & a_{13} \\ 0 & a_{22} & a_{23} \\ 0 & 0 & a_{33} \end{bmatrix}$$

The matrices "a" and "b" below are examples of zero matrices.

$$a = \begin{bmatrix} 3 & 0 & 0 \\ 5 & 4 & 0 \\ 1 & 8 & 7 \end{bmatrix} \quad OR \quad a = \begin{bmatrix} 8 & 3 & 14 \\ 0 & 1 & 2 \\ 0 & 0 & 9 \end{bmatrix}$$

A matrix in which all the entries above the main diagonal are zero is called an upper triangular matrix, and the matrix in which all the entries below the main diagonal are is called a lower triangular matrix.

A.2.9 Vector Representation of Matrix

Rows and columns of a matrix of $m*n$ dimensions can be represented in the form of vectors. The rows of an $m*n$ matrix can be represented as:

$$A = [a_{i1}, a_{i2}, a_{i3} \ldots, a_{in}] \text{ where } i = [1, 2, 3 \ldots n]$$

Similarly, the columns of a matrix of $m*n$ dimensions can be represented in the form of a column vector as follows:

$$a = \begin{bmatrix} a_{11} \\ a_{21} \\ a_{31} \\ \vdots \\ a_{im} \end{bmatrix}$$

A.2.10 Determinant

The determinant is the process of associating a single element with a matrix. With the help of a determinant, we can find that either a matrix can be inverted or not. If a matrix has a nonzero determinant, it can be inverted; however, a matrix cannot be inverted if it has a zero determinant. The determinant of a 2×2 square matrix can be found by the following method:

$$A = \begin{bmatrix} a_{11} & a_{12} \\ a_{21} & a_{22} \end{bmatrix}$$

$$\text{Det } A = |A| = a_{11}a_{11} - a_{11}a_{11}$$

For example, consider the following matrix:

$$a = \begin{bmatrix} 5 & 7 \\ 4 & 3 \end{bmatrix}$$

$$\text{Det } A = |A| = 5 * 3 - 4 * 7$$

$$\text{Det } A = |A| = 15 - 28$$

$$\text{Det } A = |A| = -13$$

The determinant of a "3*3" matrix can be determined as follows:

$$a = \begin{bmatrix} a_{11} & a_{12} & a_{13} \\ a_{21} & a_{22} & a_{23} \\ a_{31} & a_{32} & a_{33} \end{bmatrix}$$

$$\text{Det } A = |A| = a_{11} * \det \begin{bmatrix} a_{22} & a_{23} \\ a_{31} & a_{33} \end{bmatrix} - a_{12} * \det \begin{bmatrix} a_{21} & a_{22} \\ a_{31} & a_{32} \end{bmatrix} + a_{13} * \det \begin{bmatrix} a_{21} & a_{23} \\ a_{31} & a_{33} \end{bmatrix}$$

For example, consider the following matrix:

$$a = \begin{bmatrix} 8 & 1 & 0 \\ 4 & 0 & 9 \\ 7 & 3 & 5 \end{bmatrix}$$

$$\text{Det } A = |A| = 8 * \det \begin{bmatrix} 0 & 9 \\ 3 & 5 \end{bmatrix} - 1 * \det \begin{bmatrix} 4 & 9 \\ 7 & 5 \end{bmatrix} + a_{13} * \det \begin{bmatrix} 4 & 0 \\ 7 & 3 \end{bmatrix}$$

$$\text{Det } A = |A| = 8 * (0 * 5 - 3 * 9) - 1 * (4 * 5 - 7 * 9) + 0 * (4 * 3 - 7 * 0)$$

$$\text{Det } A = |A| = 8 * (0 - 27) - 1 * (20 - 63) + 0 * (12 - 0)$$
$$\text{Det } A = |A| = 8 * (-27) - 1 * (-43) + 0 * (12)$$
$$\text{Det } A = |A| = -216 + 43 + 0$$
$$\text{Det } A = |A| = -173$$

A.3 Probability

Probability deals with uncertainty and randomness. In simple words, it mentions the chance of occurring or not occurring something. Randomness here means that we cannot control the environment that can affect the output of the experiment.

In simple words, if we want to define probability, it is the likelihood of occurrence of some event where an event is something that results in the form of some finite number of outcomes. It can be calculated by dividing the favorable number of outcomes by the total number of outcomes. The formula for probability is:

$$P(A) = \frac{\text{favorable outcomes}}{\text{total outcomes}} \tag{A.6}$$

Let's explain it with the help of an example:

Suppose we want to predict that tossing a fair coin will result in a head or tail. So, it means that we will have to predict the probability of occurring of the head and tail.

Since the head is one favorable outcome (if we want to predict the probability of occurring of head) and the total number of outcomes is two (head and tail), the probability of occurring of head can be calculated as:

$$P(A) = \frac{1}{2} = 0.5$$

So there is a 50% likelihood that it will be head when a fair coin is tossed. The term "fair" here means that there is an equal chance of occurrence of all the outputs. Let's now discuss some key terms that are related to the concept of probability.

A.3.1 Experiment

An experiment is an operation or a test that is conducted to get an outcome. An example of an experiment may be tossing a coin three times.

Sample space: A sample space refers to all the possible outcomes of an experiment. For example, if we toss a coin, all the possible outcomes comprise "head" and "tail."

Favorable outcomes: The desired results or the expected outcomes of an event are called the favorable outcome, for example, if we want to know, when two dices are rolled together, what is the probability of getting the sum of five on both dices. In this case, the favorable outcomes are those when the sum of numbers on both dices results in five, i.e., (1,4), (2,3), (3,2), and (4,1).

A.3.2 Event

An event refers to one or more possible outcomes of an experiment. For example, occurring of two heads together when two coins are tossed is referred to as an event.

When the probability of occurrence of two or more events is the same, they are called equally likely events. For example, if we toss a coin, then there are equal chances of occurrence of a head and a tail; hence both are equally likely events.

A.3.3 Mutually Exclusive Events

The events where the probability of occurrence of one event is zero when the other event has occurred are called mutually exclusive events. For example, the probability of a plan taking off is zero when it is landing and vice versa, so both taking-off and landing are mutually exclusive events.

Now we will discuss some probability formulas as given below.

A.3.4 Union of Probabilities

When an event is a union of other events, the probability is calculated using the union rule as follows:

$$P(E_1 \cup E_2) = P(E_1) + P(E_2) - P(E_1 \cap E_2) \qquad (A.7)$$

A.3.5 Complement of Probability

The complement of an event means the inverse of the event. For example, the complement of the event that "there will a head when a toss is coined" will be "there will be no head when a toss is coined." Mathematically it is defined as:

$$P(\text{Not } E) = 1 - P(E) \qquad (A.8)$$

A.3.6 Conditional Probability

Sometimes we need to find out the probability of occurrence of an event when a certain other event has already occurred, for example, the probability of going out for shopping when it is raining. Here we need to find the probability of going out for shopping when the event of rain has already occurred. The probability will be calculated as:

$$P(A|B) = P(A \cap B)/P(B) \tag{A.9}$$

A.3.7 Intersection of Probabilities

When an event is the intersection of two other events, then the intersection of the probabilities can be calculated as:

$$P(A \cap B) = P(A|B)P(B) \tag{A.10}$$

A.3.8 Probability Tree

A probability tree is a visual representation of the probabilities of the outcomes when an experiment is performed. For example, when a coin is tossed, the probabilities of occurrence of heads and tails can be visually specified with the help of the following probability tree diagram shown in Fig. A.2.

The numbers on the arrows represent the probability and the leaf nodes represent the outcomes of the events. According to the above diagram, the probability of occurring of "Head" is 0.5.

A.3.9 Probability Axioms

The first axiom states that the minimum probability of an event will be zero and the maximum probability will be one. So, the probability of an event will always be equal to or greater than zero and less than or equal to one.

Fig. A.2 Probability tree

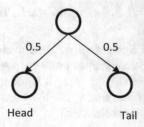

Head   Tail

Mathematically:

$$0 \leq P(A) \leq 1$$

The second axiom states that if the outcome comprises the entire sample space, the probability of the event will be equal to one. Such an event will be called a certain event, i.e., an event having the probability of "1." Similarly, an event having zero probability will never occur.

The third axiom states that the sum of probabilities of all the outcomes in a sample space is always equal to "1."

$$P(A) + P(B) + \ldots + P(N) = 1$$

Glossary

Accuracy Accuracy defines the ratio of the number of classes correctly predicted as compared to the total number of records.

Activation functions The output of the neuron is applied to the activation function. The activation functions normally convert the input value to a specific limit, e.g., from 0 to 1.

Agglomerative clustering These approaches consider the points as individual clusters and keep on merging the individual clusters on the basis of some proximity measure, e.g., the clusters that are closer to each other.

Apache Spark Apache server is one of the largest large-scale data processing tools. The tool executes applications in Hadoop clusters 100 times faster in memory and 10 times faster on disk.

Apriori algorithm The first associative algorithm that was proposed, and various further tasks in association, classification, associative classification, etc. have used this algorithm in their techniques.

Artificial neuron It is the base of the artificial neural network.

Association rules Machine learning technique that tries to find out the relationship between two items.

Autoencoder It is a feed-forward neural network based on unsupervised learning.

Bag of words It is one of the models to represent a text in numerical format for computers to understand.

Bagging It stands for "bootstrap aggregation." In bagging, weak learners are implemented in parallel that learn independently from each other.

Bias It is used for the average squared difference between the actual value and predicted value.

Binary text classification Binary text classification is a special type of single-label classification where a single text document can be assigned to a class from a set of only two classes.

Binning Also called "discretization"; can be called the inverse process of encoding. In encoding, we convert the categorical features into numerical features.

U. Qamar, M. S. Raza, *Data Science Concepts and Techniques with Applications*,
https://doi.org/10.1007/978-3-031-17442-1

Boosting It is a sequential ensemble learning method in which the weights are adjusted iteratively according to the last classification results.

Business intelligence It refers to a generic process that is useful for decision-making out of the historical information in any business, whereas data analytics is the process of finding the relationships between data to get deep insight.

Caffe (Convolutional Architecture for Fast Feature Embedding) It is an open-source deep learning library originally written in C++ with a Python interface.

Cardinality It specifies the total number of elements of a set.

Categorical features They consist of symbols to represent domain values.

Centroids They are the central points of a cluster.

Classical multidimensional scaling (MDS) It is another linear-based dimensionality reduction approach that reduces dimensions by measuring the distance or dissimilarity between two data points.

Classification Classification is one of the core machine learning techniques that use supervised learning. Classification is the problem of assigning objects to predefined groups based on their characteristics.

Cluster analysis It is the process of clustering the objects in groups on the basis of their properties and their relation with each other and studying their behavior for extracting the information for analysis purposes.

Confidence It determines the likeliness of occurrence of consequent in a transaction with a given antecedent.

Convolutional neural network (CNN) Based on convolution on one-dimensional and two-dimensional data.

Cubic polynomial regression This type of regression has variable(s) with a maximum of three degrees.

Data acquisition Data acquisition is the process of obtaining and storing the data from which the patterns are intended to be identified.

Data analytics The process of generalizing knowledge from raw data which includes various steps from storage to final knowledge extractions.

Data mining It forms the core of the entire data analytics process. It may include extraction of the data from heterogeneous sources including texts, videos, numbers, and figures.

Data science It is a multidisciplinary field that focuses on the study of all aspects of data right from its generation to processing to convert it into a valuable knowledge source.

Data Facts needed for analysis.

DBSCAN It is a density-based clustering approach.

Decision system It consists of a finite number of objects (U) and conditional attributes (C) against each object along with information about the targeted class (D).

Decision tree It is used to show the possible outcomes of different choices that are made based on some conditions.

Deep learning It comprises various other domains including machine learning, neural networks, artificial intelligence, etc. Neutral networks form a core of deep learning.

Descriptive analytics It helps us find "what happened" or "what is happening."

Diagnostics analytics Taking the analytics process a step ahead, diagnostic analytics help in analyzing "Why it happened?".

Disjoint sets Two sets A and B will be called disjoint if they do not contain any element common in both of these sets.

Divisive clustering These approaches consider all the data points as a single all-inclusive cluster and keep on splitting it. The process continues until we get singleton clusters that we cannot split further.

Dominance-based rough set approach (DRSA) It is an extension of rough set theory to consider the preference relation of attributes.

ECLAT It stands for equivalence class clustering and bottom-up lattice transversal algorithm. This is another algorithm to find the frequent patterns in the transactional database.

Elman neural network It is a type of recurrent neural network. In recurrent neural networks, the output of neurons is fed back to other neurons and the entire network works in a loop.

Empty set It is one in which there is no element.

Ensembles They are aggregate predictions made by multiple classifiers.

Entropy It defines the purity/impurity of a dataset.

Equal sets Two sets A and B will be called equal if they have an equal number of elements and all the elements are the same.

Equivalent sets Two sets A and B will be equivalent if the cardinality of both sets is the same which means that there is an equal number of elements in both sets and there is a one-to-one correspondence between the elements of the sets.

Error calculation The output of the network is matched with the actual output by calculating the mean error.

Error rate Error rate defines the ratio of incorrectly predicted classes to the total number of records.

ETL It is the process of collecting data from various homogeneous and heterogeneous sources and then applying the transformation process to prepare the data for storing in the data warehouse.

Euclidean distance It is the distance between two points that is simply the length of the straight line between the points in Euclidean space.

Event It refers to one or more possible outcomes of an experiment.

Experiment It is an operation or a test that is conducted to get an outcome.

Feature encoding It is the process of converting features into a form that is easy for data science applications to process.

Feature engineering It is the process of creating the features relevant to the underlying problem using the knowledge about the available data.

Feature extraction Feature extraction techniques, on the other hand, project current feature space to a new feature subset space.

Feature selection A process that selects features from the given feature subset without transforming or losing the information. So, we can preserve data semantics in the transformation process.

Feature It is an individual characteristic or property of anything existing in this world.

Feed forward Information is provided from one layer to the next. The output of the first layer becomes the input of the next layer.

Frequent pattern mining It deals with identifying the common trends and behaviors in datasets.

Fuzzy Apriori algorithm It is an effort to combine the fuzzy set theory with association rule mining.

Genetic algorithm One of the most common heuristics-based approaches inspired by the nature. The algorithm is based on the concept of survival of the fittest.

Hyperplane A hyperplane is a decision boundary that separates the two or more decision classes.

In-warehouse data transformation This is a slight modification from the conventional method. Data is extracted from the source and moved into the data warehouse and all the transformations are performed there.

Inverse document frequency It specifies how common a word is in a document.

Isomap It is a dimensionality reduction for handling nonlinear data structures.

Jordan neural network It is similar to the Elman neural network. The only difference is that context neurons take input from the output layer instead of the hidden layer.

K-Means It is based on prototype-based clustering, i.e., they create prototypes and then arrange objects in clusters according to those prototypes. K-Means define prototypes in terms of the centroids which are normally the mean points of the objects in clusters. K-Medoid, on the other hand, defines the prototypes in terms of medoids which are the most representative points for the objects in a cluster.

Keras It is a free and open-source Python library used for neural network development and evaluation within machine learning and deep learning models.

KNIME An open-source platform that lets you analyze and model data. Through its modular data pipeline concepts, KNIME provides a platform for reporting and integration of data.

Lift It is the ratio of the probability of consequent being present with the knowledge of antecedent over the probability of consequent being present without the knowledge of antecedent.

Linear polynomial regression This type of regression has variable(s) with a maximum of one degree.

Linear regression It is the relationship between one independent variable and one dependent variable.

Locally linear embedding (LLE) It is one of the unsupervised approaches which reduces dimensions while conserving geometric features of the nonlinear dataset.

Long Short-Term Memory (LSTM) It is an advance form of neural network that lets the network persist the information.

Loss function It is a function that is used in machine learning to find the performance of a model.

Lumify Lumify is a powerful big data tool to develop actionable intelligence. It helps find the complex relationships in data through a suite of analytic options, including graph visualizations, and full-text faceted search.

Manhattan distance It specifies the difference between the vectors.

matplotlib It is a comprehensive Python library for data visualization and making interactive graphs and animations.

Matrix It is an arrangement of elements in rows and columns.

Microsoft Excel One of the most common tools used for organizational data processing and visualizations. The tool is developed by Microsoft and is part of the Microsoft Office suite.

Multi-label text classification In multi-label text classification, each text document can be assigned more than one class from the given set of classes.

Multiple linear regression It is a relationship of more than one independent variables with one dependent variable.

Multistage data transformation This is a conventional method where data is extracted from the source and stored in intermediate storage, transformation is performed, and data is moved to the data warehouse.

Naïve Bayes It is a probabilistic classifier that predicts the class of data based on the previous data by using probability measures.

Named entity recognition (NER) It is the process of identifying the named entities from the text. The identified entities can then be used for further processing.

Natural language processing (NLP) It is the branch of computer science that deals with the processing of text in natural language so that computers could understand and process it.

Null transaction It is a special type of transaction that does not contain any itemset under consideration.

Numerical attributes They represent numbers.

NumPy It is an open-source Python programming language library.

OpenCV It is a huge open-source Python library developed for image processing, machine learning, and computer vision. It can be used for detecting various objects, faces, or even handwriting from different sources of inputs (i.e., images or videos).

OpenRefine As per their official statement, OpenRefine (previously Google Refine) is a powerful tool for working with messy data: cleaning it, transforming it from one format into another, and extending it with web services and external data.

Orthogonal vector If the dot product of two vectors is zero, the vectors will be called orthogonal vectors.

pandas It is an open-source Python library that is used for machine learning tasks and data analysis. It is built on top of NumPy and is responsible for preprocessing datasets for machine learning.

Part of speech tagging It is one of the important tasks of natural language processing. In POS tagging, each token is categorized in terms of its part of speech.

Pattern recognition It is the process of reading the raw data and matching it with some already available data to see if the raw data has the same patterns as we have applied.

Power set It is the collection of all the possible subsets of a set.

Predictive analytics Predictive analytics, as the name indicates, helps in making predictions.

Prescriptive analytics Prescriptive analytics, as the name implies, prescribes the necessary actions that need to be taken in case of a certain predicted event.

Principal component analysis (PCA) It is a well-known multivariate statistical dimensionality reduction process used for data analysis.

Probability It deals with uncertainty and randomness. In simple words, it mentions the chance of occurring or not occurring something.

PyCaret It is an open-source modular machine learning library based on the popular Classification And REgression Training (CARET) R package.

Python Python is another programming language that is most widely used for writing programs related to data analytics.

PyTorch It is an open-source deep learning library developed by the Facebook AI research group and it is based on Python and a C language framework, Torch.

QlikView Offers in proc data processing, thus enhancing efficiency and performance. It also offers data association and data visualization with compressed data.

Quadratic polynomial regression This type of regression has variable(s) with a maximum of two degrees.

Quick Reduct (QR) It is a rough set-based feature selection that used the forward feature selection technique.

R It is another important programming language used for statistical analysis and data visualization.

RapidMiner Mostly used for predictive analytics, the tool can be integrated with any data source including Excel, Oracle, SQL Server, etc.

Recurrent neural network (RNN) It is just like a conventional neural network; however the only difference is that it remembers the previous state of the network as well.

Regression tree It is a machine learning model that is used to partition the data into a set of smaller groups and then solve each smaller group recursively.

Regression In regression, we try to find the value of dependent variables from independent variables using the already provided data; however, a basic difference is that the data provided here is in the form of real values.

Regularization It is a technique used in machine learning to increase the generalization by decreasing high variance and avoiding overfitting.

Rough set theory It is a set theory-based tool for data reduction and classification tasks.

Scientific Python (SciPy) It is an open-source library based on NumPy and supports advance and more complex operations such as clustering, image processing, integration, etc.

scikit-learn It is an open-source and robust to use library for supervised and unsupervised machine learning algorithms.

Scrapy It is an extensive free and open-source web scraping and crawling library.

Seaborn It is an open-source enhanced data visualization and representation library based on matplotlib.

Sigmoid It is the most common activation function used in a majority of artificial neural network applications.

Simple linear regression Simple linear regression is the simplest form of regression analysis involving one dependent variable and one independent variable.

Single-label text classification When each document is assigned only one class from the given class set, then this is called single-label text classification.

Splice Machine Splice Machine is a scalable SQL database that lets you modernize your legacy and custom applications to be agile, data-rich, and intelligent without modifications.

Splunk Splunk is a specialized tool to search, analyze, and manage machine-generated data.

Subsets A set A will be called a subset of another set B if every element of A is also present in set B.

Supervised learning Use labeled data for learning.

Support It defines how frequently an itemset appears in a transaction.

Support vector machines It is used for classifying the data in N-dimensional space. The aim of the classifier is to identify the maximal margin hyperplane that classifies the data.

Supremum distance It is the distance between two vectors that is equal to the maximum difference that exists between any two axes of the points.

Tableau Public It is free software that can connect to any data source and create visualizations including graphs, charts, and maps in real time.

Talend Talend is a powerful tool for automating big data integration.

Telemedicine It is the process of contacting the doctor and physician using advance technologies without involving personal physical visits.

TensorFlow It is a Python library for fast numerical computing developed and maintained by Google and released under an open-source license.

Term frequency-inverse document frequency (TF-IDF) TF-IDF gives the importance of a term in a document.

Term frequency It specifies a specific term "t" is the ratio of the number of occurrences of t and the total number of terms in the sentence.

Test dataset Used to test the model.

The holdout method It reserves a certain amount of the labeled data for testing and uses the remainder for training.

Theano It is an optimizing compiler and numerical computation Python library that is exclusively designed for machine learning.

Training dataset Training dataset, as the name implies, is required to train our model, so that it could predict with maximum accuracy for some unknown data in the future.

Transformation-based techniques Those methods which perform dimensionality reduction at the cost of destruction of the semantics of the dataset.

Unsupervised learning Used unlabeled data for learning.

Variety The number of types of data.

Vector It is an arrangement of elements in the form of a column.

Velocity Data is generated at an immense rate.

Veracity Refers to bias and anomalies in big data.

Volume We deal with petabytes, zettabytes, and exabytes of data.

WEKA It stands for Waikato Environment for Knowledge Analysis. It is an open-source tool for performing different machine learning and data mining tasks.

Printed in the United States
by Baker & Taylor Publisher Services